Advances in Spatial Science

For further volumes:
http://www.springer.com/series/3302

Yee Leung

Knowledge Discovery in Spatial Data

Springer

Prof. Yee Leung
The Chinese University of Hong Kong
Dept. of Geography &
Resource Management
Shatin, New Territories
Hong Kong SAR
yeeleung@cuhk.edu.hk

ISBN 978-3-642-26170-1 e-ISBN 978-3-642-02664-5
DOI 10.1007/978-3-642-02664-5
Springer Heidelberg Dordrecht London New York

Cover design: SPi Publisher Services

Printed on acid-free paper

Springer is part of Springer Science+Business Media (www.springer.com)

In memory of my father

Wah Sang Leung (梁華生)

To

Yuk Lee, my postgraduate advisor
My undergraduate teachers
Sau Kuen Chu, my secondary school teacher

for their initiation, stimulation and guidance in my search for geographical knowledge at various stages of my academic development

Acknowledgements

This monograph contains figures and tables based on copyright figures and tables owned and supplied by China Academic Journal Electronic Publishing House, Elsevier, IEEE, Springer, Taylor and Francis, and Wiley, and are used with their permissions. These comprise of:

Figures 1.2, 2.8–2.12, 2.27–2.34, 4.10–4.14, 5.1–5.7; Tables 4.10–4.13, 5.1 (taken from Springer)

Figures 1.1, 2.1, 2.2, 2.6, 2.7, 2.35–2.37, 5.10, 5.11, 5.14 (taken from IEEE)

Figure 6.5 (taken from China Academic Journal Electronic Publishing House)

Figures 6.6, 6.8, 6.9 (taken from Elsevier)

Figures 1.5, 6.10–6.26; Table 6.2 (taken from Wiley)

Tables 1.1, 4.17–4.28 (taken from Taylor and Francis)

I would like to thank Prof. Manfred M. Fischer who has been encouraging me to write this book for the series. I would also like to thank my research associates, particularly Profs. Z.B. Xu, W.X. Zhang, J.S. Zhang, J.H. Ma, C.L. Mei, J.S. Mi, W.Z. Wu, J.C. Luo, V. Anh and our students who have worked with me over the years to develop the methodologies discussed in this monograph. My appreciation also goes to Ms. Kilkenny Chan and Mr. Eric Wong, particularly Kilkenny, for typing and re-typing the monograph with patience and dedication.

Last but not least, my heartfelt appreciation goes to my wife, Sau-Ching Sherry, for her love and support, and my son, Hei, for giving me a pleasant diversion from work. They both make my life complete and meaningful.

Yee Leung

Preface

When I first came across the term data mining and knowledge discovery in databases, I was excited and curious to find out what it was all about. I was excited because the term tends to convey a new field that is in the making. I was curious because I wondered what it was doing that the other fields of research, such as statistics and the broad field of artificial intelligence, were not doing. After reading up on the literature, I have come to realize that it is not much different from conventional data analysis. The commonly used definition of knowledge discovery in databases: "the non-trivial process of identifying valid, novel, potentially useful, and ultimately understandable patterns in data" is actually in line with the core mission of conventional data analysis. The process employed by conventional data analysis is by no means trivial, and the patterns in data to be unraveled have, of course, to be valid, novel, useful and understandable. Therefore, what is the commotion all about? Careful scrutiny of the main lines of research in data mining and knowledge discovery again told me that they are not much different from that of conventional data analysis. Putting aside data warehousing and database management aspects, again a main area of research in conventional database research, the rest of the tasks in data mining are largely the main concerns of conventional data analysis. Model identification, model construction, and discovery of plausible hypotheses, for example, are not unique to data mining. They are, in addition to model estimation and hypothesis testing, in the agenda of conventional data analysis, such as statistics, also. Searching for clusters, looking for separating surfaces or rules for classification, mining for association rules and relationships, and detecting temporal trends or processes in data constitute the core of the knowledge discovery process that is not unique to data mining. They form the backbone of research in conventional data analysis. From this perspective, there is very little novelty in data mining and knowledge discovery.

On the other hand, if we look at the environment within which data mining and knowledge discovery is taking place, there is something genuine that is worthy of our attention. Though we have traditionally looked for patterns in data by performing clustering, classification, relational analysis, and trend or process analysis, the kinds of data that we are dealing with nowadays are quite different from that targeted by conventional data analysis methods. The sheer volume and complexity of the data that we need to handle nowadays are substantially different from that of the past. Effective discovery of knowledge hidden in data requires novel methods for accomplishing the old tasks. Therefore, it is from this perspective that the mission of data mining and knowledge discovery is justified. This field of research can actually be treated as the continuation of the mission of conventional data analysis into the information and knowledge age. And, our main objective is simply to discover knowledge in data as we have always been doing. Nevertheless, I have no problem in using the term data mining and knowledge discovery adopted by the research community as long as we know exactly what we are doing.

Following up on the literature, I also encountered the term spatial data mining and knowledge discovery. A natural question again is what it is all about, and how it is different from data mining and knowledge discovery in general. An examination of the research activities in this area again tells me that in principle it is more or less similar to that of the general field. The major difference is that data in spatial data mining are mostly geo-referenced and much more complex. And the knowledge to be discovered is often location specific and takes on geometric shapes. Space and time are the two main dimensions along which knowledge discovery is performed. Thus, there is something unique in spatial data mining and knowledge discovery that is worth looking into.

Our main goal is then to discover knowledge in spatial data. It is again in line with conventional spatial data analysis but with special emphasis placed on the nature of spatial data. The idea is to develop novel methods for spatial knowledge discovery. Whether we should call such process spatial data mining and knowledge discovery, or just simply discovery of knowledge in spatial and temporal data is just a matter of terminology. It all involves the discovery of spatial structures, processes, and relationships from spatial and temporal data. As data mining and knowledge discovery have become a commonly employed collective term for such activities, it is used indiscriminately throughout this book. I would not painstakingly point out whether a method should be called a data mining and knowledge discovery method, or just a data analysis method targeting the unraveling of structures, processes and relationships in voluminous and complex spatial and temporal databases.

As a good number of text books and research monographs have been written on data mining and knowledge discovery, one needs a good justification to write another book on the topic. Given the unique features of knowledge discovery in spatial data and the burgeoning growth of research interest in this area, it is an opportune time to make a critical analysis of the field and explore directions for further research. Instead of repeating what has been written in many current books on data mining and knowledge discovery, I would like to write it from the perspective of my own research in this area. So, it is not a text book on data mining and knowledge discovery. It is not a book, like many others, that discusses all aspects of the knowledge discovery process. So there are no discussions on topics such as data warehousing, on-line analytical processing (OLAP), data query, and data mining software. There is no intention to give a comprehensive survey of the literature of the field, although state-of-the-art reviews under relevant topics are made in the book.

This book is intended to be a research monograph on methods and algorithms, conventionally called data mining methods, for the discovery of knowledge in spatial and temporal data. The majority of the methods discussed are based on our own research. So, when I discuss topics such as clustering, classification, relationships and temporal processes, algorithms in the literature are not discussed in detail. Emphasis is placed on the development of our own methods. Nevertheless, it is not difficult to see that some of our methods can, more or less, fit into the family of research methodologies on the same topics. They are developed on the foundation of mathematics, statistics, and artificial intelligence. In brief, the present monograph is not a text book for spatial data mining and knowledge discovery. It is a book for researchers and advanced graduate students who are interested or might have an interest in the methodologies for the discovery of knowledge in spatial and temporal data. The view is more personal, but it fits in with the overall picture of research in the field.

Yee Leung

Contents

List of Figures

List of Tables

Chapter 1
Introduction

1.1 On Spatial Data Mining and Knowledge Discovery

Understanding natural and human-induced structures and processes in space and time has long been the agenda of geographical research. Through theoretical and experimental studies, geographers have accumulated a wealth of knowledge about our physical and man-made world over the years. Based on such knowledge, we work for the betterment of the man-land relationship that hopefully will lead to the sustainable development of our man-land system. The quest for knowledge can mainly be summarized into two basic approaches. Based on some assumptions about the underlying mechanisms, we can infer the properties and behaviors of our systems. This is the well-known process of the search for knowledge through deduction. On the other hand, knowledge is often discovered through critical observations of phenomena in space and time. Structures and processes are unraveled by sipping through data that we gathered. With the advancement and rapid development of the information technologies, amassing huge volume of data for research is no longer a problem in any disciplines. It is particularly true in geographical studies where a continuous inflow of various types of data collected by various means is a common ground. The problem is then not having enough data but having too much and too complex a database for the discovery and understanding of structures, processes, and relationships. Useful knowledge is often hidden in the sea of data that awaits discovery.

Long before the dazzling development of advanced technologies for information collection in recent decades, extracting from and establishing knowledge in data has been a major undertaking in geographical research. Technically, geographers have been mining knowledge in data for a long time. Searching for spatial distribution of and spatial relationships between phenomena has been the center piece of spatial data analysis in all these years. The euphoria generated by the data mining and knowledge discovery community tends to convey an idea that something fundamentally new is in the making. The bandwagon effect is felt not only in geography but in many other academic disciplines and commercial circles. A scrutiny

Y. Leung, *Knowledge Discovery in Spatial Data*, Advances in Spatial Science,
DOI 10.1007/978-3-642-02664-5_1, © Springer-Verlag Berlin Heidelberg 2010

of the core research agenda of data mining and knowledge discovery, however, shows that the main objective is actually more or less in line with the purpose of conventional data analysis. Both intend to look for novel and potentially useful knowledge hidden in data through non-trivial processes. Furthermore, the main lines of research are also something that we are familiar with. Clustering, classification, association, relationship are basic things that we have been performing with data over and over again. One then wonders what all this excitement is about in data mining and knowledge discovery, and what remains to be done that has not been done in conventional data analysis. If one looks at it from this broad-brush perspective, then it is essentially not much new. An immediate conclusion is then "old wine in new bottle." However, if we look at the means by which the data mining tasks are accomplished, there are issues that need to be addressed from perspectives differing from that of the conventional data analysis.

Data mining and knowledge discovery can be regarded as a re-examination of the state-of-the-art in data analysis with the view of unraveling knowledge hidden in voluminous and complex databases that are not targeted at by conventional methods. It is in synchrony with the advancement of our ever changing information technologies that generate data of various types and structures which might not be of main concern in conventional data analysis. Furthermore, it deals with data generated by complex and dynamic systems that challenge traditional wisdom. It is in the methodological development front that makes data mining and knowledge discovery a serious undertaking. It is the kind of research that advocates a tight coupling of information technologies and data analysis methods.

1.2 What Makes Spatial Data Mining Different

Data mining and knowledge discovery has become an active research area in the commercial and scientific communities. Theories and methods have been developed over the years to mine knowledge from data. In the business circle, the interest might be the segmentation of clients from transaction data, direct marketing, stock trends, electronic commerce, etc. In the bio-medical field, the purpose of data mining might be tumor identification, drug response, patterns in gene sequences, long-range correlation in DNA sequences, classification of organisms, etc. In astronomy, the focus might be placed on the discovery of black holes, faint objects, new stars or galaxies, etc. In various fields of engineering, data mining might involve pattern recognition, robot movement, vision or voice identification, etc. A common feature of these investigations is, of course, the discovery of knowledge in massive databases.

With no exception, dealing with huge volume of data is of primary concern in knowledge discovery from spatial data. However, there are issues which are unique or of particular importance in spatial data mining. Compared to other databases, spatial data are much more complicated. In addition to the shear volume, spatial databases contain both non-spatial and geo-referenced information. Spatial data are multi-sourced, multi-typed, multi-scaled, heterogeneous, dynamic, noisy, and imperfect. Besides the concern of volume, geographers have to come up with means to handle the above problems in order to be able to mine effectively and efficiently knowledge hidden in spatial data. In terms of knowledge, it is often spatially relevant. Geographers are particularly keen to find out whether or not a structure, a process, or a relationship is spatially independent. Specifically, we look for autocorrelation in spatial data. Therefore, knowledge mining in spatial data is much more complicated than general data mining and knowledge discovery.

There is always the spatial dimension that needs to be taken into consideration. It goes without saying, the consideration of the temporal aspect further complicates the task.

To pave the road for the discussion in the remaining part of this book, the issues of knowledge, data, and tasks for knowledge discovery in spatial data are first examined in the sections to follow.

1.3 On Spatial Knowledge

Knowledge is extensively examined throughout human history. Philosophers have spent their lifetimes examining what knowledge is about. There is no intention to engage in this philosophical discussion in here. The notion of knowledge is relatively concrete and practical in our spatial data mining task. It simply refers to structures, processes, and relationships that exist in space and time. They can be natural or man-made. They are, however, knowledge that is essential in scientific investigations and real-life applications.

In the context of structure, the kind of knowledge we intend to unravel from spatial data might be natural land covers in remotely sensed images, hot spots of infectious disease, distribution patterns of seismic belts, geometric spatial objects of regular or irregular shapes in vector-based data, and hierarchical concept structures hidden in relational databases. This type of knowledge is generally static in nature and appears as intrinsic clusters or outliers in data sets.

With reference to process, the kind of knowledge we are referring to is the underlying mechanism that generate specific structures in space and time. Generating processes of various time series are targets of the knowledge discovery task. They can be processes manifested as regular or irregular trends of air pollution, occurrences of extreme temperature, out breaks of natural disasters, and patterns of spread of epidemics. Such knowledge is dynamic in nature and appears as trends, cycles, or spikes in temporal data.

In terms of relationship, the kind of knowledge that is intended to detect from data is associations and causal relations in space. Spatial interdependence of phenomena manifested by spatial autocorrelation in data is a common piece of knowledge one intends to detect from data. Metric, directional and topological relations are also targets to be unraveled from data. Association rules depicting co-existence of spatial phenomena are another form of knowledge depicting certain kinds of spatial relations.

Thus, knowledge to be mined from spatial and temporal databases comprises concrete structures, processes, and relationships of spatial phenomena. They are embedded or hidden in spatial data whose retrieval requires non-trivial data mining methods. This is the kind of knowledge we are referring to throughout our discussion in this monograph. It is essentially the kind of knowledge meant by the majority of the spatial data mining tasks.

1.4 On Spatial Data

As discussed above, spatial knowledge discovery encounters data that are rather unique in structure and form. They constitute tasks that are different from those in general data mining. They also create difficulties in unraveling structures, processes, and relationships in spatial and temporal data. I first discuss in brief the nature of spatial data that needs to be taken into consideration in the development of appropriate methods for knowledge mining.

1. Volume. With the advancement of information technologies, we have and we will amass voluminous data covering the spatial and temporal dimensions. The number of attributes characterizing certain phenomena can be in hundreds or even thousands. In hyperspectral analysis, images obtained by AVIRIS, HYDICE, and Hyperion range from 0.4 to 2.45 micrometers at 224 bands, 0.4 to 2.5 micrometers at 210 bands, and 0.4 to 2.5 micrometers at 220 bands with 30 meter spatial resolution, respectively. A time series can cover a time span of hundreds and millenniums of years. Discovering knowledge in this shear volume of data is a daunting task for data mining methodologies.
2. Multi-source. Development of space information technologies has enabled us to collect data from a variety of sources. Geo-referenced data can be obtained from a great variety of satellite sensors, digital map scanning, radar devices, aircrafts, and global positioning systems. Different devices collect information with different formats and measurements. Spatial and temporal coverage also varies accordingly. To discover knowledge in data needs to deal with single and multi-sourced data. Data compatibility and fusion need to be entertained by the data mining methods.
3. Multi-scale. Spatial observation with various scales has long been a tradition in geographical studies. Capturing and representing phenomena in paper maps with different scales is a common practice in order to reveal spatial structures in various details. Such practice has then been carried over to digital maps commonly used nowadays. In the capturing of spatial phenomena via remote sensing technologies, images of various resolutions are again a common place in the study of remotely sensed images. Along the temporal dimension, data of various scales are usually encountered in the study of trends and dynamics of spatial processes. They can range from seconds, minutes, days, months, and years. How to discover knowledge in data with multiple spatial and temporal scales is thus a challenge in the development of appropriate data mining techniques.

4. Multi-type. Due to the multi-source nature of spatial data, they come in with various types. Knowledge discovery might need to be carried out with raster-based data such as images collected by satellites with various sensors; vector-based data such as points, lines and polygons; or object-oriented data arranged in specific hierarchical structures. Furthermore, data can be geo-referenced and/or non-spatial. The challenge then rests on whether appropriate methods can be developed for the unraveling of knowledge from single-type and multi-type spatial data.
5. Imperfection. It is generally difficult to collect perfect information about our complex spatio-temporal systems. Spatial data are generally imperfect in the sense that they might be incomplete, even though they can be precisely measured. Sometimes information is fuzzy with imprecise characterization. Under the spell of the chance factor, structures and processes that the data represent might be random following a probabilistic distribution or a stochastic process. Captured in different scales, data are granular with varying roughness. Moreover, missing values and noise generally exist in spatial databases. In brief, imperfection can be due to randomness and/or fuzziness and/or roughness and/or noise and/or incompleteness. This constitutes the complexity of knowledge discovery in spatial data.
6. Dynamic. It goes without saying that spatial systems are ever changing. Therefore, we need to collect data that can reflect the phenomena or processes over time. Temporal data can be in discrete time, like most of the time series data, or in continuous time. How to unfold the hidden processes from temporal data is thus high in the agenda of spatial knowledge discovery.

To sum up, geographers deal with databases which are generally much more complex than those in other disciplines. Data with anyone of the above characteristics is already a challenge in itself. The co-existence of a few will further complicate the task of spatial knowledge discovery. Compounding on the complexity of spatial databases, complicated spatial relationships embedded in the data further complicate the process of knowledge discovery. It is under such high level of complexity we need to develop methods to unravel intrinsic and useful structures, processes, and relationships.

1.5 Basic Tasks of Knowledge Discovery in Spatial Data

Though there are a great variety of knowledge discovery tasks, they can be grouped under several basic categories. In terms of the kind of knowledge to be discovered, our tasks are essentially (1) clustering, (2) classification, (3) association/relationship, and (4) process.

Clustering is the basic task by which structures can be discovered as clusters in spatial data. This task searches for data that form clusters of similar features in a natural way. The whole idea is to unravel without any presumptions data clusters representing spatial structures in compact forms. Spatial knowledge represented by

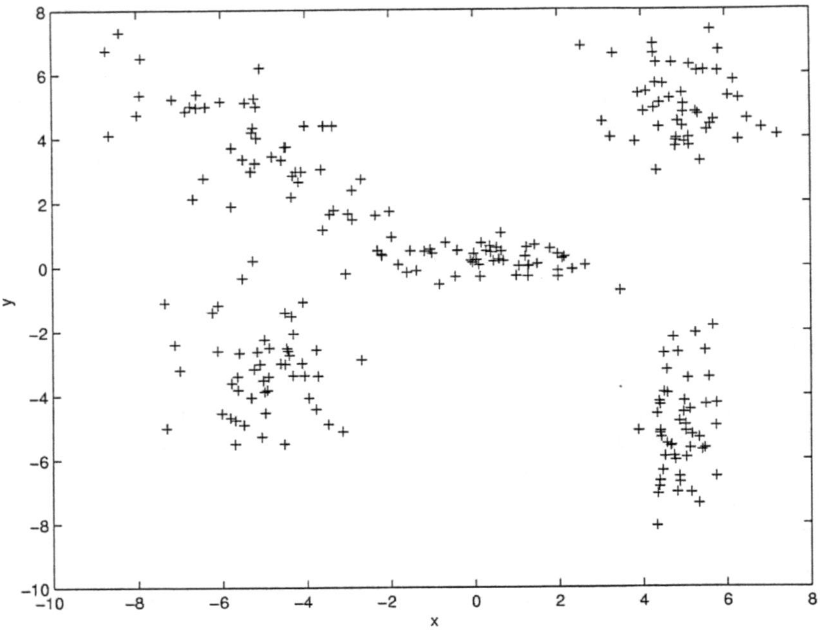

Fig. 1.1 How many clusters are there?

Fig. 1.2 How many seismic belts are there?

the clusters might be land covers, seismic zones, epidemic hot spots, etc. Figure 1.1 is a plot of a data set. The natural question is "How many clusters are there?" (The answer is provided in Sect. 2.2.5). Figure 1.2 is a distribution of earthquakes in a region over a certain time period. The curiosity is "Do they form seismic belts? If yes, how many?" (The solution can be found in Sect. 2.2.7). Thus, in the task of clustering, the general interest is to discover the partitioning or covering of space by clusters representing certain structures of a spatial phenomenon.

Classification, on the other hand, intends to discover in data hypersurfaces that can classify spatial objects into pre-specified classes. This piece of knowledge can

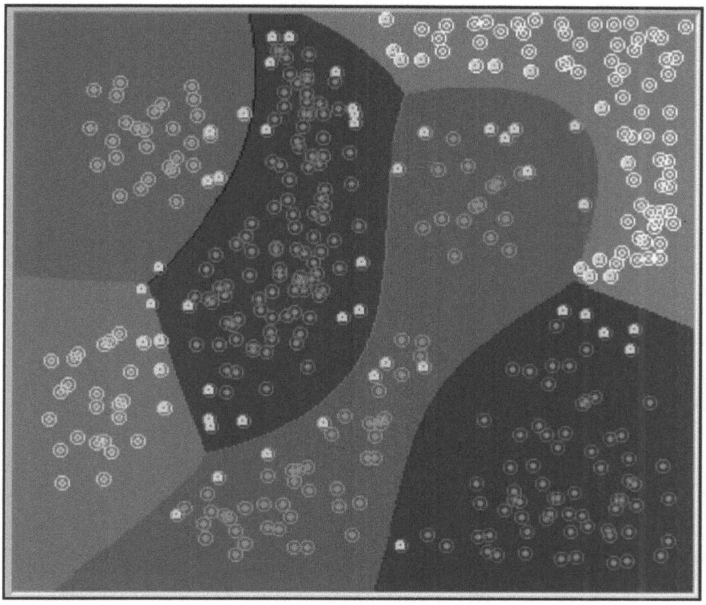

Fig. 1.3 How can the classes be best separated?

be a separating function that can classify objects into classes according to their similarities in characteristics. Figure 1.3, for example, depicts a multiple-class problem in which the best separating surface for the relevant classes is sought (The result of this data mining task is given in Sect. 3.2). It can also be a set of classification rules stipulating how different spatial objects can be assigned to pre-specified classes. Table 1.1, for instance, is a summary of a multispectal data set from which the best classification rules for some specific classes are to be discovered for the classification of remotely sensed images. (The classification rule set is induced in Sect. 4.5.5.). The task can help us to look for the most appropriate way to classify spatial phenomena with minimal error.

In terms of association, the knowledge discovery task aims at the identification of patterns of spatial association of certain phenomena. It particularly looks for spatial dependence of data indicating the dependence or independence of distribution of phenomena over space. Figure 1.4, for example, is a non-stationary distribution of mean minimal temperature over a span of over 40 years uncovered from a spatial database. The issue is to find an appropriate method for the discovery of such spatially autocorrelated distribution. (The solution is discussed in Sect. 5.3.3). In terms of relationship, the purpose is to search from data functional representations of causal relationships among spatial phenomena. Of special interest is the local and global issue of spatial association or relationships. It is important to detect whether or not there are significant local variations in a distribution.

With respect to process, the task of knowledge discovery is to unravel underlying processes that generate time series manifesting the dynamics of certain spatial

Table 1.1 What are the optimal classification rules for the data?

Land cover	No. of samples	Spectral band															
		Green (a_1)				Red (a_2)				NIR (a_3)				SWIR (a_4)			
		Mean	Variance	Min	Max	Mean	Variance	Min	Max	Mean	Variance	Min	Max	Mean	Variance	Min	Max
Water (u_1)	60	68.45	20.05	60	77	56.13	37.13	44	68	24.25	8.70	19	30	10.73	8.37	5	16
Mudflat (u_2)	60	79.22	1.43	77	81	77.13	3.64	74	80	56.47	4.80	53	60	12.17	13.60	5	19
Residential land (u_3)	60	85.02	36.08	74	97	90.40	171.36	65	116	84.92	129.54	63	107	81.57	172.32	56	107
Industrial land (u_4)	30	146.20	2,300.79	51	242	174.73	2,590.20	73	276	156.30	597.94	108	205	154.23	631.22	104	204
Vegetation e (u_5)	60	55.60	1.26	54	57	40.05	1.71	38	42	116.38	133.66	94	139	34.95	10.90	29	41

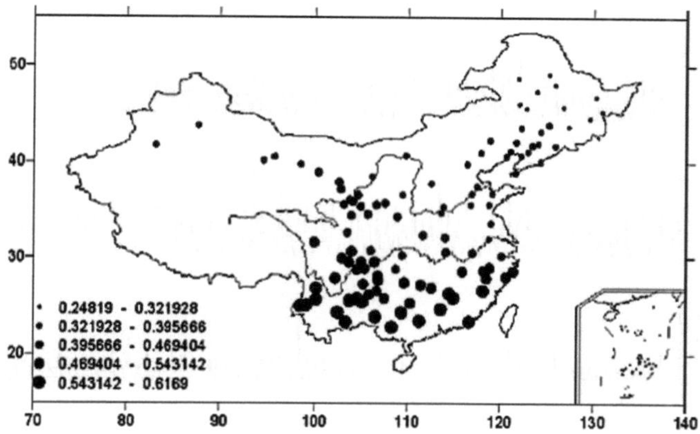

Fig. 1.4 Is the distribution of mean minimal temperature over 40 years spatially autocorrelated?

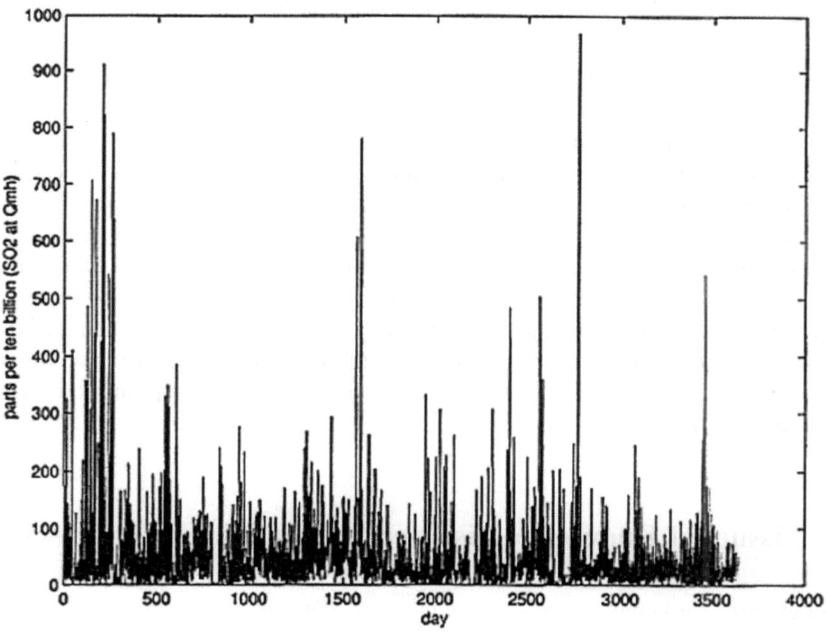

Fig. 1.5 What is the generating process of this maximum daily concentrations of SO_2?

phenomena. It looks for trends, cycles or irregular occurrences over the time horizon. The idea is to discover functional forms of such processes. Figure 1.5 is a temporal distribution of the maximum daily concentration of SO_2 recorded at a air quality monitoring station. It might be of interest to find out the form of the process that drives such a time series. (The data model that generates such intermittent distribution is found in Sect. 6.3.2). Figure 1.6 depicts the monthly runoffs recorded

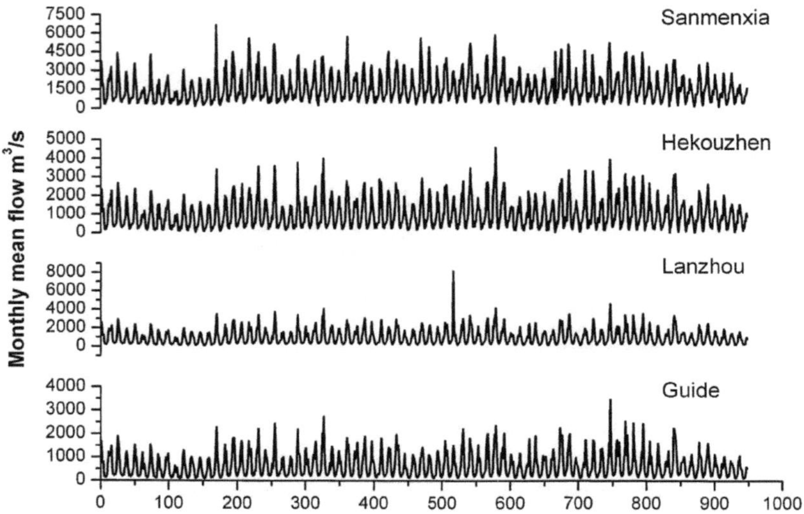

Fig. 1.6 What are the scaling behaviors of these runoffs series?

at four hydrolic stations of a river over time. Hydrologists might want to discover changes of runoffs, particularly the high and low flows, over different time scales. (The low-flow and high-flow cycles of various temporal scales are discovered in Sect. 6.2.5)

Therefore, the majority of the spatial knowledge discovery activities center on the above main tasks. Purpose wise, they are more or less in line with the tasks in general data mining and knowledge discovery. The key point is, however, to identify issues peculiar to spatial knowledge discovery under these tasks. The following section is a discussion of the issues involved in the development of appropriate knowledge discovery methodologies.

1.6 Issues of Knowledge Discovery in Spatial Data

Though tasks for knowledge discovery in spatial data are in principle similar to that of ordinary data, there are fundamental issues that need to be attended to in spatial data mining. They are highlighted as follows:

1. Scale. As pointed out above, spatial data are generally measured in different scales. How to discover knowledge in multi-scaled data is thus an important issue in spatial data mining. Of particular interest is whether phenomena under study is scale invariant, self-similar, or with long range dependence. That is, when scale varies, would there be a regularity of variation in structures, processes, or relationships? Another notion of scale is the smoothing scale. Noise or immaterial irregularities often exist in spatial data. They may deter us from

discovering spatial knowledge embedded in the noisy environment. The idea then is to smoothen out noise or irregularities by varying the scale throughout the knowledge discovery process so that spatial structures surface naturally, and genuine outliers can be automatically identified.
2. Heterogeneity. The multi-source nature of spatial information leads to the problem of knowledge discovery with mixed data types. Spatial data can be real-valued, integer-valued, categorical, geo-referenced, fuzzy, or granular. The key issue is the way knowledge should be discovered from any of these data types or a mixture of different data types.
3. Uncertainty. Spatial databases generally contain uncertainty. To put it the other way around, spatial data are generated by some process of uncertainty so that hidden knowledge might only be discovered with certain level of credibility. Uncertainty can be due to randomness, fuzziness, data granularity, or any of their combinations. Therefore, appropriate methods need to be developed to discover knowledge under different uncertain situations. Furthermore, missing values have to be appropriately handled throughout the knowledge discovery process.
4. Spatial nonstationarity. Space is the pillar of research in geographical studies. Structures, processes, and relationships always contain the spatial dimension. The key issue is whether they are global or local. In other words, are structures, processes, relationships stationary over space? If yes, then the knowledge discovered is global. Otherwise, there are significant local variations over space. The task of knowledge discovery is thus the identification and differentiation of local and global phenomena in spatial and temporal data. Spatial independence and spatial autocorrelation are also things that we need to determine in data.
5. Scalability. Computational cost is an issue in discovering knowledge in voluminous and complex spatial databases. To be effective and efficient, data mining methods have to be scalable so that computational cost will not be greatly increased when the database become larger and complex. It is particularly important when scale is explicitly considered in the knowledge discovery process.

Therefore, regardless of whether we are looking for intrinsic clusters, separation surfaces, classification rules, relationships, or time-varying processes, we need to bear in mind the above issues in order to be instrumental. Depending on the knowledge that is targeted at and the nature of the database in which knowledge is to be discovered, one may need to account for some or all of the above problems in a single task.

1.7 Methodological Background for Knowledge Discovery in Spatial Data

Due to the complexity of the problems involved, different methods may be required to accomplish different tasks of knowledge discovery in different types of spatial data. As mentioned in the preface, the majority of the methods discussed in the book come from our own research. They are formulated on top of a wide spectrum of mathematical and statistical methods that are too lengthy to discuss here as a background. It will either be too brief to be instrumental or too lengthy to be possible under the limit of space. In place of having a separate section or chapter for such background knowledge, I choose to give some introductory or complementary notes of such background knowledge as a prelude to the development of relevant methods at appropriate places of the book. This, however, may lay the burden on readers who might not be familiar with the background. On the other hand, this may lead or stimulate readers to explore different realms of mathematics, statistics, artificial intelligence, etc., that form the foundation for the methodological development of knowledge discovery in spatial data. I hope that a suitable trade off between breadth and depth has been made in here.

Throughout, the upper case bold letters, e.g. **A**, denote matrices, and the lower case bold letters, e.g. **a**, denote vectors. Other special symbols are explained in the text wherever necessary.

1.8 Organization of the Book

Chapter 2 aims at the discovery of spatial structures that appear as natural clusters in spatial data. Methods for identifying clusters and patterns of clustering are discussed with particular emphasis placed on the incorporation of scale and treatment of noise in the data mining process. Discovery of clusters under imprecision and mixture data are also scrutinized. Knowledge discovery through spatial classification is examined in Chaps. 3 and 4. Statistical and semi-statistical methods are developed in Chap. 3 for the identification of separation surfaces and extraction of classification rules from data. This discussion will takes us from the classical Bayesian and logistic-regression approaches to the support vector machines that are based on statistical learning theory. To be able to perform classification on spatial data that may not follow any probability distributions, non-statistical paradigms, that are largely algorithmic, for the discovery of separation surfaces or classification rules are discussed in Chap. 4. Developing classifiers for different data types and their mixtures is of main concern in the discussion. Scale and data imperfection are again the issues addressed in particular. For the discovery of spatial relationships, the local and global issue is examined in details in Chap. 5. Non-stationarity of spatial associations and relationships is unraveled by statistical and other novel approaches. In Chap. 6, our discussion concentrates on the discovery of the generating processes for time series data. Special attention is paid to their scaling behaviors. Issues such as self-similarity, multifractality, and long range dependence are addressed. The book is then concluded with a summary and outlook for further research in Chap. 7. Future challenges of knowledge discovery in spatial data are also outlined. Throughout our discussion in this monograph, numerical examples and real-life applications are employed to substantiate the conceptual arguments. For simplicity, all theoretical results are stated without proofs, but references are provided for readers to follow up on more indepth discussions.

Chapter 2
Discovery of Intrinsic Clustering in Spatial Data

2.1 A Brief Background About Clustering

A fundamental task in knowledge discovery is the unraveling of clusters intrinsically formed in spatial databases. These clusters can be natural groups of variables, data-points or objects that are similar to each other in terms of a concept of similarity. They render a general and high-level scrutiny of the databases that can serve as an end in itself or a means to further data mining activities. Segmentation of spatial data into homogenous or interconnected groups, identification of regions with varying levels of information granularity, detection of spatial group structures of specific characteristics, and visualization of spatial phenomena under natural groupings are typical purpose of clustering with very little or no prior knowledge about the data. Often, clustering is employed as an initial exploration of the data that might form natural structures or relationships. It usually sets the stage for further data analysis or mining of structures and processes.

Clustering has long been a main concern in statistical investigations and other data-heavy researches (Duda and Hart 1974; Jain and Dubes 1988; Everitt 1993). It is essentially an unsupervised learning, a terminology used in the field of pattern recognition and artificial intelligence, which aims at the discovery from data a class structure or classes that are unknown a priori. It has found its applications in fields such as pattern recognition, image processing, micro array data analysis, data storage, data transmission, machine learning, computer vision, remote sensing, geographical information science, and geographical research. Novel algorithms have also been developed arising from these applications. The advancement of data mining applications and the associated data sets have however posed new challenges to clustering, and it in turn intensifies the interest in clustering research. Catering for very large databases, particularly spatial databases, some new methods have also been developed over the years (Murray and Estivilli-Castro 1998; Miller and Han 2001; Li et al. 2006). To facilitate our discussion, a brief review of the clustering methods is first made in this section.

Y. Leung, *Knowledge Discovery in Spatial Data*, Advances in Spatial Science,
DOI 10.1007/978-3-642-02664-5_2, © Springer-Verlag Berlin Heidelberg 2010

These are two basic approaches to perform clustering: hierarchical clustering and partitioning clustering. With reference to some criteria for merging or splitting clusters on the basis of a similarity or dissimilarity/distance measure, hierarchical clustering algorithms produce, via an agglomerative or divisive manner, a dendrogram which is a tree showing a sequence of clustering with each being a partition of the data set. According to the structure adopted, hierarchical clustering can be further categorized into nested hierarchical clustering and non-nested hierarchical clustering. In nested hierarchical clustering, each small cluster fits itself in whole inside a larger cluster at a merging scale (or threshold) and every datum is not permitted to change cluster membership once an assignment has been made. In non-nested hierarchical clustering, a cluster obtained at small scale may divide itself into several small parts and fits these parts into different clusters at the merging scale and, therefore, each datum is permitted to change its cluster membership as the scale varies. The single-link (nearest-neighbor) algorithms (Hubert 1974; Dubes and Jain 1976), the complete-link (farthest-neighbor) algorithms (Johnson 1967; Hubert 1974), and the average-link (average-neighbor) algorithms (Ward 1963) are typical nested hierarchical clustering algorithms. The single-link method is more efficient but is sensitive to noise and tends to generate elongated clusters. Complete link and average link methods give more compact clusters but are computationally more expensive. On the other hand, the algorithms proposed in (Taven et al. 1990; Wilson and Spann 1990; Miller and Rose 1996; Blatt et al. 1997; Roberts 1997; Waldemark 1997) generate non-nested hierarchical clusterings.

Early hierarchical clustering algorithms such as AGENS (agglomerative nesting) and DIANA (divisa analysis) (Kaufman and Rousseeuw 1990) are under the curse of dimensionality and do nor scale well for large data sets because of the difficulties in deciding on the merge or split points. To handle large data sets, BIRCH (balanced iterative reducing and clustering using hierarchies) obtains clusters by compressing data into smaller sub-clusters (Zhang et al. 1996). The algorithm appears to be linearly scalable and gives reasonably good-quality clustering. Clusters are spherical in shape but they may not be natural clusters. By combining random sampling and partitioning, CURE (clustering using representatives) merges clusters via the concepts of representative objects and shirking factor (Guha et al. 1998). It is relatively robust to outliers (objects in non-dense regions) and can identify clusters with non-spherical shapes and large variance. Somewhat similar to CURE, CHAMELEON employs the concepts of interconnectivity and closeness to merge clusters (Karypis et al. 1999). The algorithm appears to be more effective than CURE in identifying clusters with arbitrary shapes and varying density. The advantage of hierarchical clustering algorithms is that it is more versatile. They give a series of clusterings along some scales. The time complexity for agglomerative algorithms is $O(n^2 \log n)$ and the space complexity is $O(n^2)$, where n is the number of objects. The disadvantage of hierarchical clustering is that it is often difficult to determine at which level the clustering gives the optimal clusters essential to an investigation.

Differing from the hierarchical approach, partitioning algorithms give only a single partition of a data set. The majority of such algorithms partition a data set into clusters through the minimization of some suitable measures such as a cost function. The K-means method, FORGY, ISODATA, WISH (MacQueen 1967; Anderberg 1973; Ball and Hall 1976; Dubes and Jain 1976), and Fuzzy ISODATA (Bezdek 1980), for examples, are essentially based on the minimization of a squared-error function. The K-means methods use the mean value of the objects in a cluster as the cluster center. Its time complexity is O(nkt), where n is the number of objects, k is the number of clusters, and t is the number of iterations. That is, for fixed k and t, the time complexity is O(n). Thus, it is essentially linear in the number of objects and this becomes its advantage. However, the K-means method is sensitive to initial partition, noise, and outliers (objects whose removal improves significantly the tightness of the clusters), and it cannot discover clusters of arbitrary shapes. By using the most centrally located object (medoid) in a cluster as the cluster center, the K-medoid is less sensitive to noise and outliers but in the expense of a higher computational cost. PAM (partitioning around medoids) is an earlier K-medoid method that uses a complex iterative procedure to replace k cluster centers (Kaufman and Rousseeuw 1990). The computational complexity in a single iteration is $O(k(n\text{-}k)^2)$. Thus, the algorithm is very costly for large data sets. To deal with large volume of data, CLARA (clustering large application) takes multiple samples of the whole data set and applies PAM to each sample to give the best clustering as the output (Kaufman and Rousseeuw 1990). The computational complexity for each iteration becomes $O(ks^2+k(n\text{-}k))$, where s is the sample size. So, the success of CLARA depends on the sample chosen. Good-quality clustering will not be achieved if the samples are biased. To better combine PAM and CLARA, CLARANS (clustering large applications based upon randomized search) is constructed to search only the subset of a data set but not confining itself to any sample at any time (Ng and Han 1994). The process is similar to searching a graph as if every one if its nodes are potential solutions. The algorithm attempts to search for a better solution by replacing the current one with a better neighbor in an iterative manner. Though CLARAN appears to be more effective than PAM and CLARA, its computational complexity is roughly $O(n^2)$. Furthermore, it assumes that all objects to be clustered are stored in the main memory. It should be noted that most of the partitioning methods cluster objects on the basis of the distance between them. It actually constitutes the expensive step of the algorithms. Since the minimization problems involved are generally NP-hard and combinatorial in nature, techniques such as simulated annealing (Kirpatrick et al. 1983), deterministic annealing (Rose et al. 1990), and EM (expectation maximization) algorithms (Celeux and Govaert 1992) are often utilized to lower the computational overhead. Moreover, most of the existing algorithms can only find clusters which are spherical in shape.

In addition to the hierarchical and partitioning approaches, there are other clustering methods such as the graph theoretic methods (Leung 1984; Karypis et al. 1999), the density-based methods (Banfield and Raftery 1993), the grid-based methods (Wang et al. 1997; Sheikholeslami et al. 1998), the neural network methods (Kohonen 1982), the fuzzy sets methods (Bezdek 1980; Leung 1984), and the evolutionary methods (Al-Sultan and Khan 1996). The graph theoretic methods often convert the clustering problem into a combinatorial optimization problem that is solved by graph algorithms or heuristic procedures. The density-based methods generally assume a mixture of distributions, with each cluster belonging to a specific distribution, for the data. Their purpose is to identify the clusters and the associated parameters. The grid-based methods impose a grid data structure on the data space in order to make density-based clustering more efficient. They however suffer from the curse of dimensionality as the number of cells in the grid increases. Neural network models generally perform clustering through a learning process. The self-organizing map, for example, can be treated as an on-line version of k-means with competitive learning. The fuzzy sets methods solve clustering problems where an object can belong to multiple clusters with different degrees of membership. The fuzzy c-means algorithm and fuzzy graph method are typical examples. The evolutionary methods are stochastic multi-point search algorithms that can be employed to solve clustering problems involving optimization. The basic principle is to devise an evolutionary strategy so that global optimal clustering can be obtained by evolving a population of clustering structures with some evolutionary operators. To achieve good quality clustering, hybrid approaches are often used in applications. In any case, all of these methods generate either the hierarchical or partitioning clustering. They can, in a sense, be fitted under either one of the frameworks.

Due to the complexity and size of the spatial databases, clustering methods should be efficient in high dimensional space (though spatial clustering is often of low dimensions), explicit in the consideration of scale, insensitive to large amount of noise, capable of identifying useful outliers, insensitive to initialization, effective in handling multiple data types, independent to a priori or domain specific knowledge (except for application specific data mining), and able to detect structures of irregular shapes. Conventional clustering algorithms often fail to fulfill these requirements. Whilst it is difficult to develop an ideal method that can meet all of these requirements, it is important to construct algorithms so that they can entertain them as much as possible. Since each method has certain assumptions about the data, it is generally impossible to determine the best clustering algorithm across all circumstances. An algorithm may be best for one problem or data set but may not perform as well for another problem or data set. A thorough understanding of the problem that needs to be solved is the first step towards the selection of the appropriate algorithm.

In the remaining part of this chapter, a detailed examination of some clustering methods that we, with the view of satisfying some of the requirements specified above, have developed to solve particular classes of clustering problems over the years. In Sect. 2.2, scale space filtering is introduced as a method of hierarchical clustering for the discovery of natural clusters in spatial data. Incorporation of scale and treatment of noise, which are essential in spatial data analysis, are explicitly dealt with in the discussion. In Sect. 2.3, fuzzy relational data clustering is described as a method of partitioning clustering. The emphasis is again on the introduction of scale and robustness against noise. Similar to scale space filtering in hierarchical clustering, unidimensional scaling examined in Sect. 2.4 attempts to provide an answer to the issues of sensitivity to initialization, presupposition of a cluster number, and difficulty of solving global optimization problem commonly encountered in partitioning clustering. To solve the problem of mixture distributions in highly noisy environment, a method of mixture decomposition clustering is introduced in Sect. 2.5 to discover natural clusters in spatial data. In Sect. 2.6, the concept of convex hull is introduced to detect clusters in exploratory spatial data analysis.

2.2 Discovery of Clustering in Space by Scale Space Filtering

In pattern recognition and image processing, human eyes seem to possess a singular aptitude to group objects and find important structures in an efficient and effective way. Coding of continuities that occur in natural images was a main research area of the Gestalt school in psychology in the early twentieth century. With respect to spatial data mining, one can argue that continuity in scale/resolution in natural images is analogous to continuity in space. Partitioning of spatial structures in scale is a fundamental property of our visual system. Thus a clustering algorithm simulating our visual processing may facilitate the discovery of natural clusters in spatial databases in general and images in particular.

Based on this view, Leung et al. (2000a) propose a scale space filtering approach to clustering. In this approach, a data set is considered as an image with each datum being a light point attached with a uniform luminous flux. As the image is blurred, each datum becomes a light blob. Throughout the blurring process, smaller blobs merge into larger ones until the whole image contains only one light blob at a low enough level of resolution. If each blob is equated to a cluster, the above blurring process will generate a hierarchical clustering with resolution being the height of a dendrogram. The blurring process is described by scale space filtering which models the blurring effect of lateral retinal interconnection through the Gaussian filtering of a digital image (Witkin 1983, 1984; Koenderink 1984; Babaud et al. 1986; Hummel and Moniot 1989). The theory in fact sheds light on the way we cluster data, regardless of whether they are digital images or raw data. It also renders a biological perspective on data clustering.

The proposed approach has several advantages. (1) The algorithms thus derived are computationally stable and insensitive to initialization. They are totally free from solving difficult global optimization problems. (2) It facilitates the formulation of new cluster validity checks and gives the final clustering a significant degree of robustness to noise in the data and change in scale. (3) It is more robust where hyper-ellipsoidal partitions may not be assumed. (4) It is suitable for the preservation of the structure and integrity of the outliers, peculiarities in space, which should not be filtered out as noise in the clustering process. (5) The patterns of clustering are highly consistent with the perception of human eyes. (6) It provides a unified generalization of the scale-related clustering algorithms derived in various fields.

Scale space theory is first described in brief in the discussion to follow. It is then extended to solve problems in data clustering.

2.2.1 *On Scale Space Theory for Hierarchical Clustering*

Consider a two-dimensional image given by a continuous mapping $p(\mathbf{x}) : \boldsymbol{R}^2 \to \boldsymbol{R}$. In scale space theory, $p(\mathbf{x})$ is embedded into a continuous family $P(\mathbf{x}, \sigma)$ of gradually smoother versions of it. The original image corresponds to the scale $\sigma = 0$ and increasing the scale should simplify the image without creating spurious structures. If there are no prior assumptions which are specific to the scene, then it is proven that one can blur the image in a unique and sensible way in which $P(\mathbf{x}, \sigma)$ is the convolution of $p(\mathbf{x})$ with the Gaussian kernel, i.e.,

$$P(\mathbf{x}, \sigma) = p(\mathbf{x}) * g(\mathbf{x}, \sigma) = \int p(\mathbf{x} - \mathbf{y}) \frac{1}{(\sigma^2 2\pi)} e^{-\frac{\|\mathbf{y}\|^2}{2\sigma^2}} d\mathbf{y}, \tag{2.1}$$

where $g(\mathbf{x}, \sigma)$ is the Gaussian function $g(\mathbf{x}, \sigma) = \frac{1}{(\sigma\sqrt{2\pi})^2} e^{-\frac{\|\mathbf{x}\|^2}{2\sigma^2}}$, σ is the scale parameter, $(\mathbf{x}, \sigma)$-plane is the scale space and $P(\mathbf{x}, \sigma)$ is the scale space image.

For each maximum $\boldsymbol{y} \in \boldsymbol{R}^2$ of $p(\mathbf{x})$, we define the corresponding light blob being a region specified as follows:

$$\boldsymbol{B}_y = \left\{ \mathbf{x}_0 \in \boldsymbol{R}^2 : \lim_{t \to \infty} \mathbf{x}(t, \mathbf{x}_0) = \mathbf{y} \right\}, \tag{2.2}$$

where $\mathbf{x}(t, \mathbf{x}_0)$ is the solution of the gradient dynamic system

$$\begin{cases} \dfrac{d\mathbf{x}}{dt} = \nabla_x p(\mathbf{x}) \\ \mathbf{x}(0) = \mathbf{x}_0. \end{cases} \tag{2.3}$$

In what follows, $\mathbf{y}$ is referred to as the *blob center* of $\boldsymbol{B}_\mathbf{y}$. All blobs in an image produce a partition of $\boldsymbol{R}^2$ with each point belonging to a unique blob except the boundary points.

Let $p(\mathbf{x}) = g(\mathbf{x}, \sigma)$, which contains only one blob for $\sigma > 0$. As $\sigma \to 0$, this blob concentrates on a light point defined as

$$\delta(\mathbf{x}) = \lim_{\sigma \to 0} g(\mathbf{x}, \sigma) = \frac{1}{(\sigma\sqrt{2\pi})^2} e^{-\frac{\|\mathbf{x}\|^2}{2\sigma^2}}. \tag{2.4}$$

Mathematically, such a function is called a δ function or a generalized function.

A light point at $\mathbf{x}_0 \in \boldsymbol{R}^2$ in an image is defined as a δ function situated at $\mathbf{x}_0$, i.e., $\delta(\mathbf{x} - \mathbf{x}_0)$, which satisfies

$$g(\mathbf{x}, \sigma) * \delta(\mathbf{x} - \mathbf{x}_0) = g(\mathbf{x} - \mathbf{x}_0, \sigma), \tag{2.5}$$

where g is the Gaussian function. From (2.5) we can see that if we blur a light point, it becomes a light blob again.

In our everyday visual experience, blurring of an image leads to the erosion of structure: small blobs always merge into large ones and new ones are never created. Therefore, the blobs obtained for images $P(\mathbf{x}, \sigma)$ at different scales form a hierarchical structure: each blob has its own survival range of scale, and large blobs are made up of small blobs. The survival range for a blob is characterized by the scale at which the blob is formed and the scale at which the blob merges with others. Each blob manifests itself purely as a simple blob within its survival range of scale.

Such blurring process can be related with the process of clustering. If $p(\mathbf{x})$ is a probability density function from which the data set is generated, then each blob is a connected region containing a relatively high density probability separated from other blobs by a boundary with relatively low density probability. Therefore, each blob is a cluster, and all blobs together produce a partition of a data space which provides a clustering for the data set with known distribution $p(\mathbf{x})$.

For a given data set $\mathbf{X} = \{\mathbf{x}_i \in \mathbf{R}^2 : i = 1, \cdots, N\}$, the empirical distribution for the data set X can be expressed as

$$\hat{p}emp(\mathbf{x}) = \frac{1}{N}\sum_{i=0}^{N} \delta(\mathbf{x} - \mathbf{x}_i). \tag{2.6}$$

The image corresponding to $\hat{p}emp(\mathbf{x})$ consists of a set of light points situated at the data set, just like a scattergram of the data set. When we blur this image, we get a family of smooth images $P(\mathbf{x}, \sigma)$ represented as follows:

$$P(\mathbf{x}, \sigma) = \frac{1}{N}\sum_{i=1}^{N} \frac{1}{(\sigma\sqrt{2\pi})^2} e^{-\frac{\|\mathbf{x}-\mathbf{x}_i\|^2}{2\sigma^2}}. \tag{2.7}$$

The family $P(\mathbf{x}, \sigma)$ can be considered as the Parzen estimation with Gaussian window function. At each given scale σ, the scale space image $P(\mathbf{x}, \sigma)$ is a smooth distribution function so that the blobs and their centers can be determined by analyzing the limit of the solution $\mathbf{x}(t, \mathbf{x}_0)$ of the following differential equation:

$$\begin{cases} \dfrac{d\mathbf{x}}{dt} = \nabla_{\mathbf{x}} P(\mathbf{x}, \sigma) = \dfrac{1}{\sigma^2 N}\displaystyle\sum_{i=1}^{N} \dfrac{(\mathbf{x}_i - \mathbf{x})}{(\sigma\sqrt{2\pi})^2} e^{-\frac{\|\mathbf{x}-\mathbf{x}_i\|^2}{2\sigma^2}} \\ \mathbf{x}(0) = \mathbf{x}_0 \end{cases}. \tag{2.8}$$

Remark 2.1. Treatment of Noise. When a distribution $p(\mathbf{x})$ is known but contains noise or is indifferentiable, we can also use scale space filtering method to erase the spurious maxima generated by the noise. In this case, the scale-space image is

$$P(\mathbf{x},\ \sigma)\ =\ p(\mathbf{x})\ *\ g(\mathbf{x},\ \sigma)\ =\ \int \frac{p(\mathbf{y})}{\left(\sigma\sqrt{2\pi}\right)^2} e^{-\frac{\|\mathbf{x}-\mathbf{y}\|^2}{2\sigma^2}} d\mathbf{y}, \tag{2.9}$$

and, the corresponding gradient dynamical system is given by

$$\begin{cases} \dfrac{d\mathbf{x}}{dt}\ =\ \nabla_{\mathbf{x}} P(\mathbf{x},\ \sigma)\ =\ \displaystyle\int \frac{p(\mathbf{y})(\mathbf{y}-\mathbf{x})}{\left(\sigma\sqrt{2\pi}\right)^2 \sigma^2} e^{-\frac{\|\mathbf{x}-\mathbf{y}\|^2}{2\sigma^2}} d\mathbf{y} \\ \mathbf{x}(0)\ =\ \mathbf{x}_0 \end{cases}. \tag{2.10}$$

When the noise in $p(\mathbf{x})$ is an independent white noise process, (2.9) provides an optimal estimate of the real distribution.

Thus, instead of clustering the data by the underlying distribution $p(\mathbf{x})$, the scale space method clusters data according to a gradient dynamic system generated by $P(\mathbf{x}, \sigma)$ for each $\sigma > 0$. By considering the data points falling into the same blob as a cluster, the blobs of $P(\mathbf{x}, \sigma)$ at a given scale produce a pattern of clustering. In this way, each data point is deterministically assigned to a cluster via the differential gradient dynamical equation in (2.8) or (2.10), and the method thus renders a hard clustering result. As we change the scale, we get a hierarchical clustering. A detailed description of the clustering procedure and the corresponding numerical implementations are given in the discussion to follow.

2.2.2 Hierarchical Clustering in Scale Space

In scale space clustering, we use the maxima of $P(\mathbf{x}, \sigma)$ with respect to x as the description primitives. Our discussion is based on the following theorem:

Theorem 2.1. *For almost all data sets, we have: (1) 0 is a regular value of* $\nabla_{\mathbf{x}} P(\mathbf{x}, \sigma)$, *(2)* as $\sigma \to 0$, *the clustering obtained for* $P(\mathbf{x}, \sigma)$ *with* $\sigma > 0$ *induces a clustering at* $\sigma = 0$ *in which each datum is a cluster and the corresponding partition is a Voronoi tessellation, i.e., each point in the scale space belongs to its nearest-neighbor datum, and (3) as* σ *increases from* $\sigma = 0$, *there are* N *maximal curves in the scale space with each of them starting from a datum of the data set.*

We know that the maxima of $P(\mathbf{x}, \sigma)$ are the points satisfying

$$\nabla_{\mathbf{x}} P(\mathbf{x},\ \sigma)\ =\ 0. \tag{2.11}$$

Therefore, 0 being a regular value of $\nabla_{\mathbf{x}} P(\mathbf{x}, \sigma)$ means that: (1) all maxima form simple curves in the scale space, and (2) we can follow these curves by numerical continuation method (Allgower and Georg 1990).

Remark 2.2. Initialization. In terms of the criterion for cluster centers (i.e., maximizing $P(\mathbf{x}, \sigma)$), there is a unique solution at small scale with N centers (each maximum is the blob center of the corresponding cluster) and hence the method is independent of initialization.

2.2.2.1 Nested Hierarchical Clustering

The construction procedure of a nested hierarchical clustering based on the scale-space image is as follows:

1. At scale $\sigma = 0$, each datum is considered as a blob center whose associated data point is itself.
2. As σ increases continuously, if the blob center of a cluster moves continuously along the maximal curve and no other blob center is siphoned into its blob, then we consider that the cluster has not changed and only its blob center moves along the maximal curve. If an existing blob center disappears at a singular scale and falls into another blob, then the two blobs merge into one blob and a new cluster is formed with the associated data points being the union of those of the original clusters.
3. Increase the scale until the whole data set becomes one single cluster. This stopping rule is well-defined because we have only one blob in the data space when scale is large enough.

A hierarchical clustering dendrogram can thus be constructed with scale as height. Such a hierarchical clustering dendrogram may be viewed as a regional tree with each of its node being a region so that data falling within the same region form a cluster. Therefore, the nested hierarchical clustering thus constructed provides a partition of the data space. In one dimensional case, such a regional tree is in fact an interval tree.

2.2.2.2 Non-Nested Hierarchical Clustering

Nested hierarchical clustering has been criticized for the fact that once a cluster is formed, its members cannot be separated subsequently. Nevertheless, we can construct a non-nested hierarchical clustering which removes such a problem. In a non-nested hierarchical clustering, we partition the data set $X = \{\mathbf{x}\}$ at a given scale by assigning a membership to each datum $\mathbf{x}_0 \in \mathbf{X}$ according to (2.2). This process is similar to the way we perceive the data set at a given distance or a given resolution. Clusters obtained at different scales are related to each other by the cluster center lines. As σ changes, a non-nested hierarchical clustering is obtained since each datum may change its membership under such a scheme. The evolution of the cluster centers in the scale-space image may be considered as a form of dendrogram. By Theorem 2.1 we know that 0 is a regular value of $\nabla_{\mathbf{x}}P(\mathbf{x}, \sigma)$ for almost all data sets. This means that cluster centers form simple curves in the scale space which can be computed through the path which follows the solutions of the equation $\nabla_{\mathbf{x}}P(\mathbf{x}, \sigma) = 0$ by the numerical continuation method.

Non-nested hierarchical clustering is more consistent with that obtained by human eyes at different distances or different resolutions, while nested hierarchical clustering has more elegant hierarchical structure.

2.2.2.3 Numerical Solution for Gradient Dynamic System

In the proposed clustering method, clusters are characterized by the maxima of $P(\mathbf{x}, \sigma)$ and the membership of each datum is determined by the gradient dynamical system in (2.8) or (2.10). Since the solution of the initial value problem of either equation cannot be found analytically, some numerical methods must be used. If the Euler difference method is used, the solution of (2.8) or (2.10), $\mathbf{x}(t, \mathbf{x}_0)$, is then approximated by the sequence $\{\mathbf{x}(n)\}$ generated in one of the following difference equations:

$$\begin{cases} \mathbf{x}(n+1) = \mathbf{x}(n) + h\nabla_{\mathbf{x}}p(\mathbf{x}(n), \sigma) = \mathbf{x}(n) + \dfrac{h}{\sigma^2 N}\displaystyle\sum_{i=1}^{N} \dfrac{(\mathbf{x}_i - \mathbf{x}(n))}{\left(\sigma\sqrt{2\pi}\right)^2} e^{-\frac{\|\mathbf{x}(n)-\mathbf{x}_i\|^2}{2\sigma^2}}, \\ \mathbf{x}(0) = \mathbf{x}_0 \end{cases} \tag{2.12}$$

or,

$$\begin{cases} \mathbf{x}(n+1) = \mathbf{x}(n) + \dfrac{h}{\sigma^2} \displaystyle\int p(\mathbf{y})\,(\mathbf{y} - \mathbf{x}(n))\, e^{-\frac{\|\mathbf{x}(n)-\mathbf{y}\|^2}{2\sigma^2}} d\mathbf{y} \\ \mathbf{x}(0) = \mathbf{x}_0, \end{cases} \tag{2.13}$$

where h is the step length.

If the magnitude of P is scaled by the logarithmic function, the corresponding gradient dynamical system of (2.8) and (2.10) becomes

$$\frac{d\mathbf{x}}{dt} = \frac{1}{\sigma^2} \frac{\sum_{i=1}^{N} (\mathbf{x}_i - \mathbf{x})\, e^{-\frac{\|\mathbf{x}-\mathbf{x}_i\|^2}{2\sigma^2}}}{\sum_{i=1}^{N} e^{-\frac{\|\mathbf{x}-\mathbf{x}_i\|^2}{2\sigma^2}}}, \tag{2.14}$$

and,

$$\frac{d\mathbf{x}}{dt} = \frac{1}{\sigma^2} \frac{\int p(\mathbf{y})(\mathbf{y} - \mathbf{x})\, e^{-\frac{\|\mathbf{x}-\mathbf{y}\|^2}{2\sigma^2}} d\mathbf{y}}{\int p(\mathbf{y})\, e^{-\frac{\|\mathbf{x}-\mathbf{y}\|^2}{2\sigma^2}} d\mathbf{y}}, \tag{2.15}$$

and the discrete approximations to (2.12) and (2.13) then become

$$\mathbf{x}(n+1) = \mathbf{x}(n) + \frac{h}{\sigma^2} \frac{\sum_{i=1}^{N} (\mathbf{x}_i - \mathbf{x}(n))\, e^{-\frac{\|\mathbf{x}(n)-\mathbf{x}_i\|^2}{2\sigma^2}}}{\sum_{i=1}^{N} e^{-\frac{\|\mathbf{x}(n)-\mathbf{x}_i\|^2}{2\sigma^2}}}, \tag{2.16}$$

or,

$$\mathbf{x}(n+1) = \mathbf{x}(n) + \frac{h}{\sigma^2} \frac{\int p(\mathbf{y})(\mathbf{y} - \mathbf{x}(n))\, e^{-\frac{\|\mathbf{x}(n)-\mathbf{y}\|^2}{2\sigma^2}} d\mathbf{y}}{\int p(\mathbf{y})\, e^{-\frac{\|\mathbf{x}(n)-\mathbf{y}\|^2}{2\sigma^2}} d\mathbf{y}}. \tag{2.17}$$

Setting the step length $h = \sigma^2$ in (2.17), we get

$$\mathbf{x}(n+1) = \frac{\sum_{i=1}^{N} \mathbf{x}_i\, e^{-\frac{\|\mathbf{x}(n)-\mathbf{x}_i\|^2}{2\sigma^2}}}{\sum_{i=1}^{N} e^{-\frac{\|\mathbf{x}(n)-\mathbf{x}_i\|^2}{2\sigma^2}}}. \tag{2.18}$$

Such iteration can be interpreted as iterative local centroid estimation (Wilson and Spann 1990; Linderberg 1990).

When the size of the data set is large or the data are given in a serial form, we can use the stochastic gradient descent algorithm to search the blob center and determine the memberships of the data. The purpose is to find the maximum of $P(\mathbf{x}, \sigma)$ which can be represented as

$$P(\mathbf{x}, \sigma) = E\left[e^{-\frac{\|\mathbf{x}(n)-\mathbf{x}_i\|^2}{2\sigma^2}}\right], \tag{2.19}$$

where $E[\cdot]$ is the expectation of the density of the data set $\mathbf{y}$. By the theory of stochastic gradient descent algorithm, the blob center of a datum $\mathbf{x}_0$ can be obtained by the following iteration initialized at $\mathbf{x}_0$:

$$\mathbf{x}(n+1) = \mathbf{x}(n) + h^{(n)}\left(\mathbf{x}^{(n)} - \mathbf{x}(n)\right) e^{-\frac{\left\|\mathbf{x}(n)-\mathbf{x}^{(n)}\right\|^2}{2\sigma^2}}, \tag{2.20}$$

where $\mathbf{x}^{(n)}$ is the nth randomly chosen member of $\boldsymbol{X}$ or the nth datum generated according to the distribution $p(\mathbf{x})$ to be presented to the algorithm, and $h^{(n)}$ is the adaptive step length chosen as:

$$h^{(n)} = \frac{1}{1+n}. \tag{2.21}$$

The datum $\mathbf{x}_0$ is then associated with a center $\mathbf{x}^*$ if $\mathbf{x}(n)$ initialized from $\mathbf{x}_0$ converges to $\mathbf{x}^*$. In practice, $\mathbf{x}(n+1)$ is defined as a blob center if $\|\mathbf{x}(n+1) - \mathbf{x}(n)\| < \varepsilon$ or $\|\nabla_{\mathbf{x}} p(\mathbf{x}(n+1))\| < \varepsilon$, where ε is a small positive value which may vary with problems. If two centers $\mathbf{x}_1$ and $\mathbf{x}_2$ satisfy the condition $\|\mathbf{x}_1 - \mathbf{x}_2\| < \varepsilon$, then they are considered as one blob center.

To implement the proposed hierarchical clustering, we can use the path-following algorithm to trace the blob centers along the maximal curves. When a singular scale at which a blob center disappears is encountered, the new blob center is obtained by solving (2.8) or (2.10) with initial value $\mathbf{x}_0 = \mathbf{x}^*$. The new blob center is then followed by the path-following algorithm again. Alternatively, we can use the discretization of scale and an iterative scheme which works as follows:

2.2.2.4 Nested Hierarchical Algorithm

Step 1. Given a sequence of scales $\sigma_0, \sigma_1, \cdots$ with $\sigma_0 = 0$. At $\sigma_0 = 0$ each datum is a cluster and its blob center is itself. Let $i = 1$.

Step 2. Find the new blob center at σ_i for each blob center obtained at scale σ_{i-1} by one of the iterative schemes in (2.12) to (2.18). Merge the clusters whose blob centers arrive at the same blob center into a new cluster.

Step 3. If there are more than two clusters, let $i := i+1$, go to Step 2.

Step 4. Stop when there is only one cluster.

2.2.2.5 Non-Nested Hierarchical Algorithm

Step 1. Given a sequence of scales $\sigma_0, \sigma_1, \cdots$ with $\sigma_0 = 0$. At $\sigma_0 = 0$ each datum is a cluster and its blob center is itself. Let $\mathrm{i} = 1$.

Step 2. Cluster the data at σ_i. Find the new blob center at σ_i for each blob center obtained at scale σ_{i-1} by one of the iterative schemes in (2.12) to (2.18). If two new blob centers arrive at the same point, then the old clusters disappear and a new cluster is formed.

Step 3. If there are more than two clusters, let $\mathrm{i} := \mathrm{i} + 1$, go to Step 2.

Step 4. Stop when there is only one cluster.

Remark 2.3. Computation for Large Data Sets. When the size of the data set is very large, we can substitute each datum in the iterative scheme in (2.12)–(2.18) with its blob center and σ_i with $\sigma_i - \sigma_{i-1}$ in step 2 to reduce the computational cost of the above algorithm. In this case, (2.18) becomes

$$\mathbf{x}(n+1) = \frac{\sum_{j=1}^{N_i} k_j \mathbf{p}_j e^{-\frac{\left\|\mathbf{x}(n)-\mathbf{p}_j\right\|^2}{2\sigma^2}}}{\sum_{j=1}^{N_i} k_j e^{-\frac{\left\|\mathbf{x}(n)-\mathbf{p}_j\right\|^2}{2\sigma^2}}}, \tag{2.22}$$

where $\mathbf{p}_j$ is blob center j obtained at scale σ_i, N_i is the number of $\mathbf{p}_j$, k_j is the number of data points in the blob whose center is $\mathbf{p}_j$ and $\sigma = \sigma_i - \sigma_{i-1}$. Since N_i is usually much smaller than N, so the computational cost can be reduced significantly.
In practical applications, σ_i should increase according to

$$\sigma_i - \sigma_{i-1} = k\sigma_{i-1}. \tag{2.23}$$

This comes from the requirement of accuracy and stability of the representation, as proved in Koenderink (1984). In psychophysics, Weber's law says that the minimal size of the difference ΔI in stimulus intensity which can be sensed is related to the magnitude of standard stimulus intensity I by $\Delta I = kI$, where k is a constant called Weber fraction. Therefore, psychophysical experimental results may be used to propose a low bound for k in the algorithms since we cannot sense the difference between two images $p(\mathbf{x}, \sigma_{i-1})$ and $p(\mathbf{x}, \sigma_i)$ when k is less than its Weber fraction. For instance, $k = 0.029$ in (2.23) is enough in one dimensional applications because scale σ is the window length in the scale space and the Weber fraction for line length is 0.029 (Coren et al. 1994).

2.2.3 Cluster Validity Check

Cluster validity is a vexing but very important problem in cluster analysis because each clustering algorithm always finds clusters, no matter they are genuine or not, even if the data set is entirely random. While many clustering algorithms can be applied to a given problem, there is in general no guarantee that two different algorithms will produce consistent answers. They particularly do not provide answers to the following questions: (1) Do the data exhibit a predisposition to cluster? (2) How many clusters are present in the data? (3) Are the clusters real or merely artifacts of the algorithms? (4) Which partition or which individual cluster is valid? Therefore, cluster validity check should be an essential requirement of any algorithm. Besides some procedures in statistics (Theodoridis and Koutroubas 1999), one widely used strategy is to employ visual processing to examine distributions on each separate variable by ways such as histograms, nonparametric density estimates or plots of each pair of variables using scattergram. However, there is no theoretical basis for such visualization. Another strategy is to produce clustering algorithms based directly on the laws of psychology of form perception. Zahn (1971) has proposed a clustering algorithm based on the laws of Gestalt psychology of form perception. The algorithm is a graphical one which is based on the minimal spanning tree and attempts to mechanize the Gestalt law of *proximity* which says that perceptual organization favors groupings representing smaller inter-point distance. Zahn's algorithm has a strong influence on cluster analysis. Many algorithms have been developed on the basis of similar ideas. However, Zahn's algorithm is derived from Gestalt psychology laws in a heuristic way since Gestalt laws cannot be represented in an accurate computational model. This inaccuracy makes it difficult to establish a formal and efficient cluster validity check.

In scale space filtering, the questions are tackled on the basis of human visual experience: the real cluster should be perceivable over a wide range of scales. Thus, the notion of *lifetime* of a cluster is employed as its validity criterion: A cluster with longer lifetime is more valid than a cluster with shorter lifetime.

In Leung et al. (2000a), the lifetime of a cluster is used to test the "goodness" of a cluster, and the lifetime of a clustering is used to determine the number of clusters in a specific pattern of clustering.

Definition 2.1. *Lifetime of a cluster is defined as the range of logarithmic scales over which the cluster survives, i.e., the logarithmic difference between the point when the cluster is formed and the point when the cluster is absorbed into or merged with other clusters.*

Each pattern of clustering in a non-nested hierarchical clustering only consists of clusters which are formed at the same scale. A pattern of clustering in a nested hierarchical clustering, however, is a partition of the data set $\mathbf{X}$ which may consist of clusters obtained at the same scale or at different scales. In what follows, we define the lifetime for these two kinds of clustering's.

Definition 2.2. *Let $\pi(\sigma)$ be the number of clusters in a clustering achieved at a given scale σ. Suppose $\mathbf{C}_\sigma$ is a clustering obtained at σ with $\pi(\sigma) = m$. The σ-lifetime of $\mathbf{C}_\sigma$ is defined as the supremum of the logarithmic difference between two scales within which $\pi(\sigma) = m$.*

Definition 2.3. *Suppose a clustering $\boldsymbol{C}$ in a hierarchical clustering contains K clusters $\{\mathbf{C}_1, \cdots, \mathbf{C}_K\}$. Denote the number of data points in $\mathbf{C}_i$ by $|\mathbf{C}_i|$ and the lifetime of $\mathbf{C}_i$ by l_i. Then the mean lifetime of all clusters in clustering $\boldsymbol{C}$ is defined as*

$$\sum_{i=1}^{K} l_i \frac{|\mathbf{C}_i|}{|X|} \,. \tag{2.24}$$

The lifetime of clustering $\boldsymbol{C}$ is the mean lifetime of all of its clusters. If a cluster $\mathbf{C}_i$ is further divided into K_i sub-clusters $\{\mathbf{C}_{i_1}, \cdots, \mathbf{C}_{i_K}\}$, and the lifetime of $\mathbf{C}_{i_j}$ is denoted by l_{i_j}, then the mean lifetime of all its sub-clusters is defined as

$$\sum_{j=1}^{K_i} l_{i_j} \frac{|\mathbf{C}_{i_j}|}{|\mathbf{C}_i|} . \tag{2.25}$$

The use of logarithmic scale in the above definitions is based on the experimental tests in Roberts (1997) which show that (σ) decays with scale σ according to

$$\pi(\sigma) = ce^{-\beta\sigma} \tag{2.26}$$

if the data are uniformly distributed, where β is a positive constant related to the dimensionality of the data space. If a data structure exists, then $\pi(\sigma)$ is a constant over a range of scales. So the stability of $\pi(\sigma)$ can be used as a criterion to test whether the data tend to cluster, i.e., have a structure. However, β is unknown and $\pi(\sigma)$ is only allowed to take integers. From (2.26) we can see that even for a uniformly distributed data set, if β is small, $\pi(\sigma)$ will then be a constant over a wide range of scales for a small $\pi(\sigma)$. If β is large, then $\pi(\sigma)$ will also be a constant over a wide range of scales for a large σ. This makes it difficult to find the structure in the $\pi(\sigma)$ plot. However, if the data are uniformly distributed and we rescale σ by a new parameter k such that the number of clusters in the clustering obtained at the new parameter k, denoted by $\pi(k)$, decays linearly with respect to k, i.e.,

$$\pi(k) = \pi(0) - k, \tag{2.27}$$

we can easily find the structure in the plot of $\pi(k)$. The reason is that it is much simpler to test whether $\pi(k)$ decays linearly with respect to k than to test whether $\pi(k)$ decays according to (2.26) in which an unknown parameter β is involved.

Under the assumption that $\pi(k)$ decays linearly with respect to k, the relationship of k and σ can be derived as follows:

Suppose σ relates to k through a function $\sigma(k)$. Then we have

$$\pi(k) = \pi(\sigma(k)) = ce^{-\beta\sigma(k)} . \tag{2.28}$$

Under the assumption that $\pi(k)$ decays linearly with respect to k, see (2.27), we have

$$\frac{d\pi(k)}{dk} = -1 . \tag{2.29}$$

From (2.26), we obtain

$$\frac{d\pi(k)}{dk} = -c\beta e^{-\beta\sigma\frac{d\sigma}{dk}} . \tag{2.30}$$

Equations (2.29) and (2.30) imply that the new parameter k should satisfy

$$\frac{d\sigma}{dk} = \frac{1}{c\beta} e^{\beta\sigma} . \tag{2.31}$$

Solving this differential equation, we get

$$k = c\left(1 - e^{-\beta\sigma}\right) . \tag{2.32}$$

Such a scaling is an ideal one, but it contains a parameter β which is usually unknown. In practice, we take the approximation $\beta/e^{\beta\sigma} = \beta/(1 + \beta\sigma + \cdots) \approx 1/\sigma$ in (2.30) which does not contain the unknown parameter β, and this leads to the logarithmic scale

$$k = c \log \frac{\sigma}{\varepsilon} , \tag{2.33}$$

where ε is a positive constant.

The term k defined in (2.33) is called the *sensation intensity* under the Fechner's Law (Coren et al. 1994). In terms of the new parameter k, lifetime should be measured by the logarithmic scale of σ.

Once a partition has been established to be valid, a natural question that follows is "How good are the individual clusters?" The first measure of "goodness" of a cluster is naturally its lifetime: a good cluster should have a long lifetime. Associated measures are compactness and isolation of a cluster. Intuitively, a cluster is good if the distances between the data inside the cluster are small and those outside are large. *Compactness* and *isolation* of a cluster are two measures suggested for the identification of good clusters (Leung et al. 2000a). For a cluster $\mathbf{C}_i$, the measures are defined as follows:

$$isolation = \frac{\Sigma_{\mathbf{x}\in\mathbf{C}_i} e^{-\|\mathbf{x}-\mathbf{p}_i\|^2/2\sigma^2}}{\Sigma_{\mathbf{x}} e^{-\|\mathbf{x}-\mathbf{p}_i\|^2/2\sigma^2}} , \tag{2.34}$$

$$compactness = \frac{\Sigma_{\mathbf{x}\in\mathbf{C}_i} e^{-\|\mathbf{x}-\mathbf{p}_i\|^2/2\sigma^2}}{\Sigma_{\mathbf{x}\in\mathbf{C}_i} \Sigma_j e^{-\|\mathbf{x}-\mathbf{p}_j\|^2/2\sigma^2}} , \tag{2.35}$$

where $\mathbf{p}_i$ is the blob center of cluster $\mathbf{C}_i$. For a good cluster, the compactness and isolation are close to one. This measure is dependent on the scale and will be used to find the optimal scale at which the clustering achieved by non-nested hierarchical clustering is good.

Therefore, *lifetime*, *compactness* and *isolation* are three measures that can be employed to check the validity of a good cluster. A genuine cluster should be compact, isolated and have a relatively long life time. A natural clustering should be the one which contains a certain number of good clusters with high overall isolation and compactness, and stays relatively long in the scale space.

Remark 2.4. A data set invariably contains noisy data points which may be genuine outliers that carry crucial information. How to detect observations which appear to be markedly different from the rest of a data set is an important problem in many diagnostic or monitoring systems (Hawkins 1980; Barnett and Lewis 1994). Successful detection of spatial outliers is important in the discovery of peculiar patterns with significant spatial implications. In scale space clustering, we can use the number of data points in a cluster $\mathbf{C}_i$ and the lifetime of $\mathbf{C}_i$ to decide whether or not $\mathbf{C}_i$ is a genuine outlier. If $\mathbf{C}_i$ contains a small number of data and survives a long time, then we say that $\mathbf{C}_i$ is an outlier, otherwise, $\mathbf{C}_i$ is a normal cluster. Therefore, we can use the measure

$$outlierness_i = \frac{life\ time\ of\ \mathbf{C}_i}{number\ of\ data\ in\ \mathbf{C}_i} \tag{2.36}$$

to test for outliers. It means that an outlier is a well isolated group with small number of data in a large scale range. Since the method treats the data point as light point, each outlier should be a stable cluster in quite a large scale range. That is to say, an outlier generally exhibits a high degree of "outlierness." A threshold may be used to exclude outliers that are non-essential in data clustering.

2.2.4 Clustering Selection Rules

Hierarchical clustering provides us with a sequence of clustering's. Several selection rules are proposed in Leung et al. (2000a) to choose a good clustering from the sequence of clustering's in the hierarchy.

The first rule is based on the *σ-lifetime* of a clustering and it tries to find a scale at which the clustering achieved has long lifetime and high degree of compactness or isolation.

2.2.4.1 Rule I

1. Find the integer m such that the clustering obtained at σ with $\pi(\sigma) = m$ has the longest *σ-lifetime*.
2. (a) In nested hierarchical clustering, clusterings which satisfy $\pi(\sigma) = m$ are identical to each other, so we can get a unique clustering when m is obtained.
   (b) In non-nested hierarchical clustering, clusterings obtained at two scales σ_1 and σ_2 are usually different from each other even though the result $\pi(\sigma_1) = \pi(\sigma_2) = m$ is obtained. Therefore, we still need a method to find the right scale at which a good clustering can be achieved when m is fixed.

Define respectively the overall isolation and overall compactness for a clustering achieved at σ with $\pi(\sigma) = m$ as follows:

$$F^{(i)}(\sigma) = \left(\sum_{i}^{m} i^{th}\ isolation \ - \ m \right) \tag{2.37}$$

$$F^{(c)}(\sigma) = \left(\sum_{i}^{m} i^{th}\ compactness \ - \ m \right) \tag{2.38}$$

where the *i-th isolation* and *i-th compactness* are the isolation and compactness of the *i-th* cluster respectively. By maximizing $F^{(t)}$ or $F^{(c)}$ under the condition that $\pi(\sigma) = m$, we can get a σ at which a partition with maximal isolation or maximal compactness is achieved. In the general case, $\pi(\sigma) = m$ is held in an interval $[\sigma_1,\ \sigma_2]$. Therefore we can use the gradient descent method to optimize $F^{(t)}$ or $F^{(c)}$. The gradient is given by

$$\frac{df}{d\sigma} = \sum_{i=1}^{m} \nabla_{\mathbf{x}_i} F \frac{d\mathbf{x}_i}{d\sigma} \tag{2.39}$$

where F is $F^{(t)}$ or $F^{(c)}$, and $\mathbf{x}_i$ is the center of the *i-th* cluster. Knowing that each cluster center $\mathbf{x}$ is a maximal point of $p(\mathbf{x}, \sigma)$, the term $d\mathbf{x}_i/d\sigma$ can be obtained as

$$\frac{d\mathbf{x}}{d\sigma} = -[\nabla_{\mathbf{xx}} P(\mathbf{x},\ \sigma)]^{-1}\, \nabla_{\mathbf{x}\sigma} P(\mathbf{x},\ \sigma)\ . \tag{2.40}$$

Finally, we obtain a σ which is a minimal point of $F^{(t)}$ or $F^{(c)}$ and we consider that the clustering obtained at this scale is good.

The second selection rule is constructed to search for a clustering with the longest lifetime in nested hierarchical clustering. Let $\boldsymbol{\Omega}$ be the set of all clustering's in a nested hierarchical clustering. For each clustering $\mathbf{p}_i \in \Omega$, its lifetime is denoted by $l_{\mathbf{P}_i}$. The aim of the second rule is to find a clustering $\mathbf{P}_j$ such that

$$l_{\mathbf{P}_j} = \max_{\mathbf{P}_i \in \Omega} l_{\mathbf{P}_i}. \tag{2.41}$$

Since such problem is usually difficult to solve, several heuristic procedures may be used to obtain a solution. Leung et al (2000a) propose two greedy methods Rule II.1 (depth-first search) and Rule II.2 (breadth-first search) for such purpose.

The first procedure is similar to Witkin's "top-level description." It works as follows:

2.2.4.2 Rule II.1 (Maximization with Depth-First Search)

1. Initially, let $\boldsymbol{P}$ be a clustering with the whole data set as a cluster. Assign 0 as the lifetime of this unique cluster.
2. Find a cluster $\boldsymbol{C}_k$ in $\boldsymbol{P}$ whose lifetime is shorter than the mean lifetime of its children, and delete the cluster $\boldsymbol{C}_k$ from $\boldsymbol{P}$ and add all children clusters of $\boldsymbol{C}_k$ into $\boldsymbol{P}$, i.e., the new clustering $\boldsymbol{P}$ consists of the children clusters of $\boldsymbol{C}_k$ and other clusters except $\boldsymbol{C}_k$. Repeat this process until the lifetime of each cluster in $\boldsymbol{P}$ is longer than the mean lifetime of its own children.

Clustering obtained by this procedure is usually less complex, i.e., with small number of clusters.

The second procedure can also be considered as a 'longest-lifetime-first' procedure. It works as follows:

2.2.4.3 Rule II.2 (Maximization with Breadth-First Search)

1. Initialize $\boldsymbol{U}$ to be an empty set. Let $\mathbf{C} = \{\mathbf{C}_1, \mathbf{C}_2 \cdots, \mathbf{C}_K\}$ be the set of all clusters in the hierarchical clustering.
2. Pick the element $\boldsymbol{C}_k$ in $\boldsymbol{C}$ with the longest lifetime and put it into $\boldsymbol{U}$. Remove $\boldsymbol{C}_k$ and the clusters in $\boldsymbol{C}$ that are either contained in or contain $\boldsymbol{C}_k$ until $\boldsymbol{C}$ is empty. The number of elements in $\boldsymbol{U}$ is the number of clusters and $\boldsymbol{U}$ is the corresponding clustering.

2.2.5 *Some Numerical Examples*

The first example involves a two-dimensional data set with 250 data points generated by a five cluster Gaussian mixture model with different shapes. Figure. 2.1a is the data plot and Fig. 2.1b is the $\pi(k)$ plot. From Fig. 2.1b, we can observe that $\pi(k)$ has an approximately linear decrease with scale k between $0 < k < 60$, where $k = c \log(\sigma/\varepsilon)$ with $\varepsilon = 0.1$ and $c = 1/\log(1.05)$. For $k > 60$, the hidden data structure appears and $\pi(k) = 5$ has the longest σ-lifetime. Figure. 2.1c, d are respectively the overall isolation and overall compactness plots. $F^{(t)}$ and $F^{(c)}$ achieve their maxima at about $k = 67(\sigma = 2.628)$. At this scale, the clustering obtained by the non-nested hierarchical clustering algorithm is consistent with that obtained by the nested-hierarchical clustering algorithm (the corresponding clustering is shown in Fig. 2.2b). Figure 2.2a is the evolutionary plot of the blob centers obtained by the nested hierarchical algorithm. Figure 2.2b is the data partition obtained at different scales. It can be observed that the results obtained via the concept of σ-lifetime, isolation and compactness are consistent.

This is actually the solution for the cluster discovery problem (Fig. 1.1) raised in Chapter 1, section 1.5.

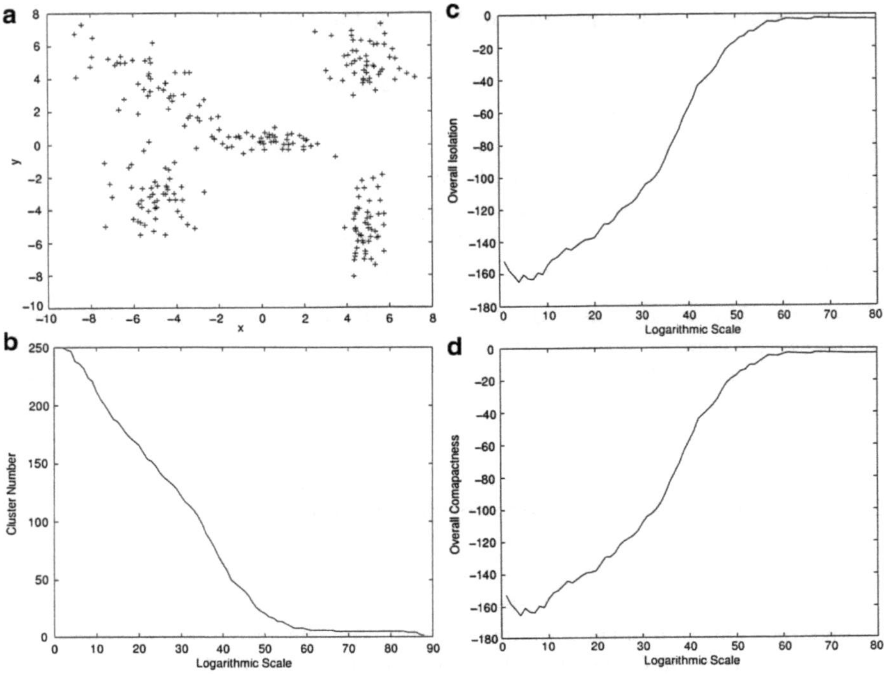

Fig. 2.1 A numerical example of scale space clustering (**a**) Plot of the data set. (**b**) Logarithmic-scale plot of the cluster number $\pi(k)$. (**c**) Logarithmic-scale plot of overall isolation. (d) Logarithmic-scale plot of overall compactness

For the sake of visualization, Fig. 2.3 depicts another two-dimensional data set with a hidden structure of $\pi(k) = 5$. At each scale we can generate the pseudo-color plot, the mesh plot and the contour plot of the scale space image. For example, Fig. 2.4a–c are respectively the pseudo-color plot, the mesh plot and the contour plot for $\sigma = 0.163$, and Fig. 2.5a–c are that for $\sigma = 1.868$. Apparently, the five clusters naturally settle in and form the natural clustering of the data at the appropriate scale.

2.2.6 Discovering Land Covers in Remotely Sensed Images

Leung et al. (2000a) apply the scale-space clustering algorithm to a real-life Landsat TM image to discover natural clusters (land covers) in multidimensional data.

It should be noted that if the data set $\mathbf{X} = \{\mathbf{x}_i \in \boldsymbol{R}^n : i = 1, \cdots, N\}$ is in the space $\boldsymbol{R}^n$, then its empirical distribution is expressed as $\hat{p}emp(\mathbf{x}) = \frac{1}{N}\sum_{i=0}^{N} \delta(\mathbf{x} - \mathbf{x}_i)$. The scale space image of $\hat{p}emp(\mathbf{x})$, $P(\mathbf{x}, \sigma)$, can be written as $P_{\mathbf{x}}(\mathbf{x}, \sigma) = \frac{1}{N}\sum_{i=1}^{k}\left(\frac{1}{\sigma\sqrt{2\pi}}\right)^{N} e^{-\frac{\|\mathbf{x}-\mathbf{x}_i\|^2}{2\sigma^2}}$, which is the convolution of $\hat{p}emp(\mathbf{x})$ with

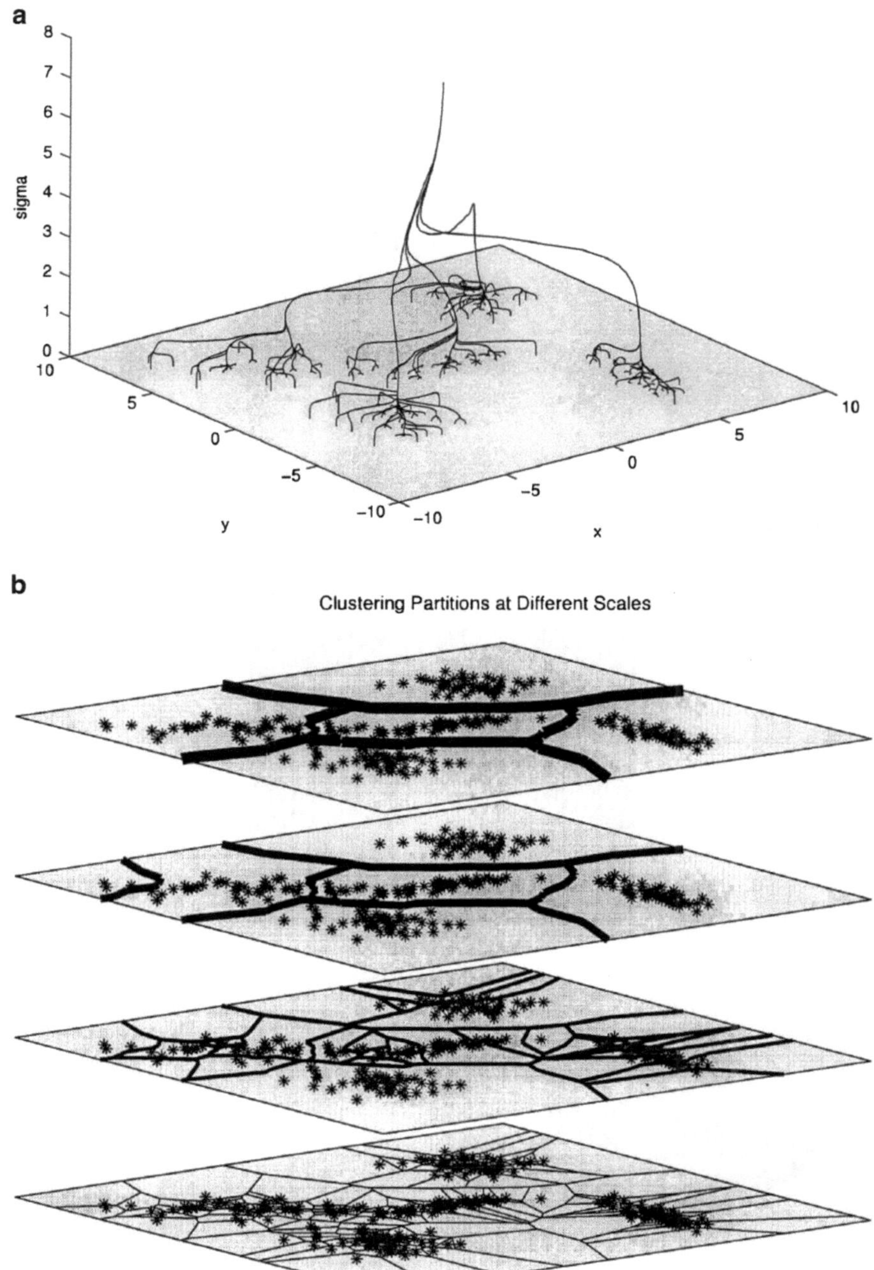

Fig. 2.2 Evolution plot of the scale space clustering in Fig. 2.1 (**a**) Evolutionary tree of cluster centers obtained by the algorithm. (**b**) The partition of the data space obtained by the nested hierarchical clustering algorithm at scales $\sigma_0 = 0$, $\sigma_1 = 0.99$, $\sigma_2 = 2.38$ and $\sigma_3 = 2.628$ (from *bottom* to *top*)

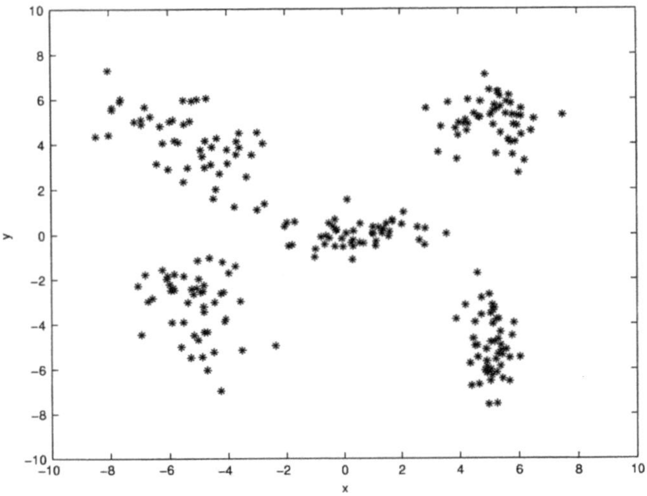

Fig. 2.3 Scatter plot of a two-dimensional data set

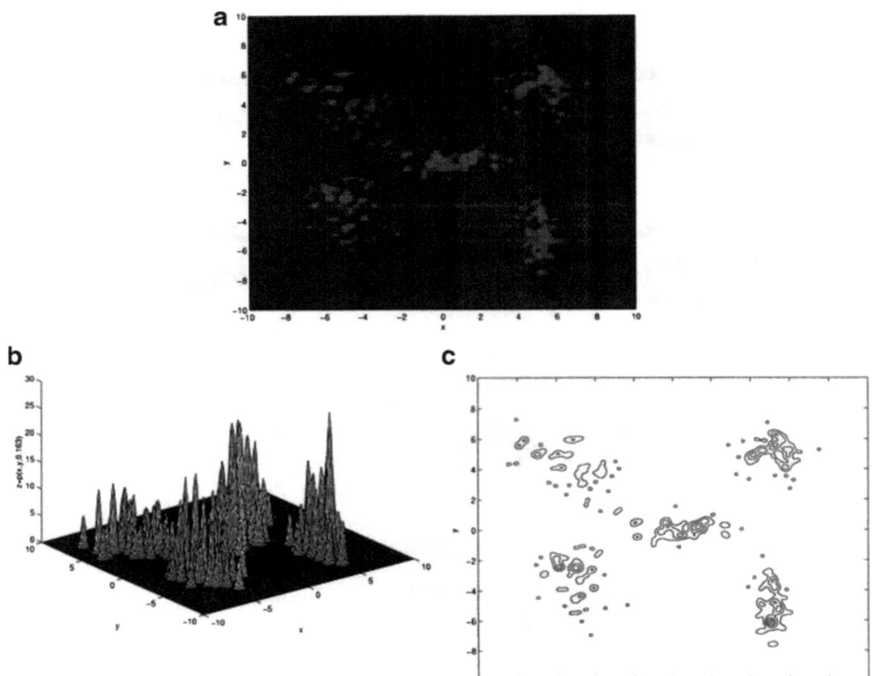

Fig. 2.4 Visualization of the scale-space image obtained from data set in Fig. 2.3 at $\sigma = 0.163$ (**a**) Scale-space image pseudo-color plot for $\sigma = 0.163$. (**b**) Mesh plot of scale-space image for $\sigma = 0.163$. (**c**) Scale-space image contour plot for $\sigma = 0.163$

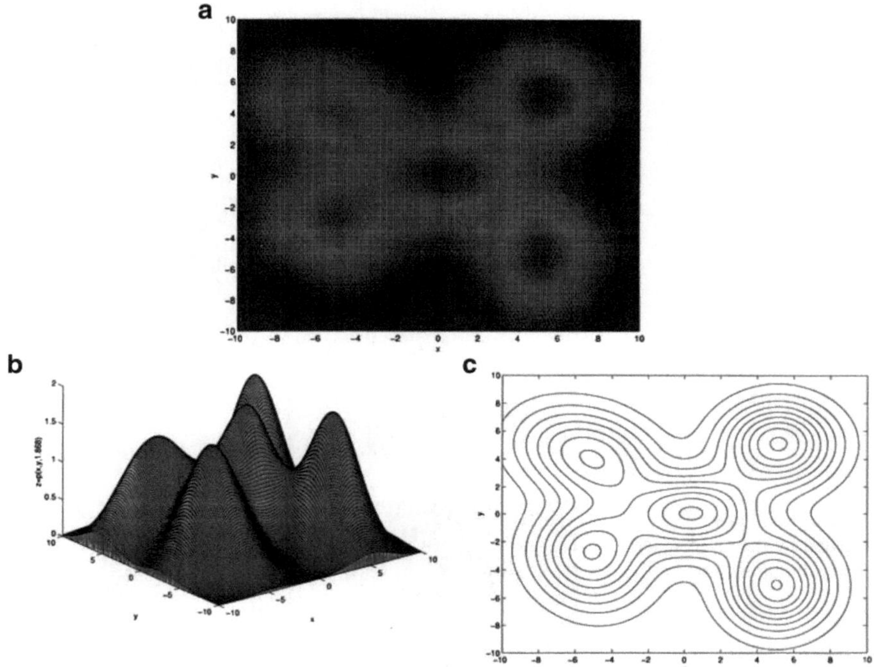

Fig. 2.5 Visualization of the scale-space image obtained from data set in Fig. 2.3 at σ=1.868 (**a**) Scale-space image pseudo-color plot for $\sigma = 1.868$. (**b**) Mesh plot of scale-space image for $\sigma = 1.868$. (**c**) Scale-space image contour plot for $\sigma = 1.868$

the Gaussian kernel $G(\mathbf{x}, \sigma) = \left(\frac{1}{\sigma\sqrt{2\pi}}\right)^N e^{-\frac{\|\mathbf{x}\|^2}{2\sigma^2}}$. Each maximum of $P(\mathbf{x}, \sigma)$ is considered as a cluster center and a point in X is assigned to a cluster via the gradient dynamic equation for $P(\mathbf{x}, \sigma)$. Since Theorem 2.1 holds in any dimension, then the scale space filtering algorithms can straightforwardly be extended to n-imensional data with slight adaptation.

The study area is Yuen Long, located in the northwest of Hong Kong, corresponding to an area of $230KM^2$ on the Hong Kong topographic maps with geographical coordinates (113°58′*E*–114°07′*E* to 22°21′*N*–22°31′*N*). The main land covers include forest, grass, rock, water, build-up area, trees, marshland, shoals, etc. They are distributed in a complex way. The Landsat TM10 image used is from 3 March 1996 with fine weather. The image size is 455 × 568 pixels. In the experiment, six bands, TM1, 2, 3, 4, 5 and 7, are utilized, i.e., the clustering is done in six dimensions.

The experiment first clusters a data set consisting of 800 pixels randomly sampled from the image and then assigns each pixel to its nearest cluster center. Figure 2.6 is the Landsat image of Yuen Long, Hong Kong, and Fig. 2.7 shows the 15 cluster solution obtained by applying the scale space clustering algorithm to the image. The 15 clusters are obtained from Rule II.2 and the outliers are deleted according to their outlierness defined in (2.36). Compared with the ground truth, the

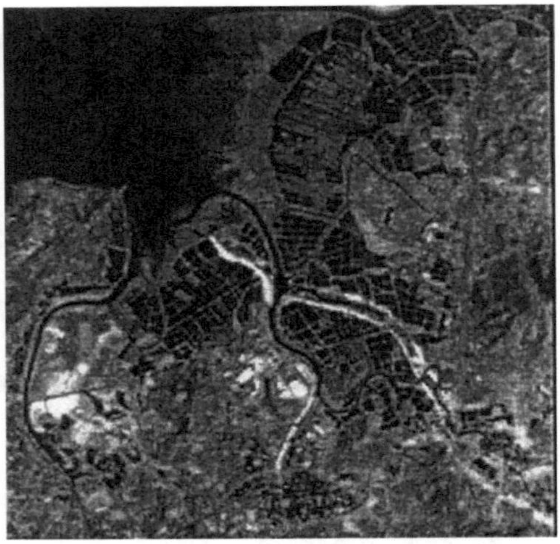

Fig. 2.6 Landsat Image of Yuen Long, Hong Kong

Fig. 2.7 Land covers revealed by the scale space clustering algorithm

scale space clustering is capable of finding the fine land covers. For example, three classes of water bodies corresponding to deep sea water, shallow seawater and freshwater of the study area have respectively been identified, while they cannot be distinguished by ISODATA method. In the experiments, it is discovered that 150–1,000 sample points are usually large enough to find the land covers contained in the image.

2.2.7 *Mining of Seismic Belts in Vector-Based Databases*

In seismology, the identification of active faults is crucial to the understanding of the tectonic pattern and the assessment of seismic risk of a specific region. In areas of strong seismic activity, major seismic faults are usually tracked by the epicenters of the seismic events. Seismic belts, by definition, are belts with dense and zonal distribution of earthquakes controlled by the tectonic belts or the geotectonic aberrance.

Seismic belts are often linear in shape because faults usually exist as wide linear features (Amorese et al. 1999). Due to the complexity of tectonic structures, perfectly linear seismic belts can hardly be found. So, methods for the discovery of seismic belts should be able to recognize features with less-than-perfect linear shape. Since seismic belts often cluster as non-spherical (ellipsoid) shape, spatial clustering algorithms need to identify such irregularly shaped structures.

Detecting all possible digital line components contained in a given binary edge image is one of the most fundamental problems in pattern recognition. Hough transform (Asano and Katoh 1996), for example, is a classical method which basically maps each point in the image space to a line in the parameter space, and counts the intersections to get the parameters of the lines in the image space. The Hough transform is, however, not suitable for detecting wide linear features such as the seismic belts (Amorese et al. 1999). Another conventional algorithm to clustering linear features is the Fuzzy C-Lines (Bezdek et al. 1981). Its basic idea is similar to ISODATA (Ball and Hall 1965), which minimizes some objective function to achieve optimal partitioning of a data set in terms of pre-specified clusters. The difference is that the centers of the clusters in Fuzzy C-Lines change from points to straight lines. The method, nevertheless, is affected by outliers (Honda et al. 2002). Seismologists have also developed several methods to search for seismic belts in databases. The collapsing method (Jones and Steward 1997), the strip method (Zhang and Lutz 1989), and the blade method (Amorese et al. 1999) are typical examples.

Though scale plays an important role in clustering, particularly for spatial databases, all of the above methods have not taken scale into consideration. Since seismic belts are natural structures which can only be detected or observed within a certain scale range, methods for the mining of such linear clusters should take scale into consideration. We particularly need to determine the appropriate spatial scale for the discovery of seismic belts, and to observe their behavior along the scale.

Mathematical morphology provides mathematical tools to analyze the geometry and structure of objects. To take advantage of such method, scale space can be constructed with several morphology filtering operators for data mining. Many attempts have been made to combine mathematical morphology with the concept of scale space or clustering. Postaire et al. (1993), for example, attempt to find the "core" of clusters with the opening and closing operators, and allocate the remainder points by the nearest neighbor method. Maragos (1989) use standard morphological opening and closing with structuring elements of varying shape and size to generate a scale space for shape representation. With increasing or decreasing scale, specific binary patterns are self-dilated or eroded and are subsequently used in the open or close operations. The scale parameter is governed by the degree of self dilation or erosion of a given pattern. In the study by Acton and Mukherjee (2000), scale space is constructed with the opening and closing operators of area morphology and the "scale space vectors" are used to perform image classification. Park and Lee (1996) have also studied the property of scale space using mathematical morphology. They point out that the scale space of one dimensional gray-scale signals based on morphological filtering satisfies causality (no new feature points are created as scale gets larger), and with the generalized concept of zero-crossing, opening and closing based morphological filtering will construct a scale space satisfying causality. Di et al. (1998), on the other hand, propose a clustering algorithm using the closing operator with structuring elements increasing iteratively in size, and use the heuristic method to find the best number of clusters. They, however, do not describe their algorithm from the viewpoint of scale space, and they do not give thorough analysis on how to specify the precision of the raster image and how to remove noise to prevent it from disturbing the subsequent morphological operations.

With special reference to the work of Di et al. (1998) but adopting the scale space point of view (Leung et al. 2000a), Wang et al. (2006) propose a scale space clustering method, called Multi-scale Clustering Algorithm with Mathematical Morphology Operators (MCAMMO), for the mining of seismic belts in spatial databases. To extract linear or semi-linear features, the algorithm is further enhanced by some more morphological operations, and the algorithm is called Linear MCAMMO (L_MCAMMO). The idea of MCAMMO is to use mathematical morphology to obtain the most suitable scale to re-segment the seismic belts first. The final belts are then obtained with further processing. The procedure of MCAMMO can in brief be summarized as follows: the vector data set is first converted into a binary image data set with a grid whose precision is specified by the sorted k-dist graph (Ester et al. 1996). A pair of closing and opening operators is used to remove the noise. A scale space is then constructed by using the closing operator with structuring elements of increasing size. Through that, the connected components (the set of cells with neighborhood relationships, i.e., clusters) in the image will gradually merge into each other and become a single cluster in the end. This is essentially a binary image segmentation process, and can also be treated as a hierarchical clustering if the points under each connected component are viewed as one cluster.

The main enhancement of MCAMMO to the work of Di et al. (1998) is that it lucidly gives an effective and easy to follow solution to specify the precision of the raster data set. Based on that, noise removing becomes easier and it makes MCAMMO a robust clustering method.

To make it more effective in the mining of near linear belts such as the seismic belts, Wang et al. (2005) perform further segmentation on the data. In brief, the procedure obtains the skeletons of the segmented image at the most suitable scale with the thinning operator. It then obtains the nodes, extracts and classifies the linear (or near linear) axes, and uses such information to re-segment the image in order to obtain the final linear belts. The procedure is a specialized MCAMMO and is called the Linear MCAMMO (L_MCAMMO). Though it intends to mine linear or near linear seismic belts, it is also suitable for the mining of other linear or semi-linear features such as roads in a remote sensed image contaminated with noise.

The advantages of MCAMMO are: (1) the number of clusters does not need to be specified a priori, (2) only a few simple inputs are required, (3) capable of extracting clusters with arbitrary shapes, and (4) robust to noise.

2.2.7.1 Experiment 2.1

The data set in this experiment comes from real-life earthquake data collected in China by the Seismic Analysis and Forecasting Center (1980, 1989). The objective is to mine seismic belts from this data set. A total of 3,201 seismic events with magnitude ≥ 2.2 in the area of [34°–42°N, 106°–115°E] are extracted. Figure 2.9a shows two nearly parallel seismic belts (in broken lines) corresponding to the north segment of the North–South seismic belt (on the left) and the Shanxi seismic belt (on the right) (Fu 1997). The difficulty in mining the belts lies on the discontinuity of the dense areas in one single belt, which is hard to pick up by the single-scale clustering algorithms such as DBSCAN (Ester et al. 1996). The task can, however, be accomplished effectively and efficiently by MCAMMO.

The lifetime of the clusterings along the scale is depicted in Fig. 2.8, and the connected components, clusters, at selected scales are shown in Fig. 2.9. From Figs. 2.8 and 2.9, we can observe that the lifetime of the 2-clusters clustering is the longest, while that of the 3-clusters clustering is the second longest. By comparing the images at scale 18 which starts the 2 clusters and scale 14 which starts the 3 clusters, we can observe that the connected components in the latter image are actually closer to the true seismic belts. It indicates that 3 is the most suitable number of clusters.

This experiment indicates that clustering of the longest lifetime may not always be the best solution to every problem. We should also pay attention to clustering's whose lifetimes are relatively long, but not the longest. The scale space approach does provide such valid patterns unraveled by the concept of lifetime. Although the seismic belts can be extracted by MCAMMO, their shapes are still not very close to the "near linear" shape of the real seismic belts. In more complex situations (see Experiments 2.2 and 2.3), the differences would even be greater. To have better performance, some specializations on MCAMMO need to be made.

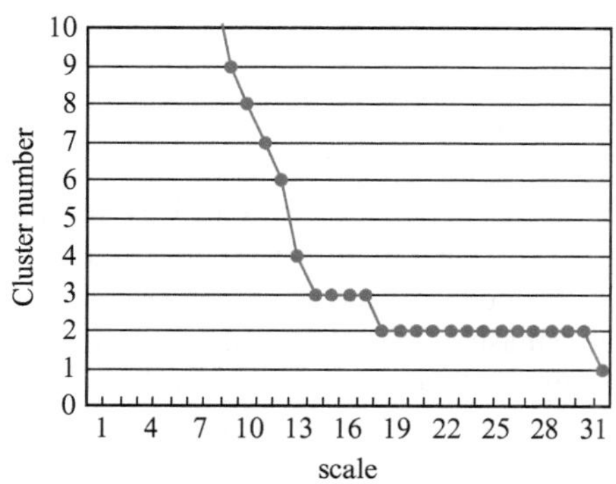

Fig. 2.8 Lifetime of the clusterings in Fig. 2.9

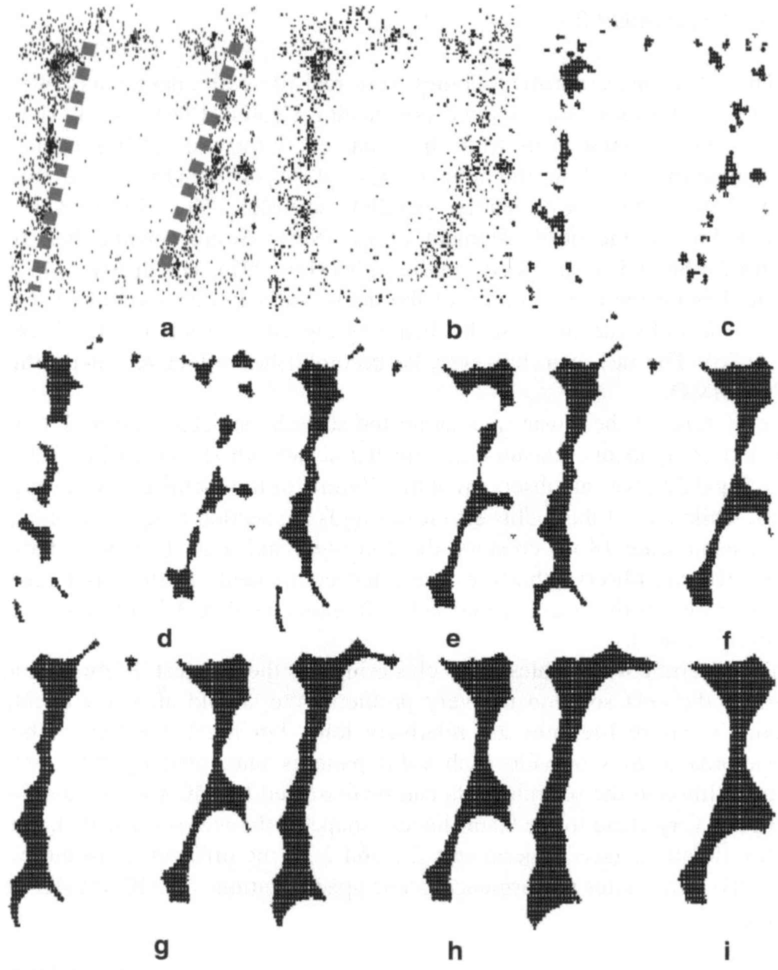

Fig. 2.9 Mining of seismic belts with MCAMMO (a) Original vector-based data set. (b) Rasterized image. (c) First scale with noises removed. (d) Scale 5. (e) Scale 10. (f) Scale 13. (g) Scale 14. (h) Scale 18. (i) Scale 25

2.2.7.2 Experiment 2.2

The image of scale 14 in Experiment 1 is re-processed with the strategy of L_MCAMMO. The skeletons of the segmented image are extracted at the most suitable scale. The nodes of the skeletons are obtained with the hit-or-miss transform. They are "smashed" to split the skeletons into the arcs which are recombined into several groups of "the longer the better" and "the straighter the better" linear (or near linear) axes. Using the information of nodes, skeletons and axes, the image is re-segmented into several linear (or near linear) belts. The belts such obtained will be very close to the true seismic belts. As a result, two linear belts are obtained which are very close to the actual seismic belts (see Fig. 2.10). This actually provides the answer to the discovery of seismic belts problem (Fig. 1.2) posed in Chapter 1, section 1.5.

2.2.7.3 Experiment 2.3

In this experiment, the test area is moved to [40–50°N, 106–115°E] to further validate the effectiveness of L_MCAMMO. There are three main seismic belts which are conglutinated with each other, with the upper one in near arch shape (Fig. 2.11a). MCAMMO is first employed to extract the most suitable image (see Fig. 2.12). It can be observed that the clustering stabilizes at scale ***9 with two clusters. Apparently, the segmented image (Fig. 2.11b) is very different from the actual seismic belts. So, by applying the L_MCAMMO, the image at scale 9 is employed to extract the skeletons (Fig. 2.11c), obtain the axes (Fig. 2.11d) and then extract the linear belts (Fig. 2.11e). Subsequently, the three longest linear belts obtained are very close to the actual seismic belts.

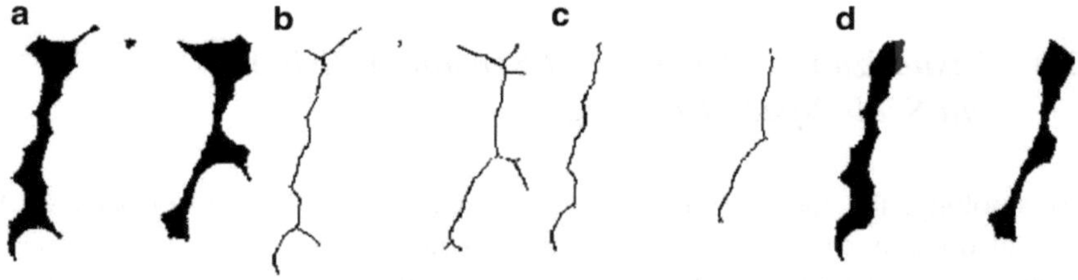

Fig. 2.10 Segmentation after specialization (**a**) Image with the longest lifetime. (**b**) Skeletons. (**c**) Axes of the two longest linear belts. (**d**) Two belts extracted

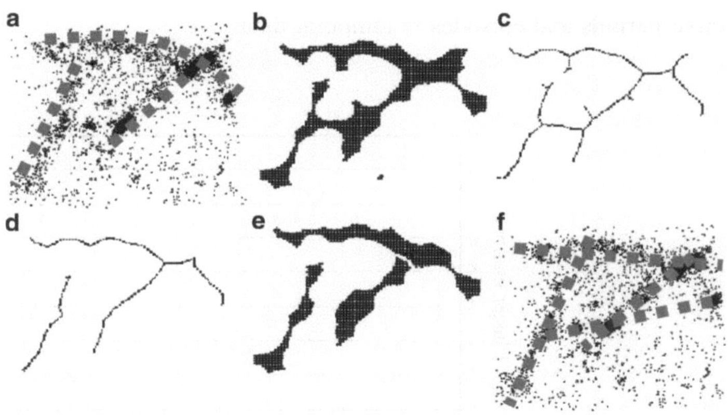

Fig. 2.11 Another seismic area (**a**) Original data set. (**b**) Image at the most suitable scale. (**c**) Skeletons. (**d**) Axes. (**e**) Linear belts. (**f**) Clustering result of Fuzzy C- Lines

As a comparison, Fuzzy C-Lines is employed to extract the belts with the same data set (Wang et al. 2003). Fuzzy C-Lines turns out to be very sensitive to noise. So, noise removal needs to be performed first. The inputs of Fuzzy C-Lines are: $m = 2$, the number of clusters = 4 (taking into account the short linear belts in the center of the image), and 100 iterations. The central lines of the final clusters and the points distributed around them are depicted in Fig. 2.11f. From this image, we find that the upper seismic belt is split apart, where as L_MCAMMO is robust to the "not very linear" clusters. Furthermore, a cluster composed by the points with very large space in-between is obtained by Fuzzy C-Lines (see the bottom-right in Fig. 2.11f), which is not very reasonable. This shows that L_MCAMMO does a better job on this data set. It should also be noted that L_MCAMMO, unlike fuzzy C-Lines, does not require the number of lines (m = 2) and the number of clusters = 4, to be pre-specified as inputs. That is what makes scale space clustering, L_MCAMMO in particular, more natural and spontaneous.

To recapitulate, MCAMMO, with the L_MCAMMO enhancement, can obtain the most suitable scale to re-segment an image, and the mining of the linear belts is completed by the re-segmentation procedure.

2.2.8 Visualization of Temporal Seismic Activities via Scale Space Filtering

In seismology, the identification of seismic active periods and episodes in the temporal domain, the seismic belts in the spatial domain, the seismic sequence and the seismic anomaly in the spatio-temporal domain can all be treated as a clustering problem. I have shown in Sect. 2.2.7 how scale space clustering can be employed to mine seismic belts in spatial data. I will show in this subsection how the clustering algorithm, together with its visualization, can be used to identify seismic active periods and episodes in temporal data.

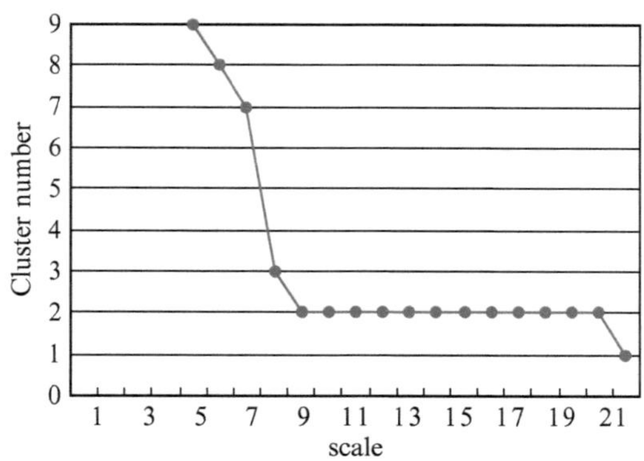

Fig. 2.12 Lifetime of the clusterings in Fig. 2.11

In a larger spatial context, the temporal sequence of strong earthquakes exhibits a certain pattern of clustering with interspersed quiescence and active periods, i.e., quasi-periodicity (Ma and Jiang 1987; Kagan and Jackson 1991; Fu and Jiang 1994). Accordingly, the regional seismic activity in the temporal domain can be segmented into the *seismic active periods* and the *seismic active episodes* on the finer temporal scale (Cao and Fu 1999). Exact and quantitative analysis of seismic active periods and episodes has important implications to the understanding and forecasting of long- and medium-term earthquakes.

Due to the complexity and unpredictability of earthquakes, as well as the difficulty in analyzing the seismic active periods and episodes, the study of seismic activities often rely on the seismologists' expertise and judgments with simple statistical indices (Matthews and Reasenberg 1988). To make the analysis more rigorous and results easier to evaluate, quantitative methods are often needed in conjunction with domain specific expertise (Kagan 1999). Cluster analysis has thus become a common approach to study seismic activities.

As discussed, clustering by scale space filtering has an intrinsic relationship with our visual system. The visualization of clustering by scale space filtering includes two phases: namely visual representation and interactive analysis.

In the first phase, the construction process of scale space clustering can naturally be visualized via a top-to-bottom tree-growing animation in two-dimensional/three-dimensional (2D/3D) views. Animation facilitates the generation of the original qualitative cognition about the clustering in the whole scale space. We can interactively set the visual properties of animation and navigate the scale space in 2D/3D view, including the rotation of a view and the one-dimensional or all-dimensional zooming of a view. This phase suits the visual representation of the scale space.

After the construction of the scale space, visualization based on the scale space and the indices for cluster validity check can assist us to interactively construct, verify and revise at any scale our cognition of the optimal clustering until the final result is obtained. The visualization techniques include the 2D/3D graphs and diagrams of indices which provide the interaction with the concrete numeric indices and customization of the visual properties. Based on the information conveyed by the indices, we can use the slider technique to select the scale of interest in free-style. The corresponding clustering result is shown by both the view of the scale space and the map or time sequence graph. Obviously this phase enables interactive analysis for obtaining the optimal result.

For illustration, I give in the following a brief description of a study on the visualization of seismic activities by scale space clustering (Qin et al. 2006).

2.2.8.1 Experimental Data

In this application, periodic seismic activity of strong earthquakes in Northern China (34–42°N, 109–124°E) is identified via the visualization of the clustering process of scale space filtering. Considering the completeness of the strong earthquake catalog (Huang et al., 1994a, b), two datasets are chosen: (1) the strong earthquakes (Ms ≥ 6.0) of 1290–2000 AD which have 71 records, and (2) the strong earthquakes (Ms ≥ 4.7) of 1484–2000 AD which have 670 records. In seismology, both Ms6.0 and Ms4.7 are lower bounds of strong seismic meanings.

2.2.8.2 Temporal Segmentation of Strong Earthquakes (Ms ≥ 6.0) of 1290–2000 AD

The scale space for the time sequence of earthquakes in this period is depicted in 2D in Fig. 2.13. The number of clusters and the indices including *lifetime*, *isolation* and *compactness* of the clustering are shown in Fig. 2.14.

The scale-space graph and indices call for special attention to the patterns appearing in both the 59–95th and the 6th scale steps (Fig. 2.14). In the 59–95th scale range, there are three clusters in the clustering with the longest lifetime, isolation and compactness. It is the seismic active period recognized through the visualization of the clustering algorithm (Fig. 2.15a). It actually corresponds to the Second, Third, and Fourth Seismic Active Periods singled out by the seismologists (Jiang and Ma 1985). The correspondence between the clustering and seismologists' results is summarized in Table 2.1.

In the 6th scale step, the number of clusters changes dramatically. The number of clusters deceases rapidly for scales preceding the 6th. After the 6th step, however, the change in clustering becomes comparatively smooth. This clustering process shows that the earthquakes, which are comparatively frequent in the time dimension preceding the 6th step, merge rapidly into clusters when the observation scale increases in this scale range. When the time scale is larger than six and seven, however, clusters are formed in more apparent isolations. Fewer clusters are formed in a relatively long scale range. The clustering result in the 6th scale step in fact corresponds to what is recognized by the seismologists as the seismic active episodes (Fig. 2.15b).

2.2.8.3 Temporal Segmentation of Strong Earthquakes (Ms ≥ 4.7) of 1484–2000 AD

Similar analysis and visualization are applied to the time sequence of strong earthquakes (Ms ≥ 4.7) of 1484–2000 AD. Based on the indices shown in Fig. 2.16, two

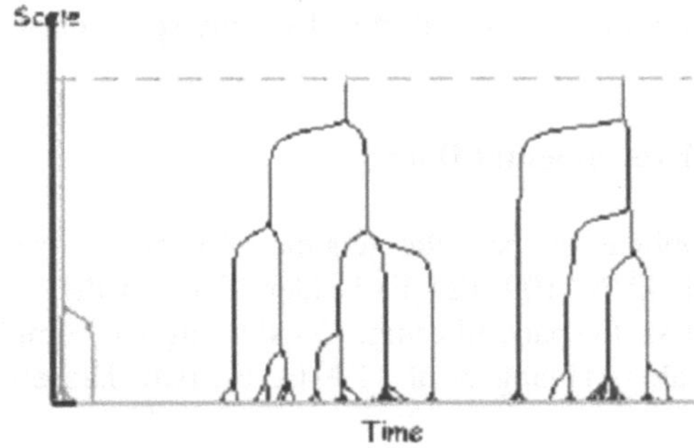

Fig. 2.13 Scale-space clustering for earthquakes (Ms ≥ 6)

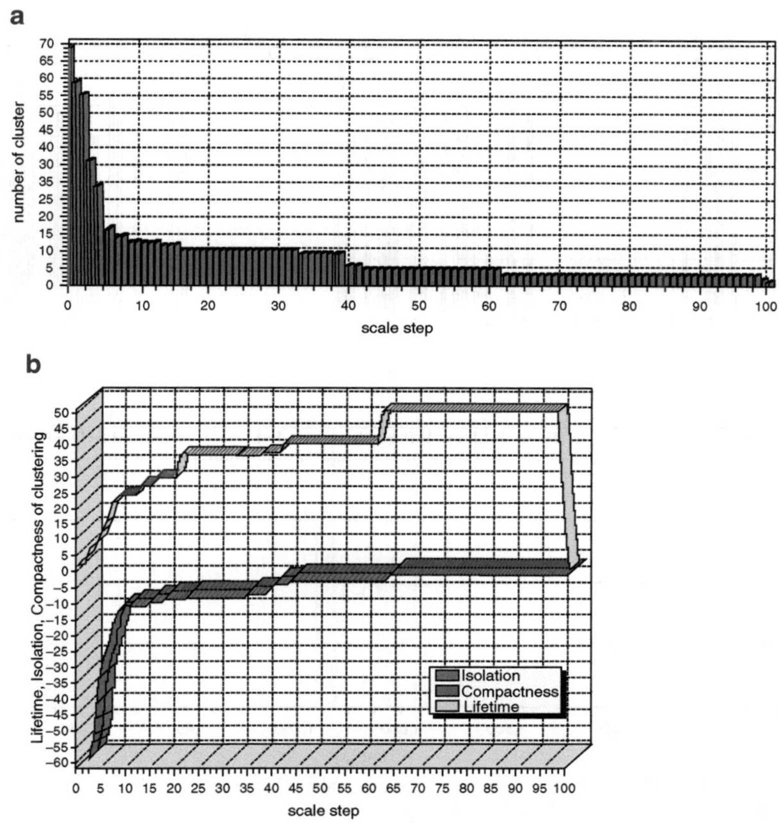

Fig. 2.14 Indices of clustering along the time scale for earthquakes (Ms $\geq$ 6.0) (**a**) number of clusters. (**b**) lifetime, isolation and compactness of the clustering

clusters are deciphered in the 74–112th scale range. They correspond well with the Third and Fourth Seismic Active Periods identified by the seismologists (Fig. 2.17a). Similar to the 1290–2000 AD situation, in the 10th scale step of this time period, we discover 18 clusters which match well with the seismic active episodes identified by the seismologists (Fig. 2.17b).

2.2.8.4 An Overall Interpretation of the Clustering Results

Table 2.1 tabulates the seismic active periods and episodes unraveled by the scale space clustering algorithm versus that of the seismologists.

It can be observed that the periods and episodes of earthquakes (Ms $\geq$ 6) and (Ms $\geq$ 4.7) obtained by scale space clustering are consistent with the results identified by the seismologists' domain specific expertise, with the exception that the episodes of

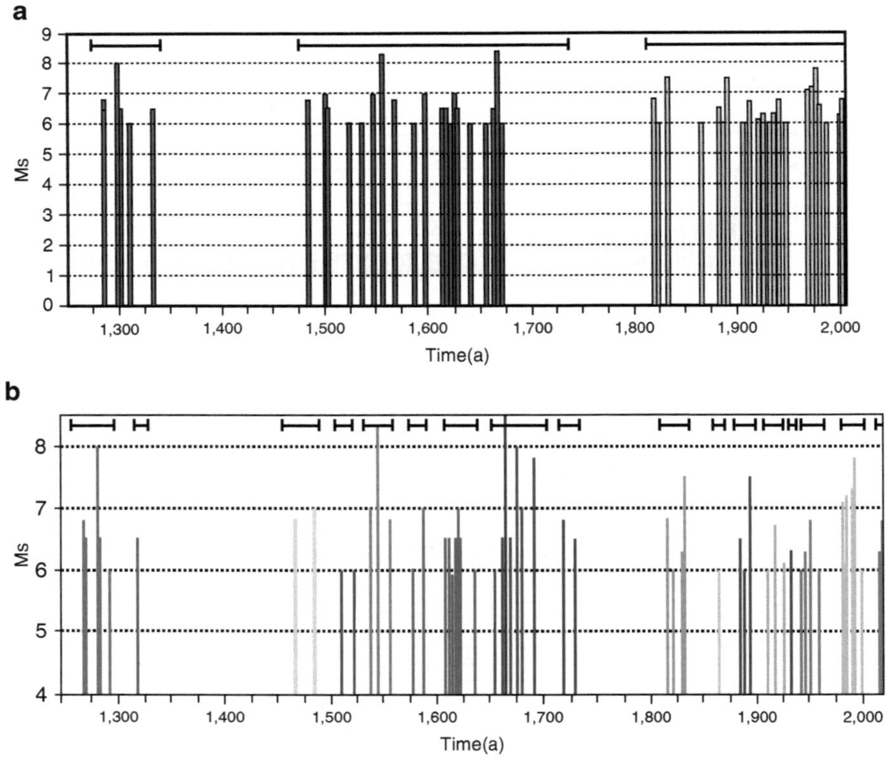

Fig. 2.15 Ms-time plot of clustering results for earthquakes (Ms $\geq$ 6) (**a**) 3 clusters in the 59–95th scale range. (**b**) 17 clusters at the 6th scale step

the Fourth Seismic Active Period recognized by the clustering algorithm is not as consistent. It seems that there is a quasi-periodicity of about 10–15 years for active episodes.

2.2.9 Summarizing Remarks on Clustering by Scale Space Filtering

1. *Lifetime* is a suitable cluster-validity criterion. This can be observed in Fig. 2.2.
2. The algorithms are robust to the variation of cluster shape which can even be non-Gaussian. This is mainly because the objective function in (2.7) is the density distribution estimate and the algorithm is a "mode-seeking" one which tries to find the dense regions. If the data consist of long and thin clusters, we can make use of the Mahalanobis distance instead of the Euclidean distance in the algorithms, and the covariance matrices can be estimated iteratively with a

Table 2.1 Seismic active periods and episodes obtained by the clustering algorithm and the seismologists

Seismologists' results				Clustering result	
Seismic active period	Seismic active episode	(Jiang and Ma 1985)	(Gu et al. 1995)	Ms≥6	Ms≥4.7
II				1290–1340 (6)	
III		1484–1730	1481–1730	1484–1730 (31)	1484–1772 (200)
IV		1815–	1812–	1815–(34)	1789–(470)
II	1 (?)			1290–1314 (5)	
	2 (?)			1337 (1)	
III	1	1484–1487	1481–1487	1484–1502 (3)	1484–1494 (12)
	2	1497–1506	1501–1506		1495–1533 (37)
	3	1522–1538	1520–1539	1524–136 (2)	
	4	1548–1569	1548–1569	1548–1568 (4)	1536–1569 (30)
	5	1578–1597	1580–1599	1587–1597 (2)	1576–1599 (28)
	6	1614–1642	1614–1642	1614–1642 (8)	1610–1633 (31)
					1638–1649 (10)
	7	1658–1683	1658–1695	1658–1695 (10)	1654–1695 (38)
	8	1695–1708			1698–1708 (3)
	9	1720–1730	1720–1730	1720–1730 (2)	1720–1746 (7)
Quiescent					1754–1772 (4)
Period					1789–1798 (6)
V	1	1815–1820	1812–1820	1815–1830 (4)	1805–1835 (26)
	2	1829–1835	1827–1835		
	3	1855–1862	1846–1863	1861 (1)	1851–1862 (11)
	4	1880–1898	1880–1893	1879–1888 (3)	1879–1893 (13)
	5	1909–1923	1909–1918	1903–1918 (4)	1898–1924 (28)
	6	1929–1952	1921–1952	1922 (1)	1929 (2)
				1929–1945 (6)	1931–1948 (15)
	7	1966–1978	1965–1976	1966–1983 (13)	1952– (369)
				1998– (2)	

(The number in parentheses is the number of earthquakes in the cluster)

particular regulation technique if too few a data is contained in a given cluster. This phenomenon can also be seen in Fig. 2.1 and the other experiments where data are of different shapes.

3. The algorithms are insensitive to outliers because outliers can easily be detected in these algorithms. From (2.7) and (2.8), we can see that the influence of one point on a given cluster center is proportional to $O\left(de^{-d^2/\sigma^2}\right)$ with d being the distance between them. When d is large, $O\left(de^{-d^2/\sigma^2}\right)$ is very small. An outlier is usually very far from the cluster centers, so it has little influence on the estimation of the cluster center. On the other hand, the normal data points are usually far away from the outlier, so they have little influence on an outlier. That is to say, an outlier can survive for a long time as a cluster. Therefore, it has a high degree of outlierness (see (2.36)) and can easily be detected.

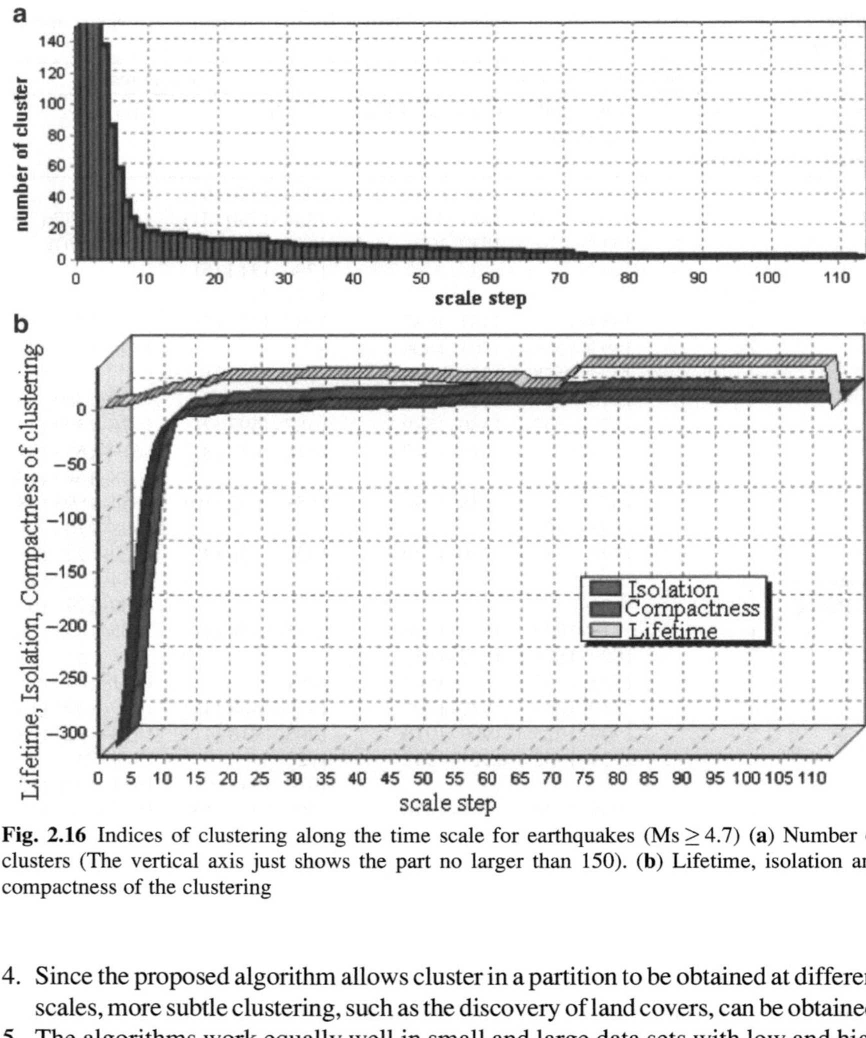

Fig. 2.16 Indices of clustering along the time scale for earthquakes (Ms ≥ 4.7) (**a**) Number of clusters (The vertical axis just shows the part no larger than 150). (**b**) Lifetime, isolation and compactness of the clustering

4. Since the proposed algorithm allows cluster in a partition to be obtained at different scales, more subtle clustering, such as the discovery of land covers, can be obtained.
5. The algorithms work equally well in small and large data sets with low and high dimensions.
6. The proposed clustering method can also be applied to the clustering of data with known distribution containing noise or being indifferentiable.
7. Several scale-based clustering algorithms have been proposed in recent years (Taven et al. 1990; Wilson and Span 1990; Wong, 1993; Chakravarthy and Ghosh 1996; Miller and Rose 1996; Waldemark 1997; Roberts 1997; Blatt et al. 1997). They are derived from very different approaches, such as estimation theory, self-organization feature mapping, information theory, statistical mechanics, and radial basis function networks. One, however, can show that these algorithms are closely related to each other, and in fact, each of these algorithms is equivalent to a special implementation of the proposed algorithm in Leung et al. (2000a).

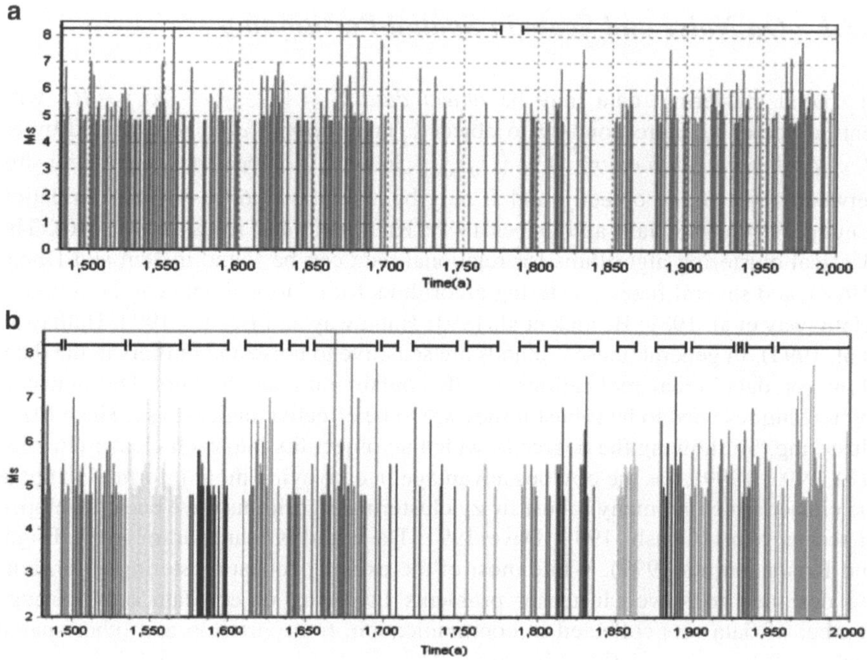

Fig. 2.17 Ms-time plot of clustering results for earthquakes (Ms $\geq$ 4.7) (**a**) 2 clusters in the 74–112th scale range. (**b**) 18 clusters at the 10th scale step

8. For further research, mechanism should be devised to separate clusters which are close to each other. Furthermore, since Gaussian scale space theory is designed to be totally non-committal, it cannot take into account any a priori information on structures which are worthy of preserving. Such a deficiency may be improved by employing more sophisticated nonlinear scale space filters or by integrating appropriate methods, such as mathematical morphology in the seismic belt experiment.

2.3 Partitioning of Spatial Data by a Robust Fuzzy Relational Data Clustering Method

As discussed in Sect. 2.1, there are two basic approaches to discover clusters in data. Scale space filtering that has just been discussed in Sect. 2.2 belongs to hierarchical clustering. To make our discussion more complete, a method for partitioning clustering, called robust fuzzy relational data clustering, is introduced in this section. Similar to scale space filtering, special attention is again paid to the issue of scale and noise in the clustering of spatial data.

2.3.1 On Noise and Scale in Spatial Partitioning

In spatial clustering, data may be *object data* $\mathbf{X} = \{\mathbf{x}_1, \mathbf{x}_2, \cdots, \mathbf{x}_N\} \in \boldsymbol{R}^s$, with feature vector $\mathbf{x}_k$ corresponding to object k, or *relational data* represented by an $N \times N$ *relational data matrix* $\mathbf{D} = (D_{ij})_{N \times N}$, in which D_{ij} measures the relationship between object i and object j, and $\mathbf{D}$ may be a similarity or dissimilarity relation (Leung 1984, 1988; Jain and Dubes 1988; Kaufmann and Rousseeuw 1990). The classical clustering algorithms for relational data can be found in Jain and Dubes (1998), and several fuzzy clustering algorithms for relational data can be found in (Hathaway et al. 1989; Bezdek et al. 1991; Hathaway and Bezdek 1994; Hathaway et al. 1994). In general, these methods are sensitive to noise and outliers in the data. However, data in real applications usually contain noise and outliers. Thus, clustering techniques need to be robust if they are to be effective under noise. Since fuzzy clustering, by showing the degree to which an object fits into each cluster (Bezdek et al. 1991, 1999), has the obvious advantage in conveying more information about the cluster structure, many robust fuzzy clustering algorithms have been developed in recent years (Ohashi 1984; Dave 1991; Dave and Krishnapuram 1997; Frigui and Krishnapuram 1999). While most of the existing robust clustering algorithms are designed to solve clustering problems involving object data only, a huge number of data sets collected in communication, transportation and other spatial analyses is however relational in nature. Therefore, it is essential to develop robust fuzzy relational data clustering algorithms for the analysis of such data type. By incorporating the concept of clustering against noise in the relational algorithms, Hathaway et al. (1994) and Sen and Dave (1998) have developed algorithms for clustering relational data contaminated by noise. Since the algorithms proposed by Ohashi (1984) and Dave (1991) are robust against noise in the object data, its relational versions are expected to be insensitive to noise in relational data. However, this approach is criticized for having only one "scale" parameter whilst in practical applications each cluster may have its own special scale. Another deficiency of the current clustering approach under noise is that a consistent method to find an appropriate value for the scale parameter is non-existent.

To be able to handle noise and scale, Zhang and Leung (2001) proposed a robust fuzzy relational clustering method by introducing multiple scale parameters into the objective function so that each cluster has its own scale space parameters. Without loss of generality, the method only considers dissimilarity relation, and the value of D_{ij} is arbitrary and no specific relations, such as positivity, reflexivity / anti-reflexivity or symmetry, are imposed on the dissimilarity matrix $\mathbf{D}$. (A fuzzy graph theoretic approach to clustering on the basis of a similarity or dissimilarity matrix resulting in hierarchical partitioning of spatial data can be found in Leung (1984)).

Based on Zhang and Leung (2001), noise clustering techniques are first briefly reviewed in this section, and a multiple-scale parameter clustering algorithm for object data containing noise is then proposed. Its relational versions are subsequently described and a new necessary condition for optimizing the corresponding objective function is stipulated. The estimation of the scale parameters and detailed description of the proposed algorithm are then made and substantiated with examples.

2.3.2 Clustering Algorithm with Multiple Scale Parameters for Noisy Data

For an object data set $\mathbf{X} = \{\mathbf{x}_1, \mathbf{x}_2, \cdots, \mathbf{x}_N\}$, we denote its cluster centers by $\mathbf{P}_v, v = 1, \cdots, k$. The fuzzy c-means algorithm (FCM) (Bezdek et al. 1999) assumes that the number of clusters c is known a priori and the goal is to minimize

$$J_{fcm} = \sum_{v=1}^{k} \sum_{i=1}^{N} (u_{iv})^m d_{iv}, \tag{2.42}$$

where $m > 1$ is fixed, d_{iv} is the squared distance from a feature point $\mathbf{x}_i$ to the cluster center $\mathbf{p}_v$, and u_{iv} is the membership of $\mathbf{x}_i$ in cluster v which satisfies:

$$v_{iv} \geq 0, \quad for \quad i = 1, \cdots, n; \quad v = 1, \cdots, k, \tag{2.43}$$

$$\sum_{v=1}^{k} u_{iv} = 1, \quad for \quad i = 1, \cdots, n. \tag{2.44}$$

The necessary conditions for local extrema of the minimization of (2.42) subject to (2.43) and (2.44) are

$$u_{iv} = \left(\sum_{w=1}^{k} \left(\frac{d_{iv}}{d_{iw}} \right)^{1/(m-1)} \right)^{-1}, \qquad i = 1, \cdots, n; \quad v = 1, \cdots, k, \tag{2.45}$$

and,

$$\mathbf{p}_v = \frac{\sum_{i=1}^{N} (u_{iv})^m \mathbf{x}_i}{\sum_{i=1}^{N} (u_{iv})^m}, \quad v = 1, \cdots, k. \tag{2.46}$$

Similar to hard c-means algorithms, fuzzy c-means algorithm is sensitive to noise and outliers. Robust clustering technique is thus introduced to make FCM less sensitive to noise. The goal of such an algorithm is to minimize

$$J_{nc} = \sum_{v=1}^{k} \sum_{i=1}^{N} (u_{iv})^m d_{iv} + \sum_{i=1}^{N} \eta \left(1 - \sum_{v=1}^{k} u_{iv} \right)^m \tag{2.47}$$

subject to

$$u_{iv} \geq 0, \qquad i = 1, \cdots, n; \quad v = 1, \cdots, k, \tag{2.48}$$

$$\sum_{v=1}^{k} u_{iv} \leq 1, \qquad i = 1, \cdots, n. \tag{2.49}$$

The necessary conditions for local extrema of the above optimization problem are

$$u_{iv} = \left(\sum_{w=1}^{k} \left(\frac{d_{iv}}{d_{iw}} \right)^{1/(m-1)} + \left(\frac{d_{iv}}{\eta} \right)^{1/(m-1)} \right)^{-1}, \qquad i = 1, \cdots, n; \tag{2.50}$$
$$v = 1, \cdots, k,$$

and,

$$\mathbf{p}_v = \frac{\sum\limits_{i-1}^{N} (u_{iv})^m \mathbf{x}_i}{\sum\limits_{i=1}^{N} (u_{iv})^m \mathbf{x}_i}, \qquad v = 1, \cdots, k. \tag{2.51}$$

It should be noted that the clustering algorithm works satisfactorily provided that an appropriate value of the scale parameter,η, is known. However, a consistent method to find a good value of η is not available. Another deficiency of clustering under noise is that only one "scale" parameter is used while in practical applications, each cluster may have its own special scale. Zhang and Leung (2001) address these problems by letting each cluster have its own scale parameter. The proposed objective function becomes

$$J_{nc} = \sum_{v=1}^{k} \sum_{i=1}^{N} (u_{iv})^m \frac{d_{iv}}{\eta_v} + \sum_{i=1}^{N} \left(1 - \sum_{v=1}^{k} u_{iv} \right)^m, \tag{2.52}$$

where u_{iv}, $i = 1, \cdots, n$; $v = 1, \cdots, k$, are membership values that need to satisfy (2.48) and (2.49). The necessary conditions for local extrema of the minimization of (2.52) subject to (2.48) and (2.49) are

$$u_{iv} = \left(\sum_{w=1}^{k} \left(\frac{d_{iv}/\eta_v}{d_{iw}/\eta_w} \right)^{1/(m-1)} + \left(\frac{d_{iv}}{\eta_v} \right)^{1/(m-1)} \right)^{-1}, \tag{2.53}$$
$$i = 1, \cdots, n; \ v = 1, \cdots, k,$$

and,

$$\mathbf{p}_v = \frac{\sum_{i=1}^{N} (u_{iv})^m \mathbf{x}_i}{\sum_{i=1}^{N} (u_{iv})^m \mathbf{x}_j}, \quad v = 1, \cdots, k. \tag{2.54}$$

Since each cluster has its own scale parameters, we can use the techniques developed in the possibilistic c-means clustering approach (Dave and Krishnapuram 1997) to estimate the scale parameters as follows:

Obtain a pilot clustering by the FCM first and then estimate η_v by

$$\eta_v = K \frac{\sum_{i=1}^{N} (u_{iv})^m d_{iv}}{\sum_{i=1}^{N} (u_{iv})^m}, \quad v = 1, \cdots, k, \tag{2.55}$$

where u_{iv} is the membership value obtained by the FCM, d_{iv} is the corresponding squared distance between $\mathbf{x}_i$ and cluster center p_v, and K is typically chosen to be 1.

Another estimate of η_v is given by

$$\eta_v = \frac{\sum_{k=1}^{N} (u_{ik})^{\geq\alpha} d_{ik}}{\sum_{k=1}^{N} (u_{ik})^{\geq\alpha}}, \quad v = 1, \cdots, k, \tag{2.56}$$

where $\alpha \in (0, 1)$ gives the crisp α-cut partition

$$(u_{ik})^{\geq\alpha} = \begin{cases} 0, & if \quad u_{ik} < \alpha, \\ 1, & if \quad u_{ik} \geq \alpha. \end{cases} \tag{2.57}$$

Based on the multiple-scale parametric objective function in (2.52), the multi-scale parametric clustering algorithm (MPCA) for noisy data is formulated as follows:

Step 1. Execute a FCM algorithm to find an initial membership values u_{iv}.
Step 2. Apply (2.55) to compute $\eta_1, \cdots, \eta_k$ based on the membership values and cluster centers obtained in step 1.
Step 3. Repeat the following sub-steps: Apply (2.54) to update p_v, Apply (2.53) to compute u_{iv}, until $\max_{iv}|u_{iv}(i+1) - u_{iv}(i)| < \varepsilon$.
Step 4. Apply (2.55) or (2.56) to compute $\eta_1, \cdots, \eta_k$ based on the membership values obtained in step 3.
Step 5. Repeat step 3 to improve d_{iv} and u_{iv}, and then stop.

In possibilistic c-means clustering, Krishnapuram and Keller (1993) have suggested the use of (2.55) in step 2 and (2.56) in step 4. However, there is no consistent method for finding an appropriate value of α for a given data set at present. Zhang and Leung (2001) propose to use (2.55) in steps 2 and 4 since the membership values obtained in step 3 are made robust by the noise clustering algorithm. Therefore, the outliers are of small membership values and they contribute very little to the estimates of η_v's.

2.3.3 Robust Fuzzy Relational Data Clustering Algorithm

The clustering algorithm for relational data containing noise is perhaps first considered by Hathaway et al. (1994), and subsequent relational versions are developed by Sen and Dave (1998). These algorithms are the robust versions of fuzzy relational data clustering algorithms and their objective function is

$$J(U,D) = \sum_{v=1}^{k} \frac{\sum_{i,j=1}^{n} (u_{iv})^m (u_{jv})^m D_{ij}}{2\sum_{j=1}^{n} (u_{jv})^m} + \eta \sum_{i=1}^{N} \left(1 - \sum_{v=1}^{k} u_{iv}\right)^m, \tag{2.58}$$

where the membership values u_{iv} are subjected to (2.48) and (2.49). The dissimilarity matrix D in these algorithms is assumed to have the following property:

$$D_{ij} \geq 0, D_{ij} = D_{ji}, \quad i \neq j \quad \text{and} \quad D_{jj} = 0. \tag{2.59}$$

It has been proved that the necessary conditions for minimizing (2.58) subject to (2.48) and (2.49) are as follows:

$$u_{iv} = \frac{(1/d_{iv})^{1/(m-1)}}{\sum_{w=1}^{k} (1/d_{iw})^{1/(m-1)} + 1/\eta^{1/(m-1)}}, \quad i = 1, \cdots, n; \quad v = 1, \cdots, k, \tag{2.60}$$

where

$$d_{iv} = \sum_{j=1}^{n} D_{ij} \left(\frac{(u_{jv})^m}{q_v} \right) - \frac{1}{2} \sum_{j,k=1}^{n} D_{jk} \left(\frac{(u_{jv})^m (u_{kv})^m}{(q_v)^2} \right), \quad i = 1, \cdots, n; \quad v = 1, \cdots, k, \tag{2.61}$$

and $q_v = \sum_{j=1}^{n} (u_{jv})^m$. When d_{iv} is negative, then u_{iv} may become negative. Therefore, there is no guarantee that the constraint in (2.48) will be satisfied. This problem can be solved by applying a "spreading" transformation proposed by Hathaway and Bezdek (1994). The spreading transformation adds a positive number β to all off-diagonal elements of $\mathbf{D}$. In fact, Hathaway and Bezdek's algorithms are derived under the condition that the relational data $\mathbf{D}$ is Euclidean, which means that $D_{ij} = \left\|\mathbf{x}_j - \mathbf{x}_i\right\|^2$ for some data set $\mathbf{X} = \{\mathbf{x}_1, \mathbf{x}_2, \cdots, \mathbf{x}_N\}$, and it has been proved that there exists a positive β_0 such that all dissimilarity matrices obtained by the spreading transformation with $\beta \geq \beta_0$ are Euclidean. When $\mathbf{D}$ is Euclidean, d_{iv} is the squared Euclidean distance between $\mathbf{x}_i$ and center of the cluster $\mathbf{p}_v$. Therefore, all d_{iv}'s are non-negative.

Zhang and Leung (2001) propose a new robust fuzzy relational data clustering algorithm with multiple-scale parameters and give an alternative approach to address the problem of negative d_{iv}. The algorithm aims at the minimization of the objective function

$$J(\mathbf{U}, \mathbf{D}) = \sum_{v=1}^{k} \frac{\sum_{i,j=1}^{n} (u_{iv})^m (u_{jv})^m D_{ij}}{2\eta_v \sum_{j=1}^{n} (u_{jv})^m} + \sum_{i=1}^{N} \left(1 - \sum_{v=1}^{k} u_{iv}\right)^m, \tag{2.62}$$

with the membership value u_{iv}, $i = 1, \cdots, n$; $v = 1, \cdots, k$, constrained by (2.48) and (2.49). In (2.62), η_v, $v \in \{1, \cdots, k\}$, is a normalization constant, called the scale parameter (which is usually a threshold used to determine which object is an outlier), and k is the given cluster number. No restriction is imposed on the dissimilarity matrix $\mathbf{D}$.

The first term in the objective function is employed to reduce the u_{iv} when object i is with high dissimilarity with other object j in cluster v, and the second term is employed to guarantee that most data should be in the meaningful clusters.

For the object data clustering problem, if $\eta_v = 1$ and $\mathbf{D}_{ij}$ is the Euclidean distance between two vectors representing object i and object j, the first term in the objective function is the general fuzzy c-means clustering objective function (Bezdek et al. 1999). Furthermore, if $\eta_1 = \cdots = \eta_k = \eta$, then the objective function in (2.62) is equivalent to the objective function in Dave (1991).

If we denote

$$\begin{aligned} d_{iv} &= \frac{1}{2}\sum_{j=1}^{n} D_{ij}\left(\frac{(u_{jv})^m}{q_v}\right) + \frac{1}{2}\sum_{j=1}^{n} D_{ji}\left(\frac{(u_{jv})^m}{q_v}\right) \\ &\quad - \frac{1}{2}\sum_{j,k=1}^{n} D_{jk}\left(\frac{(u_{jv})^m (u_{kv})^m}{(q_v)^2}\right), i = 1, \cdots, n; v = 1, \cdots, k, \end{aligned} \tag{2.63}$$

where $q_v = \sum_{j=1}^{n} (u_{jv})^m$, then we can prove that

$$u_{iv} = \frac{(1/|d_{iv}|)^{1/(m-1)}}{\sum_{w=1}^{k} (1/|d_{iw}|)^{1/(m-1)} + 1/\eta_v^{1/(m-1)}}, \qquad i = 1, \cdots, n; \quad v = 1, \cdots, k, \tag{2.64}$$

satisfies the Karush–Kuhn–Tucker conditions for optimality of the problem in (2.62) when $m-1=r_1/2r_2$ with r_1 and r_2 being odd numbers. Since each $m-1$ can be approximated by such numbers, (2.63) and (2.64) are used to estimate the membership value in the proposed algorithm for any $m \geq 1$.

If D_{ij} is the squared Euclidean distance between objects i and j, then d_{iv} is the squared distance from object i to the center of cluster v.

In the proposed algorithm, we must give the estimated value of η_v. In Zhang and Leung (2001), a fuzzy clustering is first obtained by minimizing the following objective function

$$J_1(\mathbf{U},\mathbf{D}) = \sum_{v=1}^{k} \frac{\sum_{i,j=1}^{N} (u_{iv})^m (u_{jv})^m D_{ij}}{2\sum_{j=1}^{n} (u_{jv})^m}, \tag{2.65}$$

with membership values u_{iv}, $i=1,\cdots,n;\quad v=1,\cdots,k$, constrained by (2.43) and (2.44). This objective function is a natural extension of a fuzzy relational data clustering algorithm called *FANNY* (Kaufmann and Rousseeuw 1990) and is first proposed by Hathaway et al. (1989). As discussed in the above section, we can derive a necessary condition for the optimal membership variables:

$$u_{iv} = \frac{(1/|d_{iv}|)^{1/(m-1)}}{\sum_{w=1}^{k} (1/|d_{iw}|)^{1/(m-1)}}, \qquad i=1,\cdots,n;\quad v=1,\cdots,k, \tag{2.66}$$

in which

$$\begin{aligned} d_{iv} = &\frac{1}{2}\sum_{j=1}^{n} D_{ij}\left(\frac{(u_{jv})^m}{q_v}\right) + \frac{1}{2}\sum_{j=1}^{n} D_{ji}\left(\frac{(u_{jv})^m}{q_v}\right) \\ &- \frac{1}{2}\sum_{j,k=1}^{N} D_{jk}\left(\frac{(u_{jv})^m (u_{kv})^m}{(q_v)^2}\right), \quad \text{for}\quad i=1,\cdots,n;\quad v=1,\cdots,k. \end{aligned} \tag{2.67}$$

The fuzzy relational data clustering algorithm (FRDC) based on (2.66) and (2.67) is as follows (Zhang and Leung 2001):

Step 1. Initialize the membership values $u_{iv}(0)$, taking into account constraints in (2.43) and (2.44). Let $i=0$.
Step 2. Compute d_{iv} by (2.67).
Step 3. Compute $u_{iv}(i+1)$ by (2.66).
Step 4. If $max_{i,v}|u_{iv}(i+1)-u_{iv}(i)| < \varepsilon$, then stop. Otherwise, $i=i+1$, then go to step 2.

When there is one $d_{iv}=0$, we can update u_{iv}, as proposed in the fuzzy c-means algorithms, in step 3.

Compared with other fuzzy relational data clustering algorithms, the proposed algorithm has no restrictions on the fuzzy exponent m and the data type. Therefore, it is a more general fuzzy relational data clustering algorithm.

When a fuzzy clustering is obtained by the FRDC algorithm, the obtained membership value u_{iv} is employed to estimate η_v as follows:

$$\eta_v = \frac{\sum_{j=1}^{n} (u_{jv})^m |d_{jv}|}{\sum_{j=1}^{n} (u_{jv})^m}, \quad v = 1, \cdots, k. \tag{2.68}$$

To formulate the robust fuzzy relational data clustering algorithm (RFRDC), the alternating optimization approach with three stages is employed to minimize the objective function in (2.62). In the first stage, we execute the FRDC algorithm to determine an initial membership value. In the second stage, (2.68) is applied to compute the scale parameters $\eta_1, \cdots, \eta_k$ based on the initial cluster membership values. Then (2.63) and (2.64) are employed to iteratively update the pseudo-distance and membership values until a given stopping criterion is satisfied (i.e., when the membership values u_{iv} cannot be significantly changed in two successive iterations). In the third stage, η_v is estimated on the basis of the membership values determined in the second stage. Then (2.63) and (2.64) are applied to refine d_{iv} and u_{iv}. Details of the robust fuzzy relational data clustering algorithm (RFRDC) are given as follows:

Step 1. Execute the FRDC algorithm to find the initial membership values u_{iv}.

Step 2. Apply (2.68) to compute $\eta_1, \cdots, \eta_k$ based on the membership values determined in step 1.

Step 3. Repeat the following sub-steps:
Apply (2.63) to update d_{iv},
Apply (2.64) to compute u_{iv},
until $max_{i,v}|u_{iv}(i+1) - u_{iv}(i)| < \varepsilon$.

Step 4. Apply (2.68) to compute $\eta_1, \cdots, \eta_k$ based on the membership values determined in step 3.

Step 5. Repeat step 3 to improve d_{iv} and u_{iv}, and then stop.

2.3.4 *Numerical Experiments*

2.3.4.1 A Pedagogic Example

This example involves two well separated clusters of seven points each and three noisy points (Fig. 2.18). We assume that the dissimilarity matrix **D** is Euclidean with $D_{ij} = \|\mathbf{x}_i - \mathbf{x}_j\|^2$. In this case, the sequence of partitioning membership value u_{iv} produced by the relational data clustering under noise is identical to the sequence produced by the corresponding clustering for object data under noise. The cluster centers in the experimental results can be computed by (2.46) which are listed in Table 2.2.

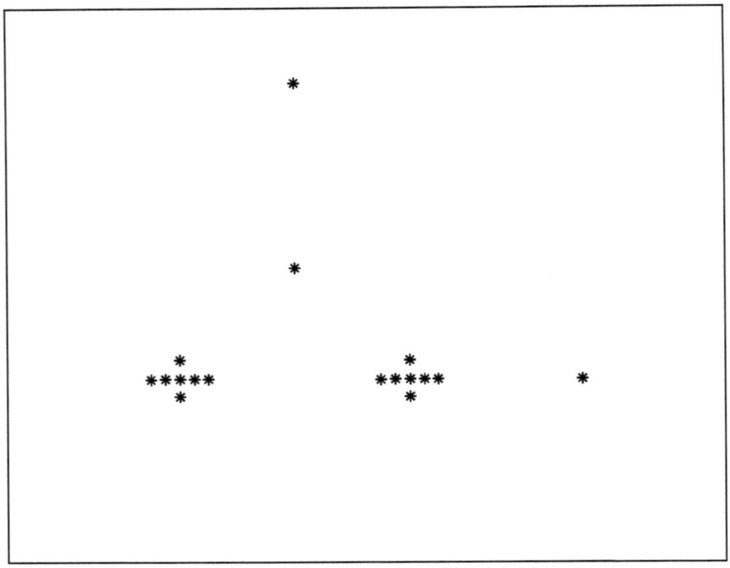

Fig. 2.18 Scatter plot of a noisy data set

Table 2.2 Cluster centers in the experiment

	cluster 1	cluster 2
real cluster centers	(60,150)	(140,150)
cluster centers obtained by noise clustering algorithm (Hathaway et al., 1994b)	(60.2724, 150.2078)	(140.3632, 150.1987)
cluster centers obtained by MPCA	(60.0002, 150.0001)	(140.0006, 150.0001)

From Table 2.2, we can see that the cluster centers found by the proposed algorithm are more precise than that of the relational noise clustering algorithm. Similar phenomena have also been observed in many other numerical experiments (Zhang and Leung 2001).

2.3.4.2 Concordance in Languages

This example is based on the real relational data from the study carried out by Johnson and Wichern (1992, Table 12.4), called "concordant first letters for numbers in eleven languages" which compares eleven European languages (English, Norwegian, Danish, Dutch, German, French, Spanish, Italian, Polish, Hungarian, and Finnish) by looking at the first letters of the first ten numbers. The words for the same number in two different languages are concordant if they have the same first letters and discordant if they do not. The following matrix of discordant first letters for numbers is used as the dissimilarity matrix **D** to cluster these languages.

$$\begin{pmatrix}
 & E & N & Da & Du & G & Fr & Sp & I & P & H & Fi \\
E & 0 & & & & & & & & & & \\
N & 2 & 0 & & & & & & & & & \\
Da & 2 & 1 & 0 & & & & & & & & \\
Du & 7 & 5 & 6 & 0 & & & & & & & \\
G & 6 & 4 & 5 & 5 & 0 & & & & & & \\
Fr & 6 & 6 & 6 & 9 & 7 & 0 & & & & & \\
Sp & 6 & 6 & 5 & 9 & 7 & 2 & 0 & & & & \\
I & 6 & 6 & 5 & 9 & 7 & 1 & 1 & 0 & & & \\
P & 7 & 7 & 6 & 10 & 8 & 5 & 3 & 4 & 0 & & \\
H & 9 & 8 & 8 & 8 & 9 & 10 & 10 & 10 & 10 & 0 & \\
Fi & 8 & 9 & 9 & 9 & 9 & 9 & 9 & 9 & 9 & 8 & 0
\end{pmatrix}$$

For $k = 2$, the results obtained by the proposed RFRDC algorithm and NERF (Hathaway and Bezdek 1994) are listed in Table 2.2, where $u_{\cdot v}$ denotes the membership value of a language in cluster v obtained by the RFRDC algorithm, $u^N_{\cdot v}$ denotes the membership value of a language in cluster v obtained by NERF, $d_{\cdot v}$ denotes the distance value obtained from (2.63) by the RFRDC algorithm.

From Table 2.3, we can observe that English, Norwegian, Danish, Dutch and German form a group, French, Spanish, Italian, and Polish form another group, while Hungarian and Finnish appear to be standing alone. This clustering result can be checked by our visual impression of the dissimilarity matrix ***D***. The advantage of the proposed approach is that it is less subjective in creating clusters and it gives the extent to which a language is in a cluster. For example, from Table 2.3, we can see that English, Norwegian, and Danish are more typical than Dutch and German in cluster 2, and Dutch is less typical than German in this

Table 2.3 Experimental results of the concordance in languages

	$u_{\cdot 1}$	$u_{\cdot 2}$	$u^N_{\cdot 1}$	$u^N_{\cdot 2}$	$d_{\cdot 1}$	$d_{\cdot 2}$
E	0.0346	0.8694	0.1548	0.8452	5.2637	1.0948
N	0.0059	0.9794	0.0488	0.9512	5.2348	0.4095
Da	0.0136	0.9602	0.1041	0.8959	4.5612	0.5492
Du	0.0622	0.3488	0.1427	0.8573	8.1486	4.2307
G	0.0901	0.4447	0.1834	0.8166	6.1874	3.2897
Fr	0.9155	0.0168	0.9542	0.0458	0.6598	4.9528
Sp	0.9499	0.0109	0.9805	0.0195	0.4923	4.6468
I	0.9858	0.0030	0.9824	0.0176	0.2572	4.6619
P	0.3495	0.1405	0.8609	0.1391	3.0218	5.5482
H	0.0589	0.1781	0.2874	0.7126	9.0909	6.7481
Fi	0.0743	0.1491	0.4172	0.5828	8.1090	7.3931

cluster. In cluster 1, Italian is the most typical one and Polish is the least typical one. While these conclusions can be drawn from the clustering results produced by the proposed RFRDC algorithm, they are not obvious in the results obtained by NERF (see Table 2.3).

2.3.4.3 Clustering of Oil Types

This example employs a real data set from Gowda and Diday (1992) for eight different types of oil. The similarity matrix obtained from that study is given as follows:

$$\begin{pmatrix} \textit{Oil Type} & o_1 & o_2 & o_3 & o_4 & o_5 & o_6 & o_7 & o_8 \\ \\ o_1 : \textit{Linseedoil} & - & & & & & & & \\ o_2 : \textit{Perillaoil} & 4.98 & - & & & & & & \\ o_3 : \textit{Coiion-seedoil} & 3.66 & 5.70 & - & & & & & \\ o_4 : \textit{Sesmaeoil} & 3.77 & 5.88 & 7.00 & - & & & & \\ o_5 : \textit{Camelia} & 3.84 & 4.70 & 6.25 & 5.90 & - & & & \\ o_6 : \textit{Oliveoil} & 3.24 & 5.30 & 6.68 & 6.37 & 6.24 & - & & \\ o_7 : \textit{Beef-tallow} & 0.86 & 2.78 & 4.11 & 3.61 & 3.48 & 4.28 & - & \\ o_8 : \textit{Lard} & 1.22 & 3.08 & 4.44 & 3.97 & 3.89 & 4.68 & 6.74 & - \end{pmatrix}$$

The dissimilarity matrix can be generated from the similarity matrix in either of the following ways:

$$D_{ij} = 1/S_{ij} - min_{r \neq t}(1/S_{rt}), i \neq j$$

or

$$D_{ij} = max_{r \neq t}(S_{rt}) - S_{ij}, i \neq j$$

and $D_{ij} = 0$ for all i. The dissimilarity matrices $\mathbf{D}_1$ and $\mathbf{D}_2$ generated respectively by the above equations are as follows:

$$\mathbf{D}_1 = \begin{pmatrix} 0 & 0.0579 & 0.1304 & 0.1224 & 0.1176 & 0.1658 & 1.0199 & 0.6768 \\ 0.0579 & 0 & 0.0326 & 0.0272 & 0.0699 & 0.0458 & 0.2169 & 0.1818 \\ 0.1304 & 0.0326 & 0 & 0 & 0.0171 & 0.0068 & 0.1005 & 0.0824 \\ 0.1224 & 0.0272 & 0 & 0 & 0.0266 & 0.0141 & 0.1342 & 0.1090 \\ 0.1176 & 0.0699 & 0.0171 & 0.0266 & 0 & 0.0174 & 0.1445 & 0.1142 \\ 0.1658 & 0.0458 & 0.0068 & 0.0141 & 0.0174 & 0 & 0.0908 & 0.0708 \\ 1.0199 & 0.2169 & 0.1005 & 0.1342 & 0.1445 & 0.0908 & 0 & 0.0055 \\ 0.6768 & 0.1818 & 0.0824 & 0.1090 & 0.1142 & 0.0708 & 0.0055 & 0 \end{pmatrix}$$

and

$$\mathbf{D}_2 = \begin{pmatrix} 0 & 0.0579 & 0.1304 & 0.1224 & 0.1176 & 0.1658 & 1.0199 & 0.6768 \\ 0.0579 & 0 & 0.0326 & 0.0272 & 0.0699 & 0.0458 & 0.2169 & 0.1818 \\ 0.1304 & 0.0326 & 0 & 0 & 0.0171 & 0.0068 & 0.1005 & 0.0824 \\ 0.1224 & 0.0272 & 0 & 0 & 0.0266 & 0.0141 & 0.1342 & 0.1090 \\ 0.1176 & 0.0699 & 0.0171 & 0.0266 & 0 & 0.0174 & 0.1445 & 0.1142 \\ 0.1658 & 0.0458 & 0.0068 & 0.0141 & 0.0174 & 0 & 0.0908 & 0.0708 \\ 1.0199 & 0.2169 & 0.1005 & 0.1342 & 0.1445 & 0.0908 & 0 & 0.0055 \\ 0.6768 & 0.1818 & 0.0824 & 0.1090 & 0.1142 & 0.0708 & 0.0055 & 0 \end{pmatrix}$$

Table 2.4 exhibits the final memberships found by the RFRDC algorithm and NERF on the dissimilarity matrices $\mathbf{D}_1$ and $\mathbf{D}_2$. The cluster number is $k = 2$. In Table 2.4, $u^{(1)}_{\cdot v}$ is the membership value produced by the proposed algorithms for dissimilarity matrix $\mathbf{D}_1$, and $u^{(2)}_{\cdot 1}$ is the membership value for $\mathbf{D}_2$; $u^{N1}_{\cdot v}$ is the membership value produced by NERF for dissimilarity matrix $\mathbf{D}_1$, and $u^{N2}_{\cdot v}$ is the membership value produced by NERF for dissimilarity matrix $\mathbf{D}_2$ (the membership value is taken from Hathaway and Bezdek 1994). It is interesting to see that in the results obtained by the RFRDC algorithm, o_2, o_3, o_4, o_5, o_6 form the fist cluster; o_7, o_8 form another cluster; o_1 seems to be alone and o_2 seems to be less typical in the first cluster. However, we cannot observe these phenomena in the clustering results obtained by NERF.

2.4 Partitioning of Spatial Object Data by Unidimensional Scaling

2.4.1 A Note on the Use of Unidimensional Scaling

In Sect. 2.3, I have introduced an algorithm for the discovery of optimal partitioning of fuzzy relational data in noisy environment. The emphasis is on the robustness to noise and the multiplicity of scale for clusters. The method falls within the realm

Table 2.4 Experimental results of clustering of oil types

	$u^{(1)}_{\cdot 1}$	$u^{(1)}_{\cdot 2}$	$u^{(2)}_{\cdot 1}$	$u^{(2)}_{\cdot 2}$	$u^{N1}_{\cdot 1}$	$u^{N1}_{\cdot 2}$	$u^{N2}_{\cdot 1}$	$u^{N2}_{\cdot 2}$
o_1	0.0619	0.0000	0.0771	0.0019	0.888	0.112	0.704	0.296
o_2	0.4914	0.0002	0.3513	0.0039	0.811	0.189	0.818	0.182
o_3	0.9998	0.0000	0.9993	0.0001	0.631	0.369	0.935	0.065
o_4	0.9874	0.0004	0.9726	0.0016	0.696	0.304	0.924	0.076
o_5	0.8329	0.0006	0.7015	0.0051	0.663	0.337	0.816	0.184
o_6	0.9687	0.0011	0.9340	0.0046	0.539	0.461	0.834	0.166
o_7	0.0001	0.8275	0.0005	0.9391	0.087	0.913	0.036	0.964
o_8	0.0002	0.8475	0.0006	0.9395	0.096	0.904	0.028	0.972

of partitioning clustering. Though it is robust and scale-based, it, similar to other partitioning methods, is sensitive to initialization and is subjected to the presupposition of a class number k.

To circumvent the sensitivity to initial seed values (if handled appropriately), the presupposition of a cluster number, and the trapping by local minima, I introduce in this section the clustering of object data by unidimensional scaling (UDS). The method is mainly developed by Guttman (1968). It has been applied to social science and medical science research (Gorden 1977; McIver and Carmines 1981), and equipped with algorithms for solving the associated global optimization problem (Pliner 1984, 1996; Simantiraki 1996; Lau et al. 1998). Our discussion in this section is based on the study by Leung et al. (2004e) on the mining of natural clusters in remotely sensed data.

2.4.2 Basic Principle of Unidimensional Scaling in Data Clustering

The basic idea of UDS is to arrange n objects on the real line so that the inter-point distances/dissimilarities can best approximate the observed distances (McIver and Carmines 1981). UDS is a relatively simple but effective algorithm. Compared with multidimensional Scaling (MDS) methods such as K-means and ISODATA, UDS is easier to understand and implement, free from the presupposition of a cluster number, insensitive to initial seed values, independent of information structure, and not limited by the feature-space dimension.

In UDS, the basis of analysis is the dissimilarity matrix. Let there be n observed objects with p dimensions:

$$\mathbf{x}_i = \left(x_{i1}, x_{i2}.....x_{ip}\right)^T, \mathbf{x}_i \in \mathbf{R}^p,\ i = 1....n. \tag{2.69}$$

Then we can establish a matrix of dissimilarities among these objects. As discussed in Leung (1984), dissimilarity between objects can be expressed by the distance between them as follows:

$$d_{ij} = \left\{ \sum_{k=1}^{p} |x_{ik} - x_{jk})|^q \right\}^{1/q},\ 1 \le q \le \infty. \tag{2.70}$$

Specifically, d_{ij} is the L1 distance or City block metric when $q = 1$, and the L2 distance or Euclidean distance when $q = 2$. Here we select the Euclidean distance as a basis of measurement.

Based on the distance measure, we can establish the $n \times n$ matrix of dissimilarity between objects as:

$$\mathbf{D} = (d_{ij}) \tag{2.71}$$

UDS attempts to map n objects $\mathbf{x}_i, i = 1....n$, from the p-dimensional space into the one dimensional coordinates $y_i, i = 1....n$, and arrange them on the real line so that their inter-point distances are as close as possible to their observed distances. That is, it arranges these coordinates in ascending order. The objective is to find the real numbers $y_1, \ldots, y_n$ by minimizing the following objective function:

$$\sigma(\mathbf{y}) = \sum_{i<j} (d_{ij} - |y_i - y_j|)^2, \quad \mathbf{y} = (y_1, y_2 \ldots y_n)^T, \; y_i \in \mathbf{R}\,. \tag{2.72}$$

The solution is however not unique. For example, the translation and reflection of $\mathbf{y}$ also give the same minimum. To overcome this shortcoming, Guttman imposes a centering constraint: $\sum_{i=1}^{n} y_i = 0$ on the above function.

As an illustration, the Guttman algorithm and Pliner algorithm are outlined as follows:

2.4.2.1 Guttman Algorithm (1968)

First, we set the initial value, y^0, for $\mathbf{y}$ as:

$$\mathbf{y}^0 = \left(y_1^0, y_2^0 \ldots y_n^0\right)^T, \tag{2.73}$$

where y^0starts with any random value, or $y_i^0 = \frac{1}{p} \sum_{k=1}^{p} x_{ik}$.

Then, the optimal estimator $\mathbf{y}$ can be obtained with an iterative algorithm using the equation below:

$$y_i^{(r+1)} = \frac{1}{n} \sum_{j=1}^{n} d_{i,j} \mathrm{sign}(y_i^{(r)} - y_j^{(r)}), \; i = 1 \ldots n, \tag{2.74}$$

where $y_i^{(r)}$ is the coordinate of object i at the r-th iteration.

The iterative process stops when $\left|y_i^{(r+1)} - y_i^{(r)}\right| < \delta$ (a small number) is met. The algorithm is fast, simple and has the self-centering property. Nevertheless, the objective function $\sigma(\mathbf{y})$ has many local minima and they increases with n. To prevent trapping by local minima, Pliner (1996) proposes a smoothing algorithm for the UDS problem.

2.4.2.2 Pliner Algorithm (1996)

The smoothing technique is employed to obtain the minimum of $\sigma_\varepsilon(\mathbf{y})$:

$$\sigma_\varepsilon(\mathbf{y}) = (1/\varepsilon^n) \int_{D(\mathbf{y},\varepsilon)} \sigma(\mathbf{x}) d\mathbf{x} \tag{2.75}$$

where $D(\mathbf{y}, \varepsilon)$ is a cube in $\boldsymbol{R}^n$ with the center in $\mathbf{y}$ and a side ε. Taking the integral in the above equation we obtain:

$$\sigma(\mathbf{y}) = \sum_{i<j} \left[(y_i - y_j)^2 - 2d_{i,j}g_\varepsilon(y_i - y_j) \right] + c \tag{2.76}$$

where c is a constant and

$$g_\varepsilon(t) = \begin{cases} t^2(3\varepsilon - |t|)/3\varepsilon^2 + \varepsilon/3, & \mathit{if}\,|t| < \varepsilon, \\ |t|, & \mathit{if}\,|t| \geq \varepsilon. \end{cases} \tag{2.77}$$

It is easy to verify that $\sigma_\varepsilon(\mathbf{y})$ is twice continuously differentiable. Taking the partial derivatives of $\sigma_\varepsilon(\mathbf{y})$ and setting them to zero, we obtain the following equation:

$$y_i^{(r+1)} = \frac{1}{n} \sum_{j=1}^{n} d_{ij} u_\varepsilon(y_i^{(r)} - y_j^{(r)}), i = 1, \ldots, n \tag{2.78}$$

where

$$u_\varepsilon(t) = \begin{cases} (t/\varepsilon)(2 - |t|/\varepsilon), & \mathit{if}\,|t| < \varepsilon, \\ \operatorname{sign}(t), & \mathit{otherwise} \end{cases}. \tag{2.79}$$

The quality of solution of both methods however depends on the initial configuration. Leung et al. (2003) propose a method for finding a good starting configuration. In general, the UDS algorithm produces a curve with obvious break off points according to the number of natural clusters in a data set (Fig. 2.20). Generally, if the differences among classes are apparent, there will be distinct step changes in the UDS curve. Consequently, we can choose the corresponding y coordinates as the natural break off points demarcating the cluster. To make the identification of break off points less judgmental, Leung et al. (2004e), propose the UDS histogram method to assist us in determining more objectively the break points in the UDS curve for the discovery of land covers in remote sensing imagery (see Sect. 2.4.4).

2.4.3 Analysis of Simulated Data

To better understand the characteristics and the performance of the UDS method, Leung et al. (2004e) perform simulation studies on three sets of artificially generated data with specific data properties (Arbia 1989).

Data set G1 is of cirque shape and includes 160 sample points, with 60 samples evenly distributed on the circumference and the rest distributed randomly around the circle center. The spatial distribution is unusual but not too complicated (Fig. 2.19a1). Data set G2 includes 200 sample points distributed randomly around two cluster centers with 100 samples each. The distribution is relatively simple and common (Fig. 2.19b1). Data set G3 consists of 200 sample points splitting into two cincture lines, with each having 100 samples. The spatial distribution is complicated (Fig. 2.19c1). The three data sets are employed to test the UDS method against the K-means classifier.

Basing on the UDS curves obtained in the three experiments (Fig. 2.20a), we can observe that Data set G1 has two obvious step changes. The coordinates at which

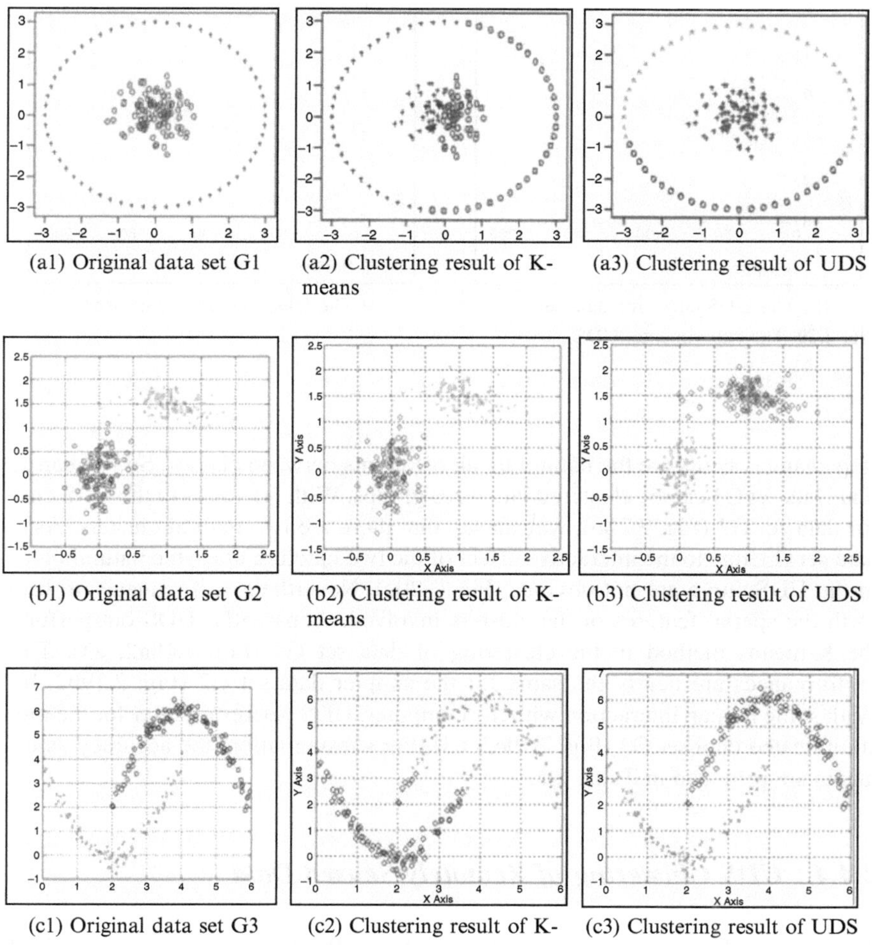

(a1) Original data set G1 (a2) Clustering result of K-means (a3) Clustering result of UDS

(b1) Original data set G2 (b2) Clustering result of K-means (b3) Clustering result of UDS

(c1) Original data set G3 (c2) Clustering result of K-means (c3) Clustering result of UDS

Fig. 2.19 Simulated Experiments of UDS clustering

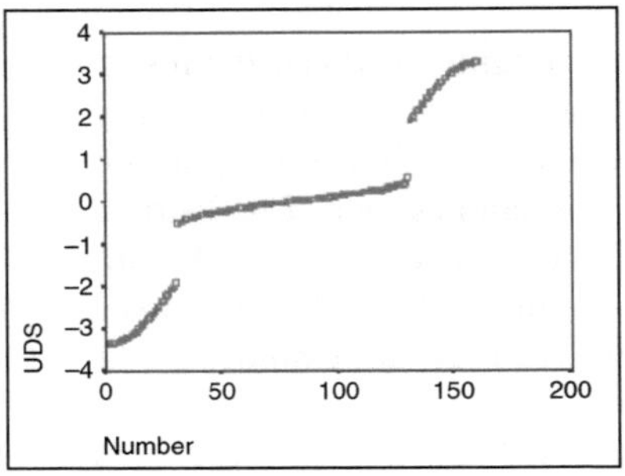

(a) The UDS curves for data set G1

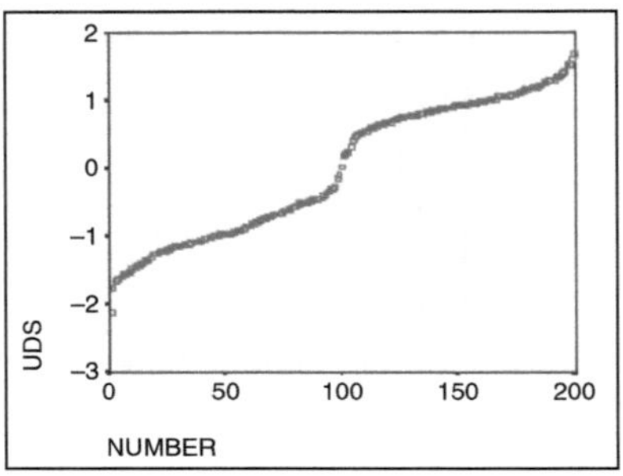

(b) The UDS curve for data set G2

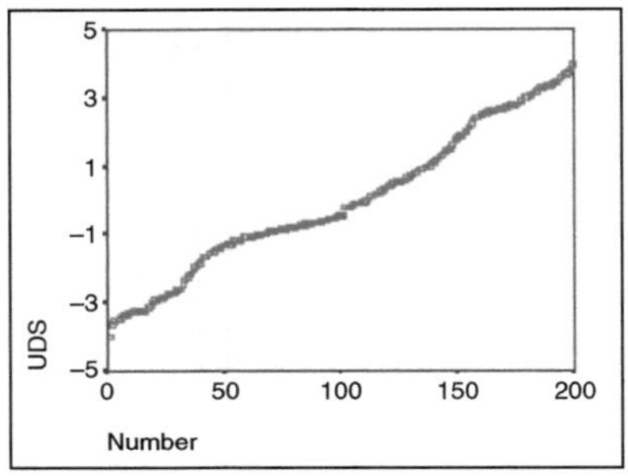

(c) The UDS curve for data set G3

Fig. 2.20 The experimental UDS curves

step change occurs are the natural break off points between classes. So, we can say that Data set G1 can be clustered into three groups. With reference to the UDS curve for data set G2 (Fig. 2.20b), we can say that there are two natural clusters. As for data set G3, due to the interactive effect of the two cincture lines, the natural breaks in the UDS curve are not obvious (Fig. 2.20c). Nevertheless, it manages to bring forth the spatial features of the clusters involved. Apparently, UDS out-performs the K-means method in the clustering of data set G1 (Fig. 2.19a2, a3). Their performances are nearly the same for the simpler data set G2 (Fig. 2.19b2, b3). With some human interaction with the computer, UDS performs better for the more complicated data set G3 (Fig. 2.19c2, c3). For substantiation, the accuracy assessments are provided in Table 2.5.

2.4.4 UDS Clustering of Remotely Sensed Data

The UDS method is customized by Leung et al. (2004e) in the analysis of a SPOT-HRV multispectral image acquired over Xinjing on August 30, 1986. The size of the original image is 3,000 × 3000 pixels with three spectral bands. It contains *sand*

Table 2.5 The error matrix of the numerical experiment

	K-Means			UDS			Total	Accuracy	
	Class	C1	C2	C1	C2	C3		K-Means	UDS
G1	C1	54	46	100	0	0	100	52.50%	76.90%
	C2	30	30	0	23	37	60		
	Total	84	76	100	23	37	160		
G2	C1	99	1	100	0		100	99.50%	100%
	C2	0	100	0	100				
	Total	99	101	100	100		200		
G3	C1	91	9	100	0		100	92%	98.50%
	C2	7	93	3	97		100		
	Total	98	102	103	97		200		

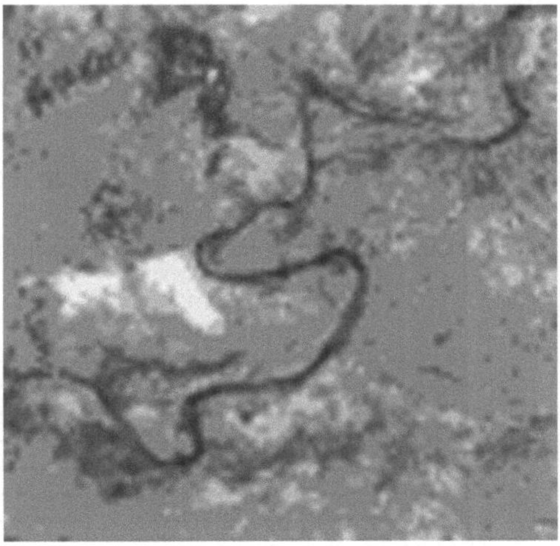

Fig. 2.21 SPOT multispectral image acquired over Xinjing

(c1) *water* (c2), and *saline area* (c3) as land covers. As usual, the image is preprocessed by filtering, stretching and geometric correction.

For pedagogy, a 100 × 100 small area (Fig. 2.21) is extracted from this image to evaluate the performance of the UDS method against the K-means and ISODATA methods.

Due to the spatial characteristics and continuity of remotely sensed data, it is necessary to take additional measures to facilitate the application of the UDS method in the clustering of remotely sensed data. First, similarity of pixels in remote sensing images should be analyzed in the multispectral space, for this case the space of 3-dimensional spectral bands. Second, the matrix of similarity D is constructed by calculating the distances between pixels in the multispectral space. Third, the ordinates y_i, $i = 1, \ldots, 10000$, are calculated and sorted in ascending order. Four, due to the continuity of ground objects in a remotely sensed image, the derived UDS curve naturally has no obvious step changes (Fig. 2.22).

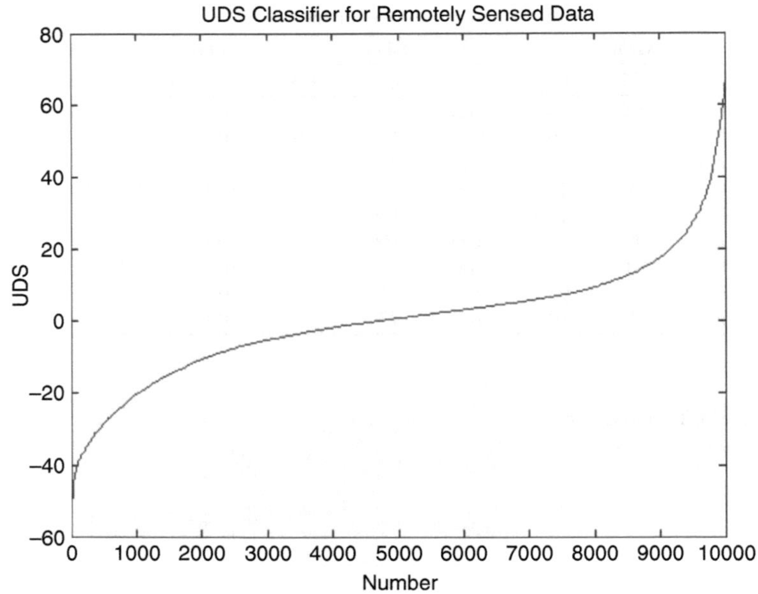

Fig. 2.22 The UDS curve obtained in the remote sensing experiment

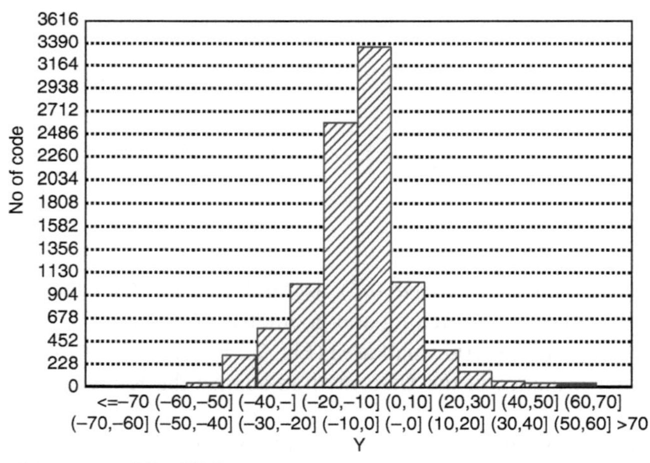

Fig. 2.23 The histogram of the UDS curve

However, we could observe an obvious ascending trend between different clusters in the UDS curve if there are natural clusters, such as this experiment. To facilitate the location of the break off points, Leung et al. (2004e) propose to plot the histogram of the UDS curve (Fig. 2.23). The abscissa is the intervals of the UDS curve and the ordinate is the number of objects in the interval.

In this experiment, $Y_i \in [-60, 70]$, and the interval length is 10. From Fig. 2.23, we can observe that the number of objects rises and drops when $Y_i = -10$ and 10 respectively. It indicates that abscissa -10 and 10 can be regarded as the natural break off points of the clusters. According to this principle, there are two break off points: (2,000, -10.7214) and (8,250, 10.7285) in the UDS curve. On that, we can partition the remotely sensed image into three clusters (Fig. 2.24), which are consistent with the real situation.

In case we cannot directly obtain the break off points, we can just adjust the interval value continually to find the optimal break off points. We can also simultaneously employ visual interpretation or other knowledge to facilitate the identification process.

As a comparison, the K-means and ISODATA methods are applied to the same image and the results are depicted in Figs. 2.25 and 2.26, respectively.

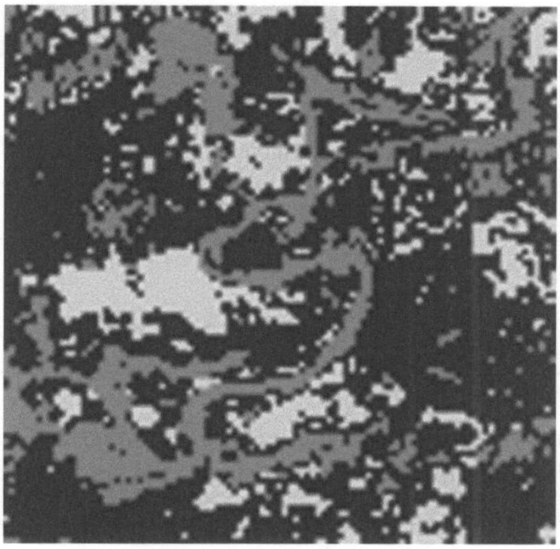

Fig. 2.24 Result obtained by the UDS method

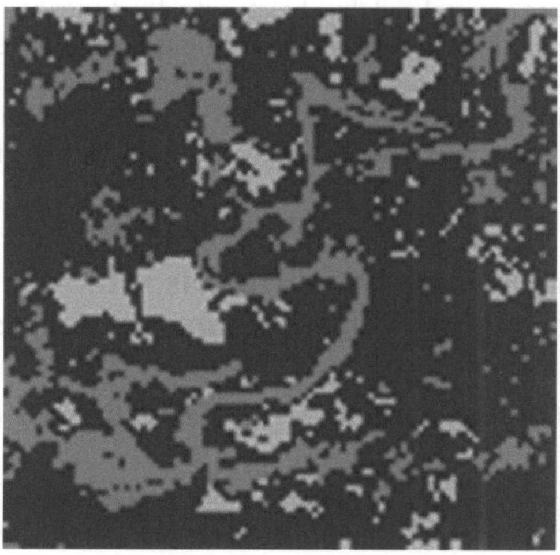

Fig. 2.25 Result obtained by the K-means method

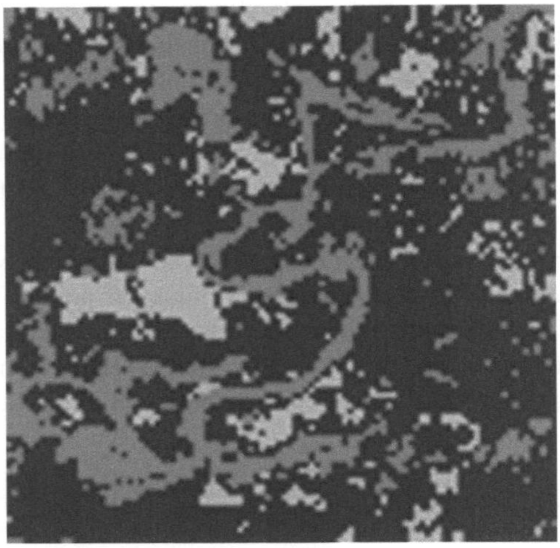

Fig. 2.26 Result obtained by the ISODATA method

Table 2.6 The error matrix of the remote sensing experiment

	K-Means			ISODATA			UDS			Total
True Class	C1	C2	C3	C1	C2	C3	C1	C2	C3	
C1	39	1	2	39	0	3	40	0	2	42
C2	7	9	0	3	13	0	3	13	0	16
C3	9	0	8	6	0	11	3	0	14	17
Accuracy	74.50%			84%			89%			

The accuracies of the three methods are summarized in Table 2.6. We can observe that the UDS method is more sensitive to the spectral features of the ground objects.

Remark 2.5. The K-means and ISODATA methods calculate the distance of a pixel from the seed and discriminate pixels according to their distances. Thus, if the seed setting is not optimal and the objects actually do not belong to the cluster determined by the seed, it may lead to undesirable results. The UDS, on the other hand, calculates the spectral distance of a pixel to all pixels rather than the seed. Therefore, the classification result of the UDS is more objective and insensitive to seed initialization. It thus leads to higher accuracy than the K-means and ISODATA.

2.5 Unraveling Spatial Objects with Arbitrary Shapes Through Mixture Decomposition Clustering

2.5.1 On Noise and Mixture Distributions in Spatial Data

A major problem in the mining of spatial objects or natural features is the rampant existence of noise and mixture distributions in large spatial databases. Thus, being able to describe distribution in the feature space with a high tolerance of noise is essential to detect successfully features in spatial data in general and remotely sensed images in particular.

Density-based method can often discover clusters of arbitrary shapes in databases consisting of noise/outliers. Unlike most partitioning methods that cluster features in terms of the distance between them, density-based methods identify clusters as dense regions interspersed by low density regions (often treated as noise/outliers) in the feature space. The DBSCAN algorithm (Ester et al. 1996) and OPTICS (Ankerst et al. 1999), for example, are early density-based methods that cluster (in a partitioning way) data on the basis of density and a set of user supplied parameters. Since both methods rely on some spatial index structures such as R^*-tree and X-tree, they are not efficient for clustering high dimensional spatial data. By using the sum of influence functions of all data points and a grid-like structure to assist the calculation of the density function and the hill-climbing procedure that identify the density attractor of each data point, the DENCLUE algorithm (Hinneburg and Keim 1998) exhibits a much better performance in cluster discovery. However, the algorithm again requires a set of parameters that need to be carefully chosen.

Due to the noise level and feature inter-mixing or overlapping, spatial features often take on the form of mixture density distributions. Thus, conventional approaches for simple distributions are inadequate for feature representation and mining in such feature spaces. Mixture density models, on the other hand, become useful for such purpose.

In a mixture model, data are assumed to follow two or more common parametric distributions mixed in varying proportions. Thus, finite mixture density models provide an important means to describe complex phenomena in relatively simple ways (Derin 1987; McLachlan and Basford 1988; Dattereya and Kanal 1990). In practice, the most important class of finite mixture densities is Gaussian (or normal) mixtures. For parametric estimation of this class of mixtures, we can usually select the expectation maximization (EM) algorithm which enables us to compute the maximum likelihood (ML) estimates of the mean vectors and covariance matrices of a Gaussian mixture distribution in an iterative manner (McLachlan and Krishnan 1997). The EM algorithms have been employed to perform spatial feature extraction, data fusion, and data mining in remotely sensed images (Bruzzone et al. 1999; Tadjudin and Landgrebe 2000).

However, the EM algorithm is severely handicapped in the estimation of suitable number of mixtures, especially when the overlapping of features exists in very noisy feature space. To overcome such difficulty, the Gaussian mixture density decomposition (GMDD) algorithm has been proposed as an effective clustering approach for data sets with mixture densities (Zhuang et al. 1996; Dave and Krishnapuram 1997). As an extension of GMDD, an effective data mining method, called regression-class mixture decomposition (RCMD), for regression relations has been developed for large data sets (Leung et al. 2001a) (see Chapter 5 for a detailed description of the approach). Within the framework of RCMD, a data set is treated as a mixture population composed of many components. Each component corresponds to a regression class defined as a subset of the data set that is subjected to a regression model. The RCMD method then extracts those regression classes in succession. In essence, a regression class reflects a kind of structure existing in the data set. Therefore, the RCMD method is more suitable for mining structured features in data sets. It can be extended to extract spatial features in remotely sensed images. Leung et al. (2006b) propose the RCMD-based feature mining model (RFMM) with genetic algorithms (GA). Through the RFMM-GA model, geometric features, represented by extended parametric models such as the linear structures, ellipsoidal structures, and more complicated parametric structures are extracted from noisy, complex and large spatial data sets. The main idea of the RFMM-GA model is to estimate effectively the parameters of the components of a mixture data set in order to find the components corresponding to the individual features. The GA is employed as a multi-point global optimization procedure to estimate efficiently the parameter sets of RFMM.

In a feature space, it is generally difficult to describe the distributions of feature sets with a common simple density distribution model since distributions of samples usually follow a mixture model. As depicted in Fig. 2.27, there are apparently three structured features in a two-dimensional space. However, the shapes of the features are so different and they overlap to some extent. Hence, the conventional density distribution model is too simple to successfully mine such features. They, however, can be appropriately unraveled by an extension of mixture density models.

As a flexible approach to density estimation, mixture density models have been applied to solve problems in a variety of disciplines. Mixture density modeling and decomposition (MDMD) (Zhuang et al. 1992, 1996; Dave and Krishnapuram 1997) can be viewed as a mixture clustering model that involves the use of robust statistics to identify individual densities more accurately and reliably. The basic flows of a MDMD algorithm are depicted in Fig. 2.28. Given a data set, the parametric distribution model corresponding to the feature to be mined is prespecified at each step. After performing estimation procedure for the parameters of the distribution model, the data subset, which is fittest to the model, is mined and taken out from the data set. The iterative decomposition procedure is completed until the whole data set is decomposed into categories of features in the mixture.

Gaussian mixtures are commonly employed to model finite mixture densities. The widespread use of Gaussian mixture densities is due to the fact that a univariate Gaussian distribution has a simple and concise representation requiring only two parameters: mean and variance. The Gaussian density is symmetric, unimodal, and isotropic, and it assumes the least prior knowledge (as measured in terms of the uncertainty or entropy of the distribution) in estimating an unknown probability density with given mean and variance. These characteristics of the Gaussian distribution along with its well-studied properties give the Gaussian mixture density models the power and effectiveness that other mixture densities can hardly surpass.

Evolved from the MDMD algorithm and the GMDD algorithm, robust regression-class mixture decomposition (RCMD) is formulated as a composition of simple

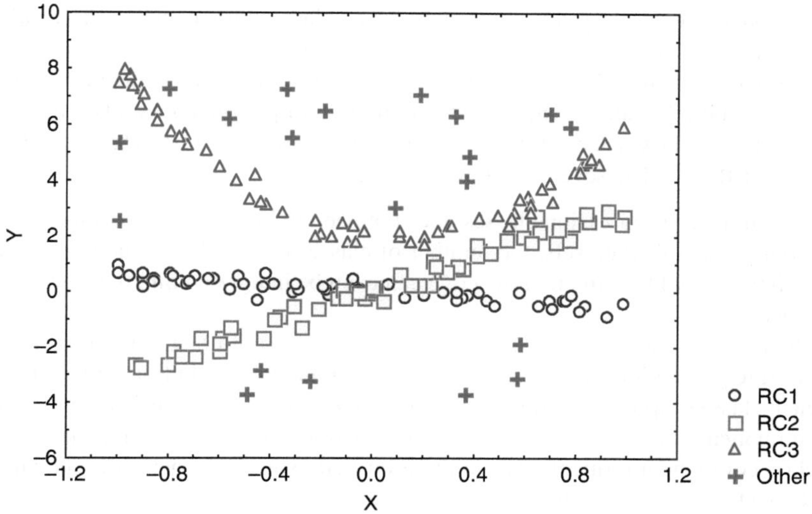

Fig. 2.27 Mixture population containing noise and genuine features

Fig. 2.28 Process of MDMD algorithm

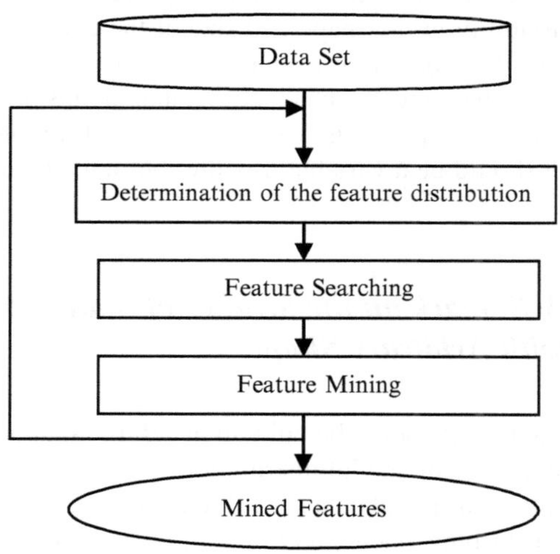

structured regression classes. With respect to a particular regression class, all data points from the other regression classes can be classified as the outlier with different statistical characterization. Thus, a mixture population can be viewed as a contaminated regression class with respect to each class component in the mixture. When all of the observations fitting a single regression class are grouped together, the remaining observations can be considered as elements of an unknown outlier set. Each class component in the mixture population can be estimated separately one at a time in an iterative fashion by using the contaminated model. The iterative estimation successively reduces the number of class components in the resulting mixture until all regression classes in the mixture are mined.

The output of the RCMD algorithm includes a number of extracted regression classes and possibly an unassigned set containing samples that do not belong to any of the detected classes. Compared to conventional statistical clustering methods, the scheme of RCMD has several distinct advantages:

1. The number of regression classes does not need to be specified a priori. Given a group of sample data sets, the number of classes can be determined one by one with the RCMD. Domain knowledge can even be integrated into the data mining process.
2. The mixtures can contain a large proportion of noise. In fact, a regression-class mixture population can be viewed as a contaminated distribution with respect to each class component in the mixture. Thus the proportion of outliers relative to a component may be large. Even in such situation, RCMD still can pick out each class component sequentially via the robust statistics approach. It is shown that RCMD can resist a large proportion of noise.
3. The estimation of parameters of each class component is virtually independent of each other. This property is derived from the search strategy adopted by the RCMD (Leung et al. 2001). However, for high dimensional feature space, it is more effective if a suitable search range can be pre-determined.
4. The variability in the shape and size of the components in the mixture is taken into consideration. In the search procedure, parameters of each component should be dynamically changed so that points identified by this component follow the corresponding distribution. Therefore, the distribution for the whole data set should be a variable mixture, but not single and fixed.

2.5.2 A Remark on the Mining of Spatial Features with Arbitrary Shapes

Spatial data mining should be built upon definite spatial analysis model and follows the true regularity of spatial distribution. The target is to discover spatial features from complicated spatial data sets. Due to complexity and uncertainty, overlapping and inter-disturbing phenomena among spatial data sets often occur. It is thus difficult to acquire substantial structure of the feature distribution which will directly affect the accuracy of the analysis and interpretability of the features unraveled. An effective way to describe complexity and uncertainty of data in statistics is mixture modeling, i.e., utilizing a mixture model of finite number of simple distributions (called components) to characterize a complicated data set.

However, features in spatial databases such as remotely sensed images may often not appear as a conventional mixture in which the distribution of each component is a density function with a fixed point as its "center" (mean), as in usual statistical distributions. The more likely situation is that they may take on a mixture whose components may contain features such as roads and rivers. It is thus inappropriate to model them by a single conventional distribution model. Many of the conventional feature mining approaches, however, are based on such a conventional distribution (especially the multivariate Gaussian distribution model) and the number of the features usually needs to be known a priori. Since the number of significant features is generally not known a priori, then the conventional single-distribution approach is not suitable to model and optimally extract features in an image. As pointed out in Richards and Xia (1999, p. 261), information classes of interest often do not appear as single distributions but rather a set of constituent spectral classes or sub-classes. So we not only need a descriptive approach to characterize spatial features in remotely sensed images but also an effective method for mining those features.

An important application of the RCMD algorithm is to identify features or classes in multi-dimensional data sets. It can easily be observed that a regression class actually corresponds to a feature so that the variable y, called the response (dependent) variable, is a function of the other variables $x_1, x_2, ..., x_p$, called the explanatory (independent) variables, i.e., $y = f(x_1, x_2, ..., x_p)$. However, in many practical situations, we may not be able to explicitly express a variable by other relevant variables. Thus, a feature may only be characterized by an equation $F(z_1, z_2, ..., z_p) = 0$ with respect to p variables $z_1, z_2, ..., z_p$, where dependent and independent variables are indistinguishable. Obviously, such a representation of features generalizes that of the regression-class framework. It is more flexible and effective in the mining of features in remotely sensed images. Leung et al. (2006b) further extend the RCMD method into the RFMM method to perform feature mining in more general situations. The RFMM method is first described in the following subsection and then the version with a genetic algorithm for more efficient performance in discussed in Sect. 2.5.4.

2.5.3 A Spatial-Feature Mining Model (RFMM) Based on Regression-Class Mixture Decomposition (RCMD)

Within the RFMM framework, the spatial feature to be mined should first be determined. The shape distribution is extended on the regression-class concepts in RCMD. Let $\mathbf{X} = \{\mathbf{x}_1, ..., \mathbf{x}_n\}$ be a p-dimensional data set, $\mathbf{x}_k = (x_{k1}, ..., x_{kp})^{\mathrm{T}} \in \boldsymbol{R}^p$, Assume that $\mathbf{x}_k$ follows a distribution $g(\mathbf{x}_k; \boldsymbol{\theta})$ with probability $1 - \varepsilon$ $(0 < \varepsilon < 1)$ and another distribution $h(\mathbf{x}_k; \boldsymbol{\theta})$ with probability ε for the outlier, where $\boldsymbol{\theta}$ is the parameter vector of the feature to be estimated. Thus, data samples are identically distributed with the common density:

$$f(\mathbf{x}_k; \boldsymbol{\theta}) = (1 - \varepsilon) \cdot g(\mathbf{x}_k; \boldsymbol{\theta}) + \varepsilon \cdot h(\mathbf{x}_k; \boldsymbol{\theta}). \tag{2.80}$$

According to the strategy of RCMD, the search process is to maximize the model-fitting function:

$$Q(\boldsymbol{\theta}) = \sum_k \log(g(\mathbf{x}_k; \boldsymbol{\theta}) + t), \tag{2.81}$$

where $t > 0$ is called *a partial model* and is selected by the method suggested in RCMD. According to the derivation in Leung et al. (2001a), the partial model t actually corresponds to the partial information about the outlier distribution $h(\mathbf{x}_k; \boldsymbol{\theta})$. Although outliers with respect to an underlying model $g(\mathbf{x}_k; \boldsymbol{\theta})$ exist inevitably in reality and the knowledge on the whole shape of the outlier distribution is usually unknown, we can approximately represent their existence by introducing a positive number t and use its value as a reduction of the information about the outlier as a whole. If the partial model t does not appear in (2.81), that is $t = 0$, then the method determined by (2.81) is the ordinary maximum likelihood (ML) method, which is not robust. However, once we have $t > 0$, the resulting method is fairly robust. Here, the shape of the distribution, controlled by the parameter $\boldsymbol{\theta}$, represents the feature structures hidden in the mixture.

As depicted in Fig. 2.29, spatial features can generally be categorized into several basic shapes, such as the simple Gaussian classes, linear structures, curvilinear structures, ellipsoidal structures, and other complicated structures integrated with domain specific knowledge. Specifically we have:

1. Simple Gaussian class (Fig. 2.29a)

The density corresponding to the Gaussian feature in a data set $\mathbf{X}$ can be expressed as:

$$g(\mathbf{x}_k; \boldsymbol{\theta}) = \frac{1}{(\sqrt{2\pi})^p \sqrt{|\Sigma|}} \exp\left(-\frac{1}{2} d^2(\mathbf{x}_k)\right), \tag{2.82}$$

$$d^2(\mathbf{x}_k) = (\mathbf{x}_k - \boldsymbol{\mu})^{\mathrm{T}} \Sigma^{-1} (\mathbf{x}_k - \boldsymbol{\mu}), \tag{2.83}$$

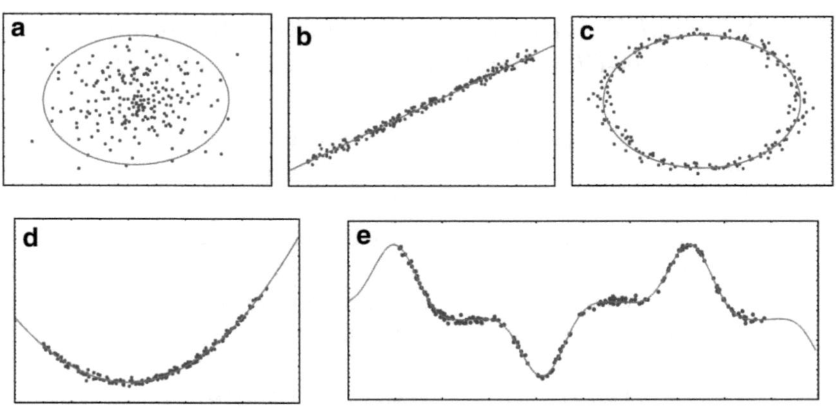

Fig. 2.29 The distributions of various spatial features (**a**) Simple Gaussian class. (**b**) Linear structure. (**c**) Ellipsoidal structure. (d) General curvilinear structure. (e) Complex structure

where $d^2(\mathbf{x}_k)$ is the square of the Mahalanobis distance, and $\boldsymbol{\Sigma}$ is the covariance matrix such that the parameter vector θ to be searched is the mean vector $\boldsymbol{\mu}$ and the covariance matrix $\boldsymbol{\Sigma}$.

2. Linear structure (Fig. 2.29b)

In multi-dimensional space, linear features can be characterized by the following distribution with parameter vector $\boldsymbol{\theta} = (\boldsymbol{\beta}^{\mathrm{T}}, \sigma)^{\mathrm{T}}$:

$$g(\mathbf{x}^k; \boldsymbol{\theta}) = \frac{1}{\sqrt{2\pi}\sigma} \exp\left(-\frac{r_k^2(\boldsymbol{\beta})}{2\sigma^2}\right), \tag{2.84}$$

where $\boldsymbol{\beta} = (\beta_0, \beta_1, \cdots, \beta_p)^{\mathrm{T}}$ is the coefficient vector of the following linear equation:

$$\beta_0 + (\beta_1, \cdots, \beta_p)\mathbf{x} = 0, \tag{2.85}$$

and r_k denotes the residuals of data $\mathbf{x}_k$ with respect to (2.85):

$$r_k = \beta_0 + (\beta_1, \cdots, \beta_p)\mathbf{x}_k, \tag{2.86}$$

σ is such that at least 98% of the points constituting the feature are contained within 3σ from the line.

3. Ellipsoidal structure (Fig. 2.29c)

In multi-dimensional space, an ellipsoidal-like structure depicted by

$$F(\mathbf{x}; \boldsymbol{\theta}) \equiv 1 - \sum_{i=1}^{p} \frac{(x_i - \beta_i)^2}{\gamma_i^2} = 0, \tag{2.87}$$

can also be considered, where $\mathbf{x} = (x_1, ..., x_p)^{\mathrm{T}}$, $\boldsymbol{\theta} = (\boldsymbol{\beta}^{\mathrm{T}}, \boldsymbol{\gamma}^{\mathrm{T}}, \sigma)^{\mathrm{T}}$ is the parameter vector, $\boldsymbol{\beta} = (\beta_1, \cdots, \beta_p)^{\mathrm{T}}$ is the location of the center point, $\boldsymbol{\gamma} = (\gamma_1, \cdots, \gamma_p)^{\mathrm{T}}$, and γ_i is the i-th semimajor-like axes of length? In this situation, its features are still characterized by (2.84), but the residuals become

$$r_k = 1 - \sum_{i=1}^{p} \frac{(x_i^k - \beta_i)^2}{\gamma_i^2}. \tag{2.88}$$

For simplicity, we consider the ellipse feature (i.e., the case $p = 2$). In this situation, the major and minor axes of the ellipse depicted by (2.87) are parallel to the coordinate axes. For more general ellipses, their equations can be transferred into (2.87) by the rotation transformation. Then the expression in (2.88) still holds.

4. General curvilinear structure (Fig. 2.29d)

For more general curvilinear structure, (2.76) is still applicable. We only need to modify the residuals r_k. As the general curve can be described by the equation $f(\mathbf{x}\,;\,\boldsymbol{\beta}) = 0$, the corresponding residuals r_k are:

$$r_k = f(\mathbf{x}_k; \boldsymbol{\beta}), \tag{2.89}$$

where f is a known function for specifying a curve and $\boldsymbol{\beta}$ is the parameter vector of the curve.

5. Complex structure (Fig. 2.29e)

Spatial features often take on a more complex shape. A simple method to represent features with complex shape is to combine simpler feature structures into an integrative one with prior knowledge. For example, a production system can be employed to determine a complex structure as follows:

$$0 = \begin{cases} f_1(\mathbf{x}), \text{if}(\mathbf{x} \in A_1), \\ f_2(\mathbf{x}), \text{if}(\mathbf{x} \in A_2), \\ \qquad \cdots \\ f_m(\mathbf{x}), \text{if}(\mathbf{x} \in A_m) \end{cases} \tag{2.90}$$

Moreover, more complicated structures or irregular structures, seemingly not being able to be parametrically represented, can be simulated by appropriate combinations of these simple parametric structures.

2.5.4 The RFMM with Genetic Algorithm (RFMM-GA)

Finding solution for the RFMM is essentially an optimization process that estimates the parameter vector θ of the feature structures. Mean squared error (MSE) is frequently employed as an optimization criterion. The disadvantages of many of the conventional optimization methods are their computational complexities and their prone to local minima. It, in particular, becomes more difficult when complex distributions integrated with domain knowledge in symbolic forms are encountered in optimization. The use of more flexible methods such as genetic algorithms (GA) is often necessary.

Genetic algorithms (GA) are highly parallel and adaptive search processes based on the principles of natural selection (Holland 1975; Goldberg 1989; Zhang and Leung 2003) (see Chap. 3 for a more formal discussion of GA). Genetic operators (namely selection, crossover and mutation) are applied to evolve a population of coded solutions (strings /chromosomes) in an iterative fashion until the optimal population is obtained. GA is thus a multi-point search algorithm which seeks the optimal solution with the highest value of a fitness function. For example, in order to solve the optimization problem in (2.81) in which $g(\mathbf{x}^k; \boldsymbol{\theta})$ is defined by (2.82) or (2.84), the function in (2.81) is selected as the fitness function to be maximized. The GA starts with a population of individuals (chromosomes) representing the parameter vector $\boldsymbol{\theta} = (\theta_1, \theta_2 \cdots, \theta_l)^{\mathrm{T}}$ which is encoded as a string of finite length. A chromosome is usually a binary string of 0's and 1's. For example, suppose the binary representation of $\theta_1, \theta_2 \cdots, \theta_l$ for 5-bit strings are 10110, 00100,. . ., 11001, respectively. Then the string $s =$ 10110 00100 ... 11001 is a binary representation of $\boldsymbol{\theta} = (\theta_1, \theta_2 \cdots, \theta_l)^{\mathrm{T}}$ and forms a one-to-one relation with $\boldsymbol{\theta}$. The q-tuple of individual strings $(s_1, ..., s_q)$ is said to be a population S in which each individual $s_i \in S$ represents a feasible solution of the problem in (2.81). The randomly generated binary strings then form the initial population to be evolved by the GA procedure, i.e., by the GA operators briefly outlined as follows:

1. Selection. It is the first operator by which individual strings are selected into an intermediate population (termed mating pool) according to their proportional fitness obtained from the fitness function. The *roulette wheel selection* technique is employed in such a way that strings with higher fitness would have higher probability to be selected for reproduction.
2. Crossover. After selection, two individuals can exchange materials at certain position(s) through the crossover operator. Crossover is a recombination mechanism to explore new solutions. The crossover operator is applied with some probability P_c. Single-point, multi-point, or uniform crossover may be employed. In practice, single-point crossover is simpler and more popular. First, individuals of the intermediate population are paired up randomly. Individuals of each pair (*parents*) are then combined, choosing one point in accordance with a uniformly distributed probability over the length of the individual strings and cutting them in two parts accordingly. The two new strings (*offspring*) are formed by the juxtaposition of the first part of one parent and the last part of the other parent.
3. Mutation. After crossover, the mutation operator is applied with uniform probability P_m. Mutation operates independently on each offspring by probabilistically perturbing each bit string. In other words, it alters the genetic code (e. g., from 0 to 1 or 1 to 0) of an individual at a certain randomly generated position. The mutation operator helps to prevent the irrecoverable loss of potentially important genetic material in an individual.

The basic procedure of the GA-based optimization for parameter estimation of the RFMM is depicted in Fig. 2.30. The GA search aims at the maximization of Q in (2.81). The parameter $\boldsymbol{\theta}$ is estimated while Q attains its maximum through the GA-based evolution. The spatial feature specified by $\boldsymbol{\theta}$ is thus successfully mined from the image.

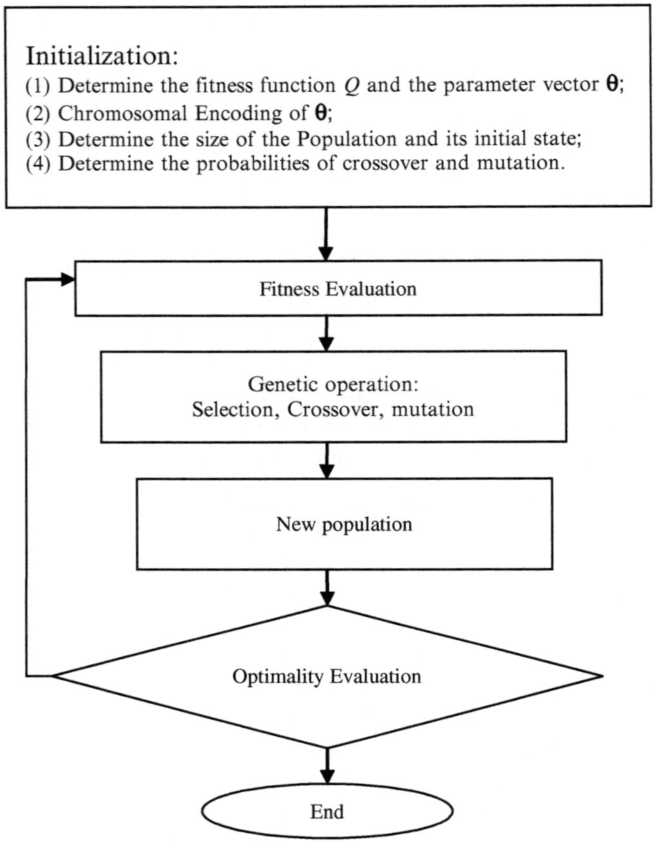

Fig. 2.30 RFMM-GA optimization algorithm

2.5.5 *Applications of RFMM-GA in the Mining of Features in Remotely Sensed Images*

For substantiation, the first two numerical experiments involve the extraction of one and two ellipsoidal features in simulated data sets contaminated with noise, and the third experiment deals with the automatic detection of linear features in a real-life remotely sensed image.

To simplify our discussion, the default set up of the RFMM-GA is specified as follows: the partial model level $t = 0.1$, $q = 300$, $P_c = 0.8$, $P_m = 0.5$.

2.5.5.1 Experiment 2.4 Ellipsoidal Feature Extraction from Simulated Data

In this experiment, the RFMM-GA is employed to extract features with ellipsoidal shape from simulated data sets contaminated with noise. It is actually a special clustering approach for estimating and extracting patterns. As shown in Fig. 2.31,

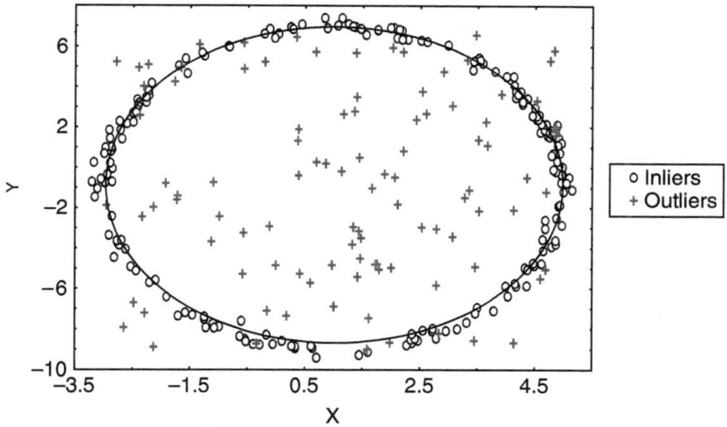

Fig. 2.31 Extraction of ellipsoidal feature

there is an ellipsoidal feature in a two-dimensional feature space with a lot of noisy points distributed randomly around it. The parameters of the true model in (2.87) are $p = 2, \beta_1 = 1 = -\beta_2, \gamma_2 = 8 = 2\gamma_1$, and σ in (2.84) is selected as 0.5. There are 300 points in Fig. 2.26, i.e., $n = 300$, in which 200 points (inliers) are generated randomly from the true model with ellipsoidal feature, and 100 points (outliers) are uniform noise. Applying RFMM-GA, the feature parameter θ can be acquired. In this experiment, the obtained parametric estimation includes the center point: $(\hat{\beta}_1, \hat{\beta}_2) = (0.999, -1.025)$; semi-major axes of length: $\hat{\gamma}_1 = 3.989$, $\hat{\gamma}_2 = 8.031$; and $\hat{\sigma} = 0.502$. With these unraveled parameters, the fitness value Q in (2.81) attains its maximum at – 124.379, and the feature is successfully mined.

2.5.5.2 Experiment 2.5 Extraction of Two Ellipsoidal Features from Simulated Data

To further illustrate the effectiveness of the RFMM-GA, this experiment is designed for the extraction of two ellipsoidal features in a data set contaminated with noise (Fig. 2.32). In the data set, 200 points come from the ellipsoidal feature characterized by the equation: $x^2/2^2 + y^2/1^2 = 1$; 200 points come from the ellipsoidal feature characterized by the equation: $(x-1)^2/1^2 + (y-5)^2/2^2 = 1$; and the other 100 points are noise. The first ellipsoidal feature in (2.87) unraveled by the RFMM-GA has the parameter estimates: $(\hat{\beta}_1, \hat{\beta}_2) = (-0.049, -0.009)$, $(\hat{\gamma}_1, \hat{\gamma}_2) = (2.023, 1.050)$, $\hat{\sigma} = 0.20$, and the fitness value Q in (2.81) attains its maximum -1952.813 at $t = 0.005$. The corresponding data points are then removed from the data set. The RFMM−GA is again applied to unravel the second ellipsoidal feature with parameters: $(\hat{\beta}_1, \hat{\beta}_2) = (1.040, 5.007)$, $(\hat{\gamma}_1, \hat{\gamma}_2) = (1.008, 2.013)$,

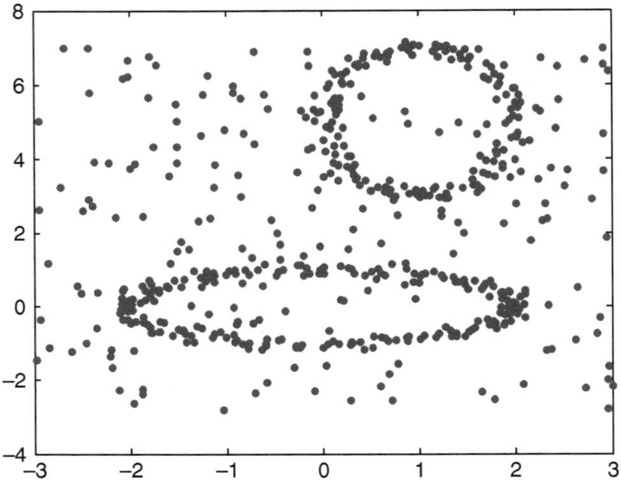

Fig. 2.32 Extraction of two ellipsoidal features

$\hat{\sigma} = 0.20$, and Q has its maximum -1277.918 at $t = 0.005$. This clearly shows that the RFMM-GA can effectively extract multiple features from noisy data sets.

2.5.5.3 Experiment 2.6 Linear Feature Extraction from a Satellite Image

A lineament in a feature space is defined as a simple or composite linear feature whose parts are aligned in a rectilinear or slightly curvilinear manner which might indicate the existence of some kind of spatial structures. Classical lineament detection methods are mainly based on gradient or Laplacian filtering which often generate a large amount of false edges and fail to link together missing occluding parts combined with the use of thresholds. Though some improvements have been achieved by applying a Hough transform to the threshold image, more recent approaches attempt to circumvent the problem by extracting the gray level in high variability through filtering techniques. Neural network models, such as adaptive resonance theory (ART), multilayer perceptron with back propagation (MLP-BP), and cellular neural networks (CNN), have also been proposed to extract connected edges (Basak and Mahata 2000; Lepage et al. 2000; Wong and Guan 2001). However, all of these approaches could only produce good results in detecting small scale edge features, but are of very limited use in the detection of linear or non-linear features, especially when lineaments have a fuzzy, gleaming, or broken appearance in aerial or satellite images (Man and Gath 1994).

Since features can be parametrically defined in the feature space under RFMM-GA, it can then provide a framework to parametrically extract spatial features from remotely sensed images. By the RFMM-GA method, the fittest linear features are successively searched and extracted from the feature space by stepwise decomposition. The RFMM-GA is supported by robust statistical technique which could reduce the interference of adjacent features and noisy points, and enable a reliable discrimination of linear features without any a priori knowledge about the number involved. Finally, linear features are mined and characterized by the associated parameters.

Figure 2.33 depicts the result of an experiment on the extraction of lineaments, defined by (2.85), by the RFMM -GA from a real-life satellite image.

Figure 2.34a depicts the original imagery of TM band 5 in another experiment located in Guangzhou, China, acquired on January 2, 1999. Three lineaments are

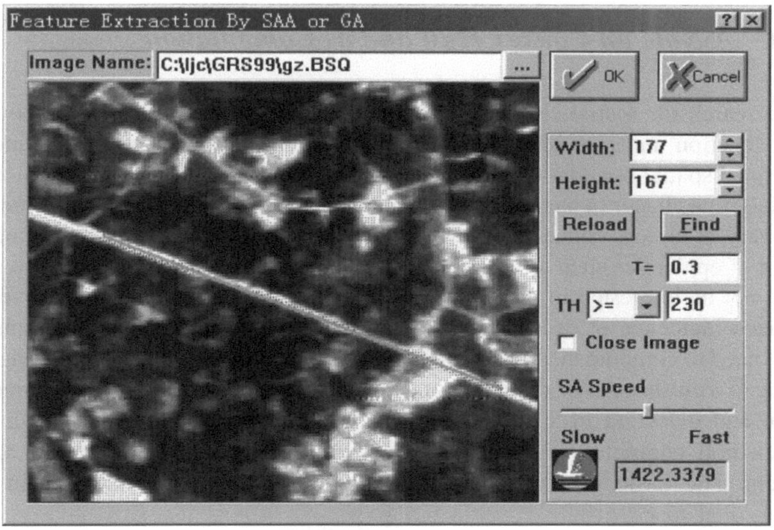

Fig. 2.33 Feature extraction system with RFMM

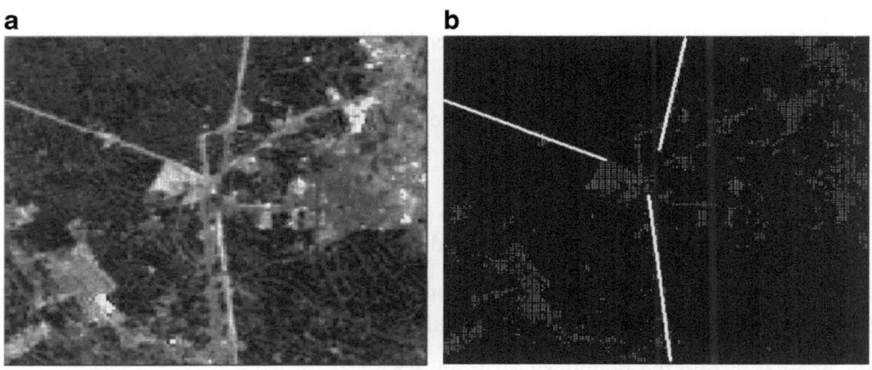

Fig. 2.34 Lineament extraction from satellite imagery (a) Original TM5 imagery (b) Results of lineament extraction

apparently three highways intersecting at a small town in the image. The features are first separated from the background with the threshold segmentation approach. Then according to the feature distribution of the lineaments, the targets are extracted by the stepwise search of the RFMM-GA. The three highways are successfully mined from the blurred imagery (Fig. 2.34b). These two experiments demonstrate that RFMM-GA provides a novel framework for feature extraction in remotely sensed images.

2.6 Cluster Characterization by the Concept of Convex Hull

2.6.1 A Note on Convex Hull and its Computation

In the search for spatial clusters, we sometimes may not have any idea about the exact location and size of a cluster, particularly in databases with undefined or ill-defined spatial boundaries. In some applications, we might just need to discover and delimit a cluster (a particular spatial concentration or hotspot) in real-time. The discovery of disease concentration, particularly the time varying concentration and spread of epidemics such as SARS and avian flu, is a typical example. Such study might not be interested in the partition of the whole data set but the discovery of localized incidence of excessive rate (Lawson 2001). Under some situations, we might need to compute the cluster diameter or to determine whether a point in space belong to a cluster. All of these tasks need a formal approach for cluster characterization and detection in spatial databases. It is proposed in here the method of convex hull computation formulated by Leung et al. (1997a) to detect spatial clusters. The basic idea is to encompass a cluster by a convex hull in high dimensional space.

To facilitate our discussion, I first give some notions of convex hulls and their computations. Let $\mathbf{S} = \{p^{(1)}, p^{(2)}, \cdots, p^{(M)}\}$ be a set of M points in $\mathbf{R}^N$. The convex hull of $\mathbf{S}$, denoted as $C(\mathbf{S})$, is the smallest convex set that contains $\mathbf{S}$. Specifically, $C(\mathbf{S})$ is a polygon in the planar case, and a polyhedron in the three-dimensional (3-D) case. In general, $C(\mathbf{S})$ can be described in terms of one of the following characteristics:

1. The *faces* of $C(\mathbf{S})$, or equivalently, the boundary of $C(\mathbf{S})$ denoted as $\mathrm{Bound}(\mathbf{S})$.
2. The *vertex* set, $Ver(\mathbf{S})$, which is the minimum subset of $\mathbf{S}$ such that $C[Ver(\mathbf{S})] = C(\mathbf{S})$.
3. The *set of hyperplanes*, denoted as $\mathbf{H}{*}(\mathbf{S})$, by which $C(\mathbf{S})$ becomes the intersection of the closed-half spaces bounded by $\mathbf{H}{*}(\mathbf{S})$.

Thus, three types of convex hull computation problems with respect to the above-stated characteristics can be formally stated as:

Problem 1: to find the boundary set $\mathrm{Bound}(\mathbf{S})$.
Problem 2: to determine the vertex set $Ver(\mathbf{S})$.
Problem 3: to specify the hyperplanes $\mathbf{H}{*}(\mathbf{S})$.

These three problems are closely related to each other. Each of them, however, has its own concern and applications. Over the years, much effort has been devoted to develop algorithms for convex hull computation which can generally be classified into two different approaches: computing the exact convex hull (Atallah 1992; Bentley et al. 1993) and computing an approximate convex hull (Bern et al. 1992; Guibas et al. 1993).

For computing the exact convex hull by a serial computer, it has been shown that the problem can be solved in the planar and the 3-D cases with a time complexity of $O(M \log M)$ if all the points $p^{(i)}$ are given (the *off-line* problem) (Graham 1972; Preparata and Hong 1977), or with a complexity of $O(\log M)$ if the points are given one by one and the convex hull is updated after each point is added (the *on-line* problem) (Preparata 1979). Bentley et al. (1993) propose a novel algorithm that computes the convex hull in N-dimensional space in $2MN + O\left(M^{1-1/N}\log^{1/N} M\right)$ expected scalar comparisons, which represents a substantial improvement over the previous best result of $2^{N+1}NM$ (Golin and Sedgewick 1988). Wennmyr (1989) presents a neural network algorithm which computes an exact convex hull in $O(M)$ time *off-line*, and $O(\log M)$ time *on-line* in the planar case. The respective performances are $O(hM)$ and $O(M)$ in the 3-D case, where h is the number of faces in the convex hull.

There are essentially two kinds of algorithms for computing an approximate convex hull. The first kind can be classified as *robust algorithms* which compute the convex hull with imprecise computation. The basic geometric tests needed to compute the convex hull are considered unreliable or inconclusive when implemented with imprecise computations (e.g., ordinary floating-point arithmetic). Such algorithms aim at constructing a convex hull very close to containing all the points under consideration. The algorithms of this kind are often much more complicated than those for computing the exact convex hull.

The second kind of algorithms for computing an approximate convex hull is the *approximate algorithms*. The geometric tests are considered as reliable and conclusive. Such algorithms compute a convex hull that closely approximates the exact one (Bern et al. 1992). Despite of losing a certain degree of accuracy in computing the convex hull, approximate algorithms have in general very low complexity but very high computation efficiency. They would be particularly useful in applications where the speed rather than the accuracy of computing a convex hull is of major concern, or generating the exact convex hull is not necessary or impossible (e.g., when the data involved in the set of points are inherently not exact).

Leung et al. (1997a) employ a neural-network approach to develop an approximate algorithm for computing approximate convex hulls in the general N-dimensional space. It solves the *off-line* problem with a linear time complexity of $O(M)$, and the *on-line* problem with $O(1)$ time complexity. Its advantages are: First, unlike the known linear expected-time complexity algorithm (Bentley et al. 1993) which might not keep linear time complexity for the worst case (it could even be much worse), the convex hull computing neural network (CHCNN) always keep linear time complexity for any case. Second, the massively parallel processing capability of the neural network makes the derived algorithm developed to be real-time in nature. This real-time processing capability is of particular importance and is required in a wide variety of applications related to adaptive and real-time processing. For example, in the collision avoidance applications (Hwang and Ahuja 1993), a robot is to be controlled to move automatically in an environment involving variable obstacles. Assume the obstacles are polyhedrons. The problem then can be deduced to determining in real-time the condition (the control strategy) under which the two convex hulls, one being the obstacle(s) and the other the range of the robot's motion, do not intersect. This then requires the real-time construction of the related convex hulls. Third, once the neural network is implemented as a physical device, it becomes extremely direct and handy to use it in various applications, such as judging if a given point (e.g., suspected outlier) belongs to a cluster, and computing cluster diameter (e.g., extent of spread of a contagious disease) when the given points constitutes a set of samples.

2.6.2 Basics of the Convex Hull Computing Neural Network (CHCNN) Model

Let $\mathbf{n} = (n_1, n_2, \cdots, n_N)^T \in \mathbf{R}^N$ be a unit vector. For any real number α, the set H defined by

$$H = \{x \in \mathbf{R}^N : \langle \mathbf{n}, x \rangle = \alpha\} \tag{2.91}$$

is called a hyperplane and the set $\bar{H}$ defined by

$$\bar{H} = \{x \in \mathbf{R}^N : \langle \mathbf{n}, x \rangle \leq \alpha\} \tag{2.92}$$

is called a closed half-space bounded by H. In the case, the vector $\mathbf{n}$ is said to be the normal vector of H.

Given any set $\mathbf{S}$ of finite number of points in $\boldsymbol{R}^N$, $C(\mathbf{S})$ can be expressed as the intersection of a finite number of closed half-space bounded by certain hyperplanes.

A hyperplane $\mathbf{H}$ is said to be a *supporting hyperplane* of $C(\mathbf{S})$ if $\mathbf{S} \subseteq \bar{H}$ and H itself contains at least one point of S. Therefore a supporting hyperplane supports $C(\mathbf{S})$ in a specific direction. Every point in $H \cap \mathbf{S}$ is referred to as a *supporting point* of $C(\mathbf{S})$, and any intersection of $H \cap C(\mathbf{S})$ is referred to as a *face* of $C(\mathbf{S})$. Any supporting point of $C(\mathbf{S})$ clearly lies on the boundary of $C(\mathbf{S})$. A supporting point p is a vertex of $C(\mathbf{S})$ if there do not exist two different points a , b in $C(\mathbf{S})$ such that p lies on the open line segment $]a, b[$ (i.e., no $\beta \in (0, 1)$ exists such that $p = (1 - \beta)a + \beta b$).

The CHCNN generates two specific approximations of the convex hull $C(\mathbf{S})$: one is inscribed within $C(\mathbf{S})$ and the other circumscribes $C(\mathbf{S})$ in the geometric sense. These two types of approximate convex hulls are specified by the following definitions:

Definition 2.4. *A convex hull* C_1 *is said to be an inscribed approximation of* $C(\mathbf{S})$ *if* $C_1 \subseteq C(\mathbf{S})$ *and any vertices of* C_1 *are on the boundary of* $C(\mathbf{S})$. *A convex hull* C_2 *is said to be a circumscribed approximation of* $C(\mathbf{S})$ if $C(\mathbf{S}) \subseteq C_2$ *and every face of* C_2 *contains at least a vertex of* $C(\mathbf{S})$.

In Figs. 2.35 and 2.36, all convex hulls demarcated by thin lines are inscribed approximations of $C(\mathbf{S})$ and those demarcated by bold lines are circumscribed approximations of $C(\mathbf{S})$. The line with medium width represents the $C(\mathbf{S})$.

The CHCNN developed in Leung et al. (1997a) is motivated by the following observations: every vertex (say, p) of the convex hull $C(\mathbf{S})$ must be supporting point and therefore, there is a direction vector $\mathbf{n}$ in which p will maximize the inner product $\langle \mathbf{n}, p^{(i)} \rangle$ among all the $p^{(i)}$s in $\mathbf{S}$. With finite points in $\mathbf{S}$, all vertices of $C(\mathbf{S})$ can be uniquely recognized in terms of the maximization procedure with a finite number of direction vectors. The basic idea in developing the CHCNN then is to yield the vertices of $C(\mathbf{S})$ through the maximization process via a prespecified set of direction vectors.

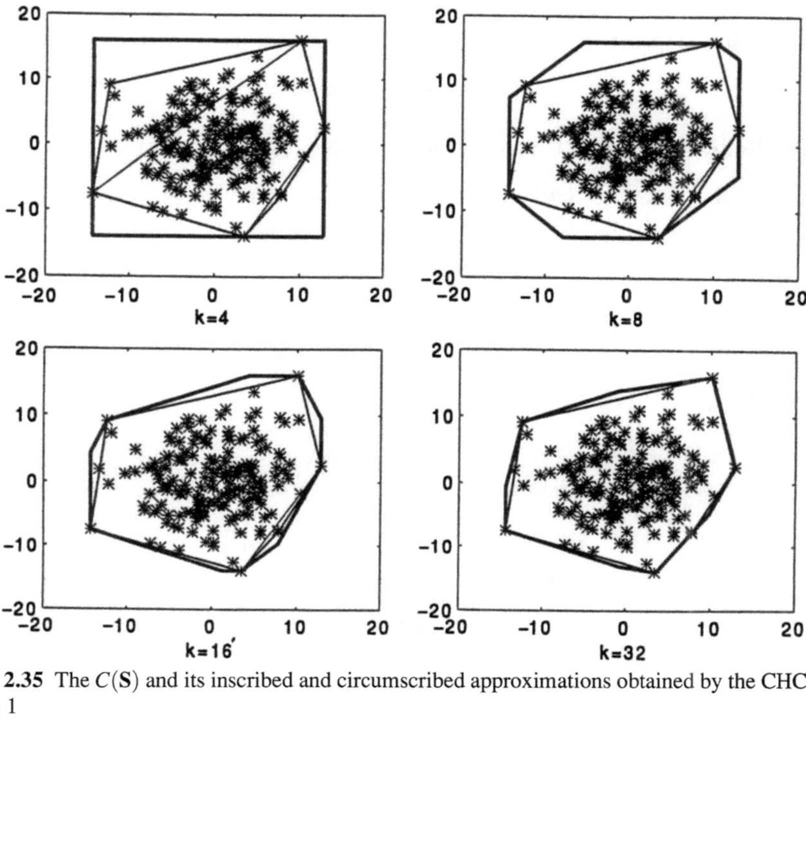

Fig. 2.35 The $C(\mathbf{S})$ and its inscribed and circumscribed approximations obtained by the CHCNN: case 1

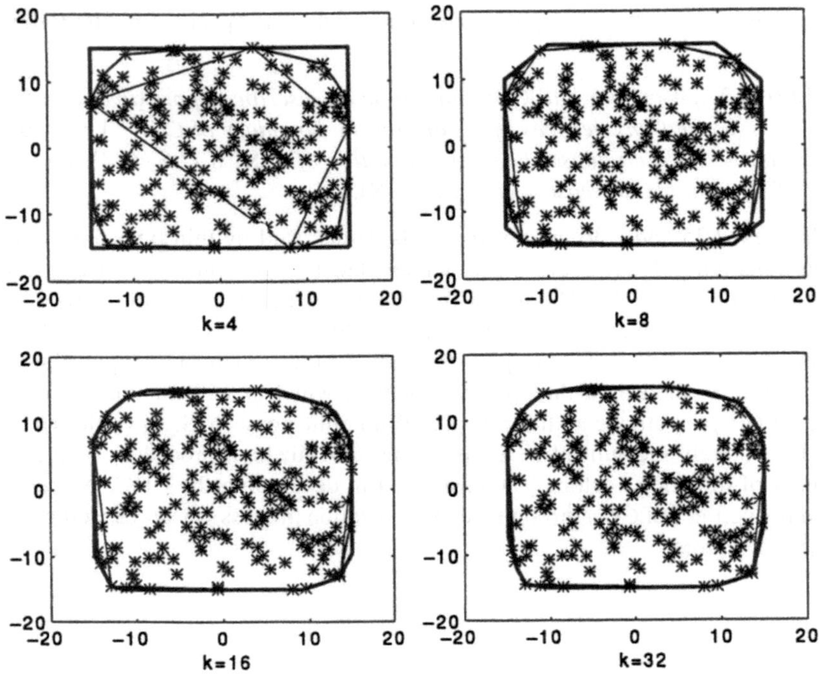

Fig. 2.36 The $C(\mathbf{S})$ and its inscribed and circumscribed approximations obtained by the CHCNN: case 2

The following Lemma Underlies the CHCNN:

Lemma 2.1. *Let* $\mathbf{U} = \{n^{(1)}, n^{(2)}, \cdots, n^{(k)}\}$ *be a given set of unit direction vectors in* $\mathbf{R}^N$. *If* $i(j) \in \{1, 2, \cdots M\}$ *is the index such that*

$$\left\langle \mathbf{n}^{(j)}, p^{[i(j)]} \right\rangle = \max_{1 \le i \le M} \left\{ \left\langle \mathbf{n}^{(j)}, p^{(i)} \right\rangle \right\} , \tag{2.93}$$

we denote

$$\theta_j = \left\langle \mathbf{n}^{(j)}, p^{[i(j)]} \right\rangle \tag{2.94}$$

$$H^{(j)} = \left\{ \mathbf{x} \in \mathbf{R}^N : \left\langle \mathbf{n}^{(j)}, \mathbf{x} \right\rangle = \theta_j \right\} \tag{2.95}$$

$$\mathbf{V}^* = \left\{ p^{[i(1)]}, p^{[i(2)]}, \cdots, p^{[i(k)]} \right\} \tag{2.96}$$

and,

$$\mathbf{H}^* = \cup\left\{H^{(j)} : j = 1, 2, \cdots, k\right\}, \tag{2.97}$$

then

1. $C_H(\mathbf{H}^*) = \cap_{j=1}^{k} \bar{H}^{(j)}$ *is a circumscribed approximation of* $C(\mathbf{S})$.
2. $C(\mathbf{V}^*)$ *is an inscribed approximation of* $C(\mathbf{S})$.

(See Leung et al. (1997) for the proof)

Lemma 2.1 indicates that for a pre-specified set of directions $\mathbf{U}$, as long as the θ_j and $p^{[i(j)]}$ defined by (2.93) and (2.94) are known, the convex hulls $C(\mathbf{V}^*)$ and $C_H(\mathbf{H}^*)$ provide respectively two approximations of $C(\mathbf{S})$ in the inscribed and circumscribed manners. Furthermore, in this case, (2) in Lemma implies that the set $\mathbf{V}^*$ defined by (2.96) offers a very good approximation to the vertex set $Ver(\mathbf{S})$, and hence to the convex hull under Problem 2. Also, (1) implies that every $H^{(j)}$ is a supporting hyperplane of the convex hull $C(\mathbf{S})$, and consequently yields an approximate solution to the convex hull under Problem 3 stated in Sect. 2.6.1.

2.6.3 The CHCNN Architecture

Given a set of k unit direction vectors $\mathbf{U} = \left\{n^{(1)}, n^{(2)}, \cdots, n^{(k)}\right\}$. According to Lemma 2.1, the aim is then to build an appropriate neural network such that after adaptive training, the network can yield the vertex set $\mathbf{V}^*$ and the hyperplanes $\mathbf{H}^*$. The network is the CHCNN shown in Fig. 2.37.

Topologically, CHCNN consists of one input layer of N neurons and one output layer of k neurons. Similar to the adaptive resonance theory (ART) developed by Carpenter and Grossberg (1987), the two layers of neurons communicate via a feedforward connection W and a feedback connection T. The input neurons are all McCulloch–Pitts type with zero threshold and linear input–output activation

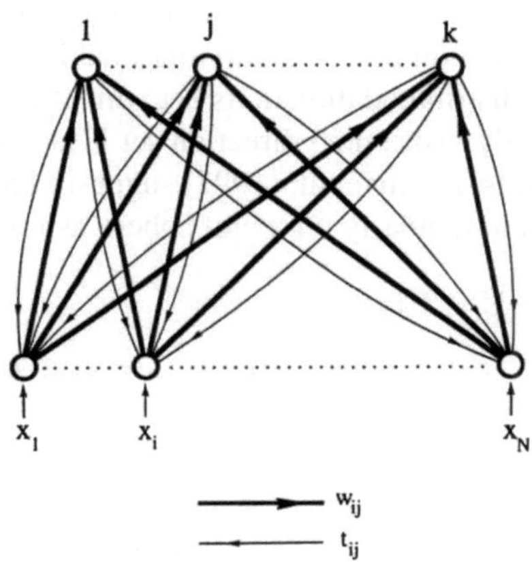

Fig. 2.37 The CHCNN architecture

function, but the output neurons all have nonzero thresholds and the hard-limiter input–output activation function defined by

$$f(x) = \begin{cases} 1, & \textit{if } x > 0, \\ 0, & \textit{if } x \leq 0. \end{cases} \tag{2.98}$$

Let w_{ij} and t_{ij} be, respectively, the feedforward and feedback connections (weights) between neuron i in the input layer and neuron j in the output layer. Let θ_j be the threshold value attached to the output neuron j. Denote

$$w^{(j)} = \left(w_{1j}\,,\, w_{2j}\,,\, \cdots\,,\, w_{Nj}\right),$$
$$t^{(j)} = \left(t_{1j}\,,\, t_{2j}\,,\, \cdots\,,\, t_{Nj}\right).$$

In the CHCNN, the feedforward connection $w^{(j)}$ is fixed as the jth prespecified direction under $\mathbf{n}^{(j)}$. The feedback connection vector $t^{(j)}$ and the threshold θ_j are trained adaptively to yield the supporting point $p^{[i(j)]}$, defined in (2.93), and the maximum value defined in (2.94) [or equivalently, the hyperplane $H^{(j)}$ defined in (2.95)], respectively. Consequently, after training, the CHCNN is capable of yielding the vertex set $\mathbf{V}^*$ and the hyperplanes $\mathbf{H}^*$ specified in Lemma 2.1.

A parameter-setting rule for $\mathbf{U}$ and a training algorithm for T and θ are as follows:

2.6.3.1 Parameter-Setting and Training Rules

The CHCNN is inherently dependent on the setting of the direction vectors $\mathbf{U}$ (which specify the direction of the supporting hyperplanes), and the training rule for adjusting the weights T (which record the supporting points) and the thresholds θ (which control the positions of the supporting hyperplanes).

1. *Setting of* $\mathbf{U}$ *and* $\mathbf{W}$. Since every supporting hyperplanes $\mathrm{H}^{(j)}$ bounds the convex hull in a given direction $\mathbf{n}^{(j)}$, so a very reasonable approximation should apply a set of uniformly distributed directions $\mathbf{U}$ to direct the hyperplanes. Nevertheless, finding a uniformly distributed direction set $\mathbf{U}$ is very difficult and complicated in high dimensions. Leung et al. (1997) suggest $\mathbf{U}$ to be specified in such a way that they distribute regularly on a unit sphere as follows:

$$\begin{aligned} \mathbf{U} &= \mathbf{U}(\alpha) \\ &= \left\{ \mathbf{n}^{(j)} : j = 1\,,\, 2\,,\, \cdots\,,\, k = 2d^{(N-1)} \right\} \end{aligned}$$

with

$$\mathbf{n}^{(j)} = \mathbf{n}^{(i_1\,,\,i_2\,,\,\cdots\,,\,i_{N-1})}$$

$$
\begin{aligned}
=\Bigg\{ & \sin\left[\left(\frac{\pi i_1}{d}\right)+\alpha_1\right]\sin\left[\left(\frac{\pi i_1}{d}\right)+\alpha_2\right]\times\cdots\times\sin\left[\left(\frac{\pi i_{N-1}}{d}\right)+\alpha_{N-1}\right],\\
& \cos\left[\left(\frac{\pi i_1}{d}\right)+\alpha_1\right]\sin\left[\left(\frac{\pi i_1}{d}\right)+\alpha_2\right]\times\cdots\times\sin\left[\left(\frac{\pi i_{N-1}}{d}\right)+\alpha_{N-1}\right],\\
& \cos\left[\left(\frac{\pi i_2}{d}\right)+\alpha_2\right]\sin\left[\left(\frac{\pi i_3}{d}\right)+\alpha_3\right]\times\cdots\times\sin\left[\left(\frac{\pi i_{N-1}}{d}\right)+\alpha_{N-1}\right],\\
& \cos\left[\left(\frac{\pi i_3}{d}\right)+\alpha_3\right]\sin\left[\left(\frac{\pi i_4}{d}\right)+\alpha_4\right]\times\cdots\times\sin\left[\left(\frac{\pi i_{N-1}}{d}\right)+\alpha_{N-1}\right],\\
& \cdots\\
& \cos\left[\left(\frac{\pi i_{(N-2)}}{d}\right)+\alpha_{N-2}\right]\times\sin\left[\left(\frac{\pi i_{N-1}}{d}\right)+\alpha_{N-1}\right],\\
& \cos\left[\left(\frac{\pi i_{N-1}}{d}\right)+\alpha_{N-1}\right]\Bigg\}
\end{aligned}
\tag{2.99}
$$

where $\alpha = (\alpha_1, \alpha_2, \cdots, \alpha_{N-1}) \in \mathbf{R}^{N-1}$, $i_1 = 1, \cdots, 2d$, $i_l = 1, \cdots, d$ for $l = 2, \cdots, N-1$, and, $j = i_1 + 2d(i_2 - 1) + 2d^2(i_3 - 1) + \cdots + 2d^{N-2}(i_{N-1} - 1)$. The variable α is a rotation parameter whose components are randomly chosen. The function of α is explained in Theorem 2.4. It is shown that with such specified direction vectors **U**, the CHCNN is always capable of yielding very accurate approximation of the convex hull $C(\mathbf{S})$. As mentioned previously, once the direction vectors **U** is specified, the feedforward connections $\mathbf{W} = \left(w^{(1)}, w^{(2)}, \cdots, w^{(k)}\right)$ are fixed as the same as **U**. That is,

$$
w^{(j)} = \left[\mathbf{n}^{(j)}\right]^T, \quad j = 1, \cdots, k.
$$

2. A Learning Rule for T and $\boldsymbol{\theta}$. The jth neuron in the output layer of the CHCNN is said to be excited by an input x if $f(\langle w, x\rangle - \theta_j) = 1$, otherwise the neuron is said to be inhibited. The feedback connections $T = \left[t^{(1)}, t^{(2)}, \cdots, t^{(k)}\right]$ and thresholds $\boldsymbol{\theta} = \left[\theta^{(1)}, \theta^{(2)}, \cdots, \theta^{(k)}\right]$ will then be adjusted according to the following learning rule.

2.6.3.2 Excited Learning Rule:

Initialize $t^{(j)}(0) = p^{(1)}$; $\theta_j(0) = \langle \mathbf{n}^{(j)}, p^{(1)}\rangle$.
Begin with $i = 1$.

Step 1. Input $p^{(i)}$, and find all neurons excited by $p^{(i)}$ in the outer layer. Denote all the excited neurons by $J(i)$.

Step 2. For every $j \in \{1, 2, \cdots, k\}$, do the following (a and b).

(a) update $t^{(j)}$ according to

$$t^{(j)}(i) = t^{(j)}(i-1) + \Delta t^{(j)}(i)$$

with

$$\Delta t^{(j)}(i) = \begin{cases} p^{(i)} - t^{(j)}(i-1), & \textit{if } j \in J(i), \\ 0, & \textit{if } j \notin J(i). \end{cases}$$

(b) update θ_j according to

$$\theta_j(i) = \theta_j(i-1) + \Delta\theta_j(i)$$

with

$$\Delta\theta_j(i) = \begin{cases} \left\langle w^{(j)}, p^{(i)} \right\rangle - \theta_j(i-1), & \textit{if } j \in J(i), \\ 0, & \textit{if } j \notin J(i). \end{cases}$$

Step 3. If $i = M$, then terminate the learning process, otherwise, go to Step 1 with $i := i + 1$.

It is shown in Leung et al. (1997a) that the learning rule can guarantee convergence to the supporting points of $C(\mathbf{S})$ within M steps. That is, the CHCNN succeeds in M-step learning as summarized by the following theorem:

Theorem 2.2.
1. *The learning algorithm of the CHCNN converges in M steps.*
2. *The CHCNN algorithm is an on-line algorithm, processing every input in a single iteration.*
3. *The trained CHCNN yields* $\mathbf{V}^*$ *and* $\mathbf{H}^*$*, such that* $C(\mathbf{V}^*)$ *is an inscribed approximation and* $C_H(\mathbf{H}^*)$ *is a circumscribed approximation of* $C(\mathbf{S})$.

From Theorem 2.2, we obtain the following conclusion:

Corollary 2.1. *The CHCNN algorithm has time complexity* $O(M)$ *for off-line* problems and $O(1)$ for *on-line problem.*

Theorems 2.3 and 2.4 below further show that $C(\mathbf{V}^*)$ and $C_H(\mathbf{H}^*)$ both actually provide very accurate approximations of $C(\mathbf{S})$.

Theorem 2.3. *Assume* $d \geq 2$, $\mathbf{V}^*$ *is the supporting point set and* $\mathbf{H}^*$ *is the supporting hyperplanes generated by the CHCNN, with the direction vectors* $\mathbf{U}$ *defined as in (2.99). Then there is a constant* K_N*, which is only dependent of N, such that*

1. $dist[C_H(\mathbf{H}^*), C(\mathbf{V}^*)] \leq K_N \mathrm{diam}(\mathbf{S}) k^{-1/(N-1)}$,
2. $dist[C(\mathbf{V}^*), C(\mathbf{S})] \leq K_N \mathrm{diam}(\mathbf{S}) k^{-1/(N-1)}$,
3. $dist[C_H(\mathbf{H}^*), C(\mathbf{S})] \leq K_N \mathrm{diam}(\mathbf{S}) k^{-1/(N-1)}$.

(See Leung et al. (1997) for the proof)

Remark 2.6. Let A and B be two subsets of $\mathbf{R}^N$. The diameter of the set A is defined by

$$\mathrm{diam}(A) = \max\{\|x - y\| : x\,,\ y \in A\}.$$

The distance of a point p to the set A, denoted by $dist(p\,,A)$, is defined by

$$dist(p\,,A) = \min\{\|p - x\| : x \in A\},$$

and the distance between A and B is defined by

$$dist\,(A\,,B) = \max\left\{\max_{x\in A} dist\,(x\,,B), \max_{y\in B} dist\,(y\,,A)\right\}.$$

The distance between two sets can serve as a measure of the difference of the sets.

Theorem 2.3 says that the $C(\mathbf{V}^*)$ and $C_H(\mathbf{H}^*)$ generated by the CHCNN approximate $C(\mathbf{S})$ with the same accuracy $O\left(k^{-1/(n-1)}\right)$, which is proportional to the number of neurons adopted in the CHCNN and is independent of the specified $\mathbf{S}$. The significance if this is twofold. First, one can determine the size of the neural network based directly on this accuracy of estimation in a given convex–hull computation application with any prespecified level of approximation. Second, it follows that the approximation accuracy of $C(\mathbf{V}^*)$ and $C_H(\mathbf{H}^*)$ can assuredly increase as k increases. Thus, any highly accurate approximation of $C(\mathbf{S})$ can be ascertained via the CHCNN. This shows further that CHCNN, as an approximate algorithm, can converge to the exact convex hull with sufficient large number of neurons.

The following theorem further explains that it is not necessary to have an infinite number of neurons in order to get an exact approximation of $C(\mathbf{S})$ via the CHCNN. Clarification of this is tightly related to another important issue: whether or not the supporting point set $\mathbf{V}$ generated by the CHCNN is a portion of $Ver(\mathbf{S})$. An affirmative answer to this question is offered in the theorem:

Theorem 2.4. *Let $\mathbf{V}^*(\alpha)$ be the set of supporting points generated by the CHCNN with the direction vectors $\mathbf{U}(\alpha)$ defined by (2.99). Then we have the following:*

1. *For almost every α in $\mathbf{R}^{N-1}$ (namely, every α except a zero measure set), $\mathbf{V}^*(\alpha)$ is a portion of $Ver(\mathbf{S})$.*
2. *There is a constant $K(\mathbf{S})$ such that, for almost every α in $\mathbf{R}^{N-1}$, $\mathbf{V}^*(\alpha) = Ver(\mathbf{S})$ whenever $k \geq K(\mathbf{S})$.*

(See Leung et al. (1997a) for the proof)

Remark 2.7. Property (2) in Theorem 2.4 shows that for any given point set $\mathbf{S}$, the CHCNN with a finite number of neurons is capable of almost always yielding the exact vertices of $C(\mathbf{S})$. Therefore, it provides an accurate solution to convex-hull problem 2. In this case, $C(\mathbf{V}^*)$ then provides an accurate solution to convex-hull Problem 1 stated in Sect. 2.6.1.

2.6.4 Applications in Cluster Characterization

2.6.4.1 Determining Whether a Point p is Inside $C(\mathbf{S})$, a Cluster.

Given a point p, check whether or not p belongs to $C(\mathbf{S})$ is a basic point-location problem in computational geometry. This naturally arises in applications such as collision avoidance problem for robot motion planning, and infection area detection problem in epidemics. The idea in this application is that instead of checking if $p \in C(\mathbf{S})$, we can check if p belongs to $C_H(\mathbf{H}^*)$, which is known to be a circumscribed approximation of $C(\mathbf{S})$. Obviously, the latter can easily be accomplished by the CHCNN. The main step are as follows:

Step 1. Input p into the neural network trained by $\mathbf{S}$.
Step 2. If there is no neuron being excited, i.e.,

$$\left\langle p\,,\,\mathbf{n}^{(i)}\right\rangle \leq \theta_i\,, \qquad i = 1\,,\,\cdots\,,\,k,$$

then, $p \in \bar{H}^{(i)}$ holds for any i. Therefore, $p \in C_H(\mathbf{H}^*)$. Otherwise, there is a neuron, denoted by j, being excited, i.e.,

$$\left\langle p\,,\,n^{(i)}\right\rangle > \theta_i\,, \qquad i = 1\,,\,\cdots\,,\,k.$$

Thus, $p \notin C_H(\mathbf{H}^*)$. This also means $p \notin C(\mathbf{S})$ since $C(\mathbf{S}) \subseteq C_H(\mathbf{H}^*)$.

It should be observed that for this application, only one iteration is required by the CHCNN to determine whether p belongs to $C_H(\mathbf{H}^*)$. This is clearly an optimal property one can expect of an on-line algorithm for dynamically changing problems such as spatial spread of epidemics.

2.6.4.2 Computing the Diameter of a Cluster S

The diameter of a set $\mathbf{S}$ is defined by

$$\mathrm{Diam}(\mathbf{S}) = \max\{\|x - y\| : x\,,\,y \in \mathbf{S}\}. \tag{2.100}$$

The problem of determining the diameter of a set **S** occurs in various applications. For instance, in clustering techniques, the "minimum diameter K-clustering" problem can be stated in the following way:

Given a set of m points in $\mathbf{R}^N$, partition them into K clusters $C_1, C_2, \cdots, C_k$ such that the maximum diameter of C_i, $i = 1, \cdots, K$, is as small as possible (Preparata and Shamos 1985).

The success of applying the CHCNN to this problem is in part due to the following well-known result:

Lemma 2.2. *The diameter of a set equals that of its convex hull, which in turn is the greatest distance between parallel supporting hyperplanes (Preparata and Shamos 1985).*

It should be noted that in the CHCNN developed in Subsection 2.6.2, if $\mathbf{n} \in \mathbf{U}$ is the pre-specified direction vector defined in (2.99), then the direction $-\mathbf{n}$ must also belong to **U**. Therefore, if $\theta(\mathbf{n})$, $\theta^-(\mathbf{n})$, $t(\mathbf{n})$, $t^-(\mathbf{n})$, $H(\mathbf{n})$, $H^-(\mathbf{n})$ respectively denote the corresponding threshold values, supporting points and supporting hyperplanes for **n** and $-\mathbf{n}$ in the CHCNN trained by **S**, then $H(\mathbf{n})$ and $H^-(\mathbf{n})$ would be parallel to each other, and the distance between them is equal to $|\theta(\mathbf{n}) - \theta^-(\mathbf{n})|$. According to Lemma 2.2, we thus can use

$$\max_{\mathbf{n}\in\mathbf{U}} \{|\theta^-(\mathbf{n}) - \theta(\mathbf{n})|\} \tag{2.101}$$

as an approximation of the diameter of **S**, However, $t(\mathbf{n})$ and $t^-(\mathbf{n})$ are both the supporting points (therefore belong to **S**), which shows Diam $(\mathbf{S}) \geq |t(\mathbf{n}) - t^-(\mathbf{n})|$ by the definition in (2.100). From the inequality $|t(\mathbf{n}) - t^-(\mathbf{n})| \geq |\theta(\mathbf{n}) - \theta^-(\mathbf{n})|$, it then follows that a more accurate approximation of Diam $(\mathbf{S})$ should be given by

$$\max_{\mathbf{n}\in\mathbf{U}} \{\, |t(\mathbf{n}) - t^-(\mathbf{n})|\,\}. \tag{2.102}$$

The advantage of this computational method is that it is not only very easy to implement but also very efficient for solving high dimensional problems. Table 2.7 shows the simulation results for diameters of a set of ten four-dimensional point sets, with all sets containing 200 points randomly chosen and the CHCNN was run with $k = 20$. In Table 2.7, $D(\mathbf{S})$ is the exact diameter of the set **S**, and

$$D_1(\mathbf{S}) = \max_{\mathbf{n}\in\mathbf{U}} \{|\theta^-(\mathbf{n}) - \theta(\mathbf{n})|\} \tag{2.103}$$

and

$$D_2(\mathbf{S}) = \max_{\mathbf{n}\in\mathbf{U}} \{|t(\mathbf{n}) - t^-(\mathbf{n})|\} \tag{2.104}$$

are the approximations defined respectively by (2.101) and (2.102).

Table 2.7 Diamater of a set S

$\mathbf{S}_i$	$D(\mathbf{S}_i)$	$D_1(\mathbf{S}_i)$	$D_2(\mathbf{S}_i)$
$\mathbf{S}_1$	131.1992	130.0681	131.1992
$\mathbf{S}_2$	143.9056	143.8920	143.9056
$\mathbf{S}_3$	137.0960	135.9434	137.0960
$\mathbf{S}_4$	138.4248	136.4933	138.4248
$\mathbf{S}_5$	144.4680	142.7688	144.4680
$\mathbf{S}_6$	135.6954	134.5078	135.6954
$\mathbf{S}_7$	136.9296	134.8861	134.8861
$\mathbf{S}_8$	146.5135	144.9885	146.5135
$\mathbf{S}_9$	149.6796	149.0020	149.6796
$\mathbf{S}_{10}$	146.6331	145.0052	146.6331

A pleasant and surprising result found in Table 2.7 is that $D_2(\mathbf{S})$ almost always yield the exact diameter of a set $\mathbf{S}$. It implies that CHCNN is highly effective and efficient in computing the diameter of a cluster.

Chapter 3
Statistical Approach to the Identification of Separation Surface for Spatial Data

3.1 A Brief Background About Statistical Classification

In spatial clustering, spatial objects are grouped into clusters according to their similarities. In terms of learning or pattern recognition, it belongs to the identification of structures/classes through an unsupervised process. In terms of data mining, it is the discovery of intrinsic classes, particularly new classes, in spatial data. It formulates class structures and determines the number of classes. I have examined in Chap. 2 the importance of clustering as a means for unraveling interesting, useful and natural patterns in spatial data. The process generally does not involve how to separate predetermined classes, or how to determine whether classes are significantly different from each other, or how to assign new objects to given classes.

Another fundamental issue of spatial knowledge discovery involves spatial classification. It essentially deals with the separation of pre-specified classes and the assignment of new spatial objects to these classes on the basis of some measurements (with respect to selected features) about them. In terms of learning or pattern recognition, it is actually a supervised learning process which searches for the decision surface separating appropriately various classes. In terms of data mining, it often involves the discovery of classification rules from the training/learning data set that can separate distinct/genuine classes of spatial objects and the assignment of new spatial objects to these labeled classes. Whether the pre-specified classes are significantly different is usually not the main concern in classification. It can be determined by procedures such as the analysis of variance in statistics.

Similar to cluster analysis, classification is a centuries old problem which even dates back to the time of Aristotle. It has been systematically studied in various disciplines over the years (Hand and Henley 1997). What reinforces its position in data analysis in general and data mining in particular is that nowadays we need to perform classification on very large data sets which may not be well-behaved in the statistical sense. Literature on the statistical approach to classification is voluminous. Depending on the nature of the underlying data, a large variety of statistical

Y. Leung, *Knowledge Discovery in Spatial Data*, Advances in Spatial Science, DOI 10.1007/978-3-642-02664-5_3,

methods has been developed for the separation of and assignment to classes. Nevertheless, we can broadly classify them into two major groups. The first group belongs to the parametric statistical classifiers. It is built directly or indirectly on the famous Bayes' rule which attempts to find an optimal classifier (separation surface, discriminant function) that minimizes a risk function stipulating the misclassification/error rate (Geisser 1982). Under the Bayes' rule, all of the information about group memberships comes from a set of conditional probabilities. Specifically, class conditional distributions and prior probabilities are estimated for each class and the Bayes' theorem is applied to obtain the posterior probability. One of the earliest parametric methods is perhaps the Fisher's linear discriminant analysis (Fisher 1936; Kranowski 1977; Das Gupta 1980). When the feature vectors follow a multivariate normal distribution and the covariance matrices are known and identical, the resulting discriminant function/separation surface is linear and is essentially the Bayes' rule. The method is based on the derivation of a low dimensional representation, a linear combination, of the multivariate feature vectors so that the classes can be best separated. It is a global model that minimizes the total error of classification by maximizing the between class distance and minimizing the within class variance. If the data are well-behaved, then it is the best rule. To accommodate for more complicated situations, the transformation can include squared and linear functions of the features. The resulting separation surface is still linear in the space spanned by the newly formed variables, but quadratic in the space spanned by the original features. The method is known as the quadratic discriminant analysis (Smith 1947). Basing on the linear and quadratic framework, extensions such as flexible, penalized and mixture discriminant analysis have also been developed to cater for special situations (McLachlan 1992; Hastie and Tibshirani 1996). By experience, linear discriminant analysis performs almost as well as the quadratic version unless the covariance matrices are substantially different. Though multivariate normality is assumed, the linear and quadratic discriminant functions are rather robust for a range of distributions in applications. Though the Bayes' rule is the best under ideal situation, it is difficult to implement because it needs a large number of conditional probabilities and prior probabilities as inputs in the estimation.

It is well known that not all data are numerical (interval-scaled) in measurement. In practice, a lot of data are categorical in nature. Furthermore, the underlying distribution of the data may not be multivariate normal and the means and covariance matrices might need to be estimated. Under these situations, parametric methods such as linear and quadratic discriminant analysis are not appropriate on the theoretical basis and non-parametric methods become necessary. Most of the non-parametric methods attempt to derive smooth estimates of the conditional probabilities by which the posterior probabilities for class assignment are estimated. In short, we make no assumption about the class probabilities. The kernel methods, the nearest neighbor methods, and the basis expansion methods are typical non-parametric procedures which usually assume that the function is locally constant. The kernel methods (Fix and Hodges 1951; Hand 1982; Wand and Jones 1995) attempt to estimate the conditional probability by assuming that a value in the feature space not only raises its probability of occurrence but also that of its surrounding values. So, a kernel is a bounded function on the feature space contributing to the local estimate of the probability density function. The basic issue is the selection of an appropriate kernel resulting in preferably simple form. Among other things, increase in the size of the data set will lead to increase of storage and computation cost. Differing from the kernel method, the nearest neighbor methods (or commonly called the k-nn method) estimate the posterior distribution as the proportions of the classes among the nearest k data points (Hart 1968; Stanfill and Waltz 1986; Aha et al. 1991). That is, the method employs the volume containing a fixed, say k, number of points as an estimator. Similar to the kernel methods, a smoothing parameter, k for this case, needs to be appropriately chosen. Storage requirement and computation cost are again major concerns in implementation. In addition to the kernel and nearest neighbor techniques, the basis expansion methods also provide a common procedure to construct non-parametric classifiers. In brief, the method expands a function by a set of suitably selected basis functions. Its general form actually belongs to the mixture model in which basis functions with parameters are combined with attached weights indicating the contributions of constituent classes to the model. Radial basis function is for example a common basis expand method used in many applications (Powell 1987). Each basis function is a function of the distance from its center. Multidimensional splines (as basis functions) yielding piecewise polynomial distributions are powerful alternatives to the radial basis function. Technically, the kernel function with components centering at corresponding data points can in fact be treated as a basis expansion method.

Apparently linear discriminant analysis and its variants attempt to construct global models for classification. They are inflexible and cannot tolerate irregularities in the separation surface. On the other hand, the class of non-parametric models is distribution-free local models giving a high degree of flexibility in accommodating local effects on the classifier. They, however, suffer from the curse of dimensionality and large sample size. In-between lies a variety of models, some are statistical and some are non-statistical, with varying degrees of flexibility. Using the dependent variable as the class indicator, logistic regression aims at approximating the class posterior probabilities via the regression framework (Berkson 1944; Anderson 1982; Hosmer and Lemeshow 1989; Collett 1991). It is employed when the log of the class likelihood ratio can be assumed to be linear. It estimates more directly the posterior probabilities. However, orthogonality no longer holds and interpretation of the coefficients is less straightforward. Besides logistic regression, additive models in more general forms have also been developed to accommodate local variations and/or situations stipulated by the parametric methods.

Placing classification in the context of statistical learning theory, support vector machines (Vapnik 1995), on the other hand, search for the optimal separating hyperplane that separates classes with maximal margins. The basic idea is to transform nonlinear problems, which are not separable, in low dimensional feature space into linearly separable problems in higher dimensional feature space. The curse of dimensionality is overcome by the use of the kernel function skill. They are particularly suitable for small sample classification problems in high dimensions. Viewing neural networks as learning machines, models such as perceptron and multilayer feedforward network become special cases of the support vector machine (Xu and Leung 2004). Support vector machines offer advantages over neural networks in terms of generalization ability, training reliability, and efficiency.

In the remaining part of this chapter, I will discuss in detail some statistical data mining methods that we have developed to unravel separating hypersurfaces or classification rules in spatial data. I will also address how different classification issues are handled under different statistical classifiers. Our discussion starts with the statistical approach to classification based on Bayes' rule. Naïve Bayes and discriminant analysis are discussed in Sects. 3.2 and 3.3. Logistic regression is then examined in Sect. 3.4. Lastly, I will discuss the development of support vector machine in Sect. 3.5.

3.2 The Bayesian Approach to Data Classification

Classification typically involves two steps: first the system is trained on a set of data and then it is used to classify a new set of unclassified cases. When the possible classes are known in advance and the system is trained on a set of classified cases, the task is termed *supervised* classification. Bayesian methods provide one of the oldest methods to perform supervised classification.

3.2.1 A Brief Description of Bayesian Classification Theory

In the context of statistics, there are two basic approaches, the informative and discriminative approach, to classification. In the informative approach, the classifier learns the class densities, while in the discriminative approach, the focus is on learning directly the class boundaries or the class membership probabilities disregard of the underlying class densities. Informative classification is done by examining the likelihood of each class producing the features and assigning objects to the most likely class. Fisher's linear discriminant analysis (LDA), hidden Markov models, and naive Bayes are typical examples. Since each class density is considered separately from the others, classifiers are relatively easy to train. Discriminative classifier such as logistic regression requires simultaneous consideration of all other classes. The classifiers are relatively more difficult to train.

In fact, data mining applications often operate in the domain of high dimensional features where the tradeoffs between informative and discriminative classifiers are especially relevant. These two types are related via the Bayes rule but often lead to different decision rules. In general, their performance depends on specific data (Rubinstein and Hastie 1997).

Denote formally the feature (or attribute) vector by $\mathbf{x} = (x_1, \ldots, x_p) \in \boldsymbol{R}^p$. A classifier can be viewed as a mapping $\alpha : \mathbf{x} \to \{1, 2, \ldots, K\}$, where K is the number of classes, that assigns class labels to observations. There is also a cost matrix $c(r,s)$, r, $s = 1, \ldots, K$, which describes the cost associated with misclassifying a member of class r to class s. The goal is to minimize the total error of classification

$$\alpha(\mathbf{x}) = \arg\min_k \sum_{m=1}^{K} c(k,m)p(y=m|\mathbf{x}). \tag{3.1}$$

In this sense, the resulting classification rule is optimal. For 0/1 loss, this reduces to classifying $\mathbf{x}$ to class k with which the posterior probability $p(y=k|\mathbf{x})$ is maximum, i.e.

$$\alpha(\mathbf{x}) = \arg\min_k p(y=k|\mathbf{x}). \tag{3.2}$$

In practice, the true density $p(\mathbf{x},y)$ is unknown and all we have available is a set of training observations $(\mathbf{x}_i, y_i)$, $i = 1, \ldots, n$. Many classification techniques seek to estimate the class posterior probabilities $p(y=k|\mathbf{x})$, because optimal classification can be achieved if these are perfectly known.

Instead of estimating the class posteriors $p(y|\mathbf{x})$ directly, informative classification methods estimate firstly the class densities $p(\mathbf{x}|y)$ and the prior probabilities $\pi_k \equiv p(y=k)$, and then gives by Bayes rule

$$p(y=k|\mathbf{x}) = \frac{p(\mathbf{x}|y=k)p(y=k)}{\sum_{m=1}^{K} p(\mathbf{x}|y=m)p(y=m)}. \tag{3.3}$$

For the Gaussian case, the optimal discrimination is the well-known Fisher's LDA. The important points with informative training are:

1. A parameter model $p_\theta(\mathbf{x}|y=k)$is often assumed for the class densities.
2. The parameters are obtained by maximizing the full log likelihood $\sum_{i=1}^{n} \log p_\theta(\mathbf{x}_i, y_i)$.
3. A decision boundary is induced.

3.2.2 *Naive Bayes Method and Feature Selection in Data Classification*

Naive Bayes (Langley and Sage 1994; John and Langley 1995; Kohavi 1996; KontKanen et al. 1998) is an informative classifier taking a specific form of the Bayesian network (Pearl 1988) which has become a common method to unravel models of interactions from large databases. The class densities assume independence among the predictors

$$
\begin{aligned}
p(\mathbf{x}|y=k) &= \prod_{j=1}^{p} p(x_j|y=k) \Rightarrow \\
\log p(\mathbf{x}|y=k) &= \sum_{j=1}^{p} \log p(x_j|y=k) = \sum_{j=1}^{p} g_{k,j}(x_j)
\end{aligned}
\tag{3.4}
$$

and is thus naive for this reason. In KontKanen et al. (1998), class densities are products of univariate Gaussians and "flexible" Gaussian kernel densities.

Let $\tilde{\mathbf{x}}$ be a subset of the set $\mathbf{x} = \{x_1, ..., x_p\}$ of all features, $\mathbf{y}_c = (y_1, ..., y_n)$ be a vector containing the values of the variable y, and $\mathbf{D}$ be the data matrix of all feature variables. According to KontKanen et al. (1998), the criterion for feature subset selection in Naive Bayes can be expressed as

$$
S(\tilde{\mathbf{x}}) \approx p(Y_c|\mathbf{D}, \hat{\boldsymbol{\theta}}, \tilde{\mathbf{x}}) = \prod_{i=1}^{n} \frac{p(y_i, \mathbf{x}_i|\hat{\boldsymbol{\theta}}, \tilde{\mathbf{x}})}{\sum_{k=1}^{K} p(y=k, \mathbf{x}_i|\hat{\boldsymbol{\theta}}, \tilde{\mathbf{x}})} \rightarrow \max!, \tag{3.5}
$$

where $\hat{\theta}$ are the maximum model parameters. The criterion means that we only select those feature subsets having the largest response probability. It should be noted that selecting variables one by one is different from selecting subsets because a single important variable is not necessarily an important subset and vice versa. The optimal subset is thought to be more advantageous for model selection (Piramuthu 1999).

3.2.3 The Application of Naïve Bayes Discriminant Analysis in Client Segmentation for Product Marketing

To promote a product such as credit cards in a spatial market, a bank needs to formulate its promotion strategy by segmenting clients' responses in its database. The purpose is to classify clients into different response classes, to characterize the class attributes (features) of clients, and to make a decision based on these attributes in order to guide the promotion strategy. The problem is apparently the discovery of classification rules for client segmentation according to their past responses (Yes and No) to similar products.

To a certain extent, the task is similar to credit granting in finance and statistics. The creditworthiness assessment method is called *credit scoring* in statistics. It aims at classifying applicants for credit into different risk categories (e.g., "high," "moderate," and "low" risk). Such a task has become increasingly important and a wide range of statistical methods has been applied to solve the problem (Hand and Henley 1997).

Leung et al. (2003a) employ the Bayesian methods to discover in given data appropriate classification rules and promotion strategies for a bank. They particularly concentrate on how to make the best out of a very undesirable data set (one with a mixture of categorical and continuous variables as well as a very large number of missing values) and evaluate how various methods perform under such a situation.

3.2.3.1 Selection and Preprocessing of Variables

The Data Set

For prototyping, the given data set consists of eight types of variables comprising basic demographics, socioeconomics, household information, shareholder variables, Finance variables, credit card payment, promotion channel and behavior. The number of variables involved is more than 50 and each variable has 16,000 records (or cases, observations). It should be noted that among all variables, 22 variables have more than 13,000 missing values (about 81% of the total records), and only 8,409 observations have no missing values with respect to all variables. Estimation and data enrichment also need to be made for some variables. The data set contains both categorical variables and continuous variables, and the difference between numerical values of some variable is very large, for instance, TOTASSET, TOTDPOS and LIQUID all have a range of at least order 10^8. In addition, the ranges for variables are quite different, as can be seen in Table 3.1 and Fig. 3.1. Such a large variability creates some difficulties in modeling. In short, the data set is highly incomplete and undesirable. However, this is the situation under which mining of classification rules has to be made. Under this circumstance, the

Table 3.1 Descriptive statistics for the bank data set

	Valid N	Mean	Minimum	Maximum	Range	Std. Dev.
AGE	15,833	36.9	0	95	95	10.
CLUSTER	16,000	21.9	0	99	99	14.
ECLASS	16,000	1.6	0	4	4	1.
MS	14,351	0.7	0	1	1	0
JOBNAT	16,000	141.4	10	990	980	205.
EDU	10,797	0.3	0	1	1	0
CAR	16,000	0.1	0	1	1	0
CHILD21	16,000	0.5	0	4	4	0
GENDER	15,919	0.5	0	1	1	0
TENURE	15,938	106.4	1	506	505	66.
PDT_AMT	16,000	4.2	0	59	59	4.
RATIO	15,938	0.1	0	2.	2.	0
CTENURE	16,000	54.6	0	213	213	41.
CINCOME	16,000	17,474.0	0	750,000	750,000	25,434.
CREDLMT	16,000	32,733.6	200	900,000	899,800	30,876.
SAVBAL	14,012	65,633.0	−3,302.	9,585,861.	9,589,163.	220,526.
LIQUID	16,000	151,280.3	−34,763,704	36,082,806	70,846,510	776,551.
HHINCOME	16,000	32,533.0	0	1,074,917	1,074,917	34,518.
TOTDPOS	16,000	157,889.4	0	60,486,605	60,486,605	938,329.
TOTASSET	16,000	182,915.5	0	139,572,840	139,572,840	1,550,294.
ROLLRATE	15,928	0.2	−209.	100	309.	3.
TOHSCPAY	15,905	43,085.3	0	3,522,293	3,522,293.	72,226.
PRICONBA	16,000	0.3	0	1	1	0
PRICONCA	16,000	0.3	0	1	1	0
RESPONSE	16,000	0.5	0	1	1	0

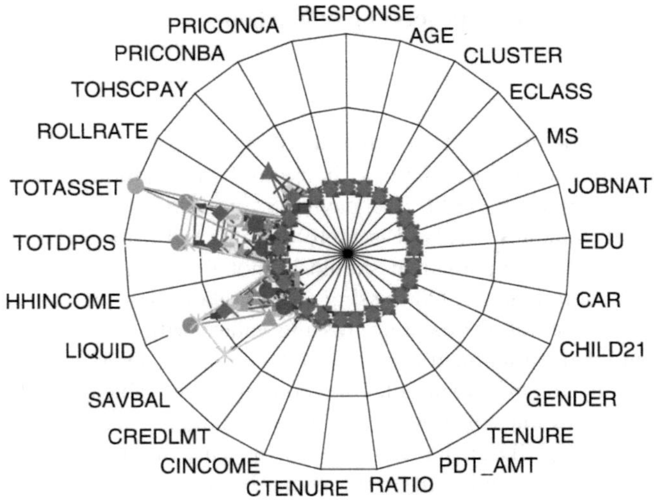

Fig. 3.1 The Radar plot for the selected variables

prerequisite to make analysis successful may rely on the selection of variables and processing of missing values.

Selection of Variables

First, we only need to consider variables with less than 8,000 missing values because theoretically we cannot provide sufficient information to characterize variables with more than 50% missing values. Twenty-five variables are thus selected at the end. Second, the selected categorical variables are coded so that they can be used in model construction. A general method to deal with categorical variables is to code the categories as indicator variables or to adopt some scaling procedure to give the categories numerical values. The details are listed in Table 3.2. Numerical variables keep their original values unchanged and are tabulated in Table 3.3. For easy understanding of the basic characteristics of all variables, their histograms are plotted in Fig. 3.2. It is hoped that at the first stage of modeling, as many variables as possible can be considered in order to maximize the information contained in the data set.

For convenience of treatment, the response variable RESPONSE and all selected (explanatory or feature) variables are denoted as y, x_1, ..., and x_{24}, respectively.

Preprocessing of Missing Values

Common approaches to handle missing data in statistical analysis are available-case analysis and complete-case analysis. The available-case analysis uses the cases in which the variable of interest is available. The available sample size is the number

Table 3.2 Selected categorical variables and their values

Variable	Name	Description	Value
y	RESPONSE	Response level for past campaign	$\begin{cases} 0, & \text{Response} = \text{Y} \\ 1, & \text{Response} = \text{N} \end{cases}$
x_1	ECLASS	Estate type	"PRE", "PRI", "HASALE," "HARENT" and "OTHERS" assign 4, 3, 2, 1, 0, respectively.
x_2	MS (Enriched)[a]	Marital status	$\begin{cases} 0, & \text{Single} = \text{Y} \\ 1, & \text{Married} = \text{N} \end{cases}$ (1,649 missing values)
x_3	EDU (Enriched)[a]	Education level	$\begin{cases} 0, & \text{College} = \text{Y} \\ 1, & \text{Non-college} = \text{N} \end{cases}$ (5,203 missing values)
x_4	CAR (Enriched)	Car ownership	$\begin{cases} 0, & \text{Not-Owner} = \text{Y} \\ 1, & \text{Owner} = \text{N} \end{cases}$
x_5	GENDER (Enriched)[a]	Sex	$\begin{cases} 0, & \text{Male} = \text{Y} \\ 1, & \text{Female} = \text{N} \end{cases}$ (81 missing values)
x_6	PRICONBA	Price conscious sense for BIA/BSA	$\begin{cases} 0, & \text{PRICONBA} = '\text{N}' \\ 1, & \text{PRICONBA} = '\text{Y}' \end{cases}$
x_7	PRICONCA	Price conscious sense for CARD	$\begin{cases} 0, & \text{PRICONCA} = '\text{N}' \\ 1, & \text{PRICONCA} = '\text{Y}' \end{cases}$
x_8	CHILD21 (Enriched)	No. of children aged < 21	(no missing values)

[a]Denotes the variables with missing values

of non-missing values. However, unequal sample sizes create practical problems. Comparative analysis across variables is difficult because different sub-samples of the original sample are used. Complete-case analysis uses cases for which all variables are present. The advantage of this approach is its simplicity, because standard statistical analysis can be applied directly to one common sub-sample of the original sample. The disadvantage is the loss of information from discarding incomplete cases. Besides the above two methods, we can also produce a complete data set by replacing missing values with, say, the mean and median.

Available-case and complete-case analyzes assume that data are missing at random. If the assumption is not satisfied, the results are unreliable with an unknown bias. The bank data set is having such a problem. Therefore, for reliable inference it is important to deal with missing values carefully. In this application, Leung et al. (2003a) employ Bayesian classification to deal with missing values and compare the result obtained by LDA.

3.2.3.2 Naive Bayes Discriminant Analysis

In Leung et al. (2003a), the "fully Bayesian" predictive inference of class memberships is made on the basis of a Naive Bayes model built from the data set. The results are represented at three levels of details: general, group-wise and individual.

Table 3.3 Selected numerical variables

Variable	Name	Description	No. of missing values
x_9	AGE (Enriched)[a]	Age of client	167
x_{10}	CLUSTER	Estate cluster	×
x_{11}	JOBNAT (Enriched)	Job nature	×
x_{12}	TENURE[a]	No. of months ago for the first account opened in H.S.	62
x_{13}	PDT_AMT	No. of active products held by each client in latest month	×
x_{14}	RATIO[a]	Ratio of no, of active products to Tenure	62
x_{15}	CTENURE	No. of months ago for the first credit card account opened	×
x_{16}	CINCOME (Enriched)	Client's enriched Income	×
x_{17}	CREDLMT	Total credit card limit for customer	×
x_{18}	SAVBAL[a]	Total saving account balance	1,988
x_{19}	LIQUID	Liquidity = SAVBAL + Unused OD + CREDLMT	×
x_{20}	HHINCOME	Estimated household income	×
x_{21}	TOTDPOS	Total deposit in KSCUSTW – total passbook and statement gold (PGAMT)	×
x_{22}	TOTASSET	Total asset = Total investment portfolio (INVPORT) + TOTDPOS	×
x_{23}	ROLLRATE[a]	Rollover rate of credit card	72
x_{24}	TOHSCPAY	Total Credit Card payment amount of HASE credit card	95

[a]Denotes variables with missing values

1. By (3.5) the most probable feature subset is: $\tilde{\mathbf{x}} = \{x_1, x_3, x_5, x_9, x_{11}, x_{13}, x_{15}, x_{16}, x_{19}, x_{23}\}$={ECLASS, EDU, GENDER, AGE, JOBNAT, PDT_AMT, CTENURE, CINCOME, LIQUID, ROLLRATE}.
2. General classification accuracy. It can be estimated that using the selected feature variables, 58.2% of the classifications will be correct. This estimation is based on the following external leave-one-out cross-validation procedure: using the selected predictor variables, we build 16,000 models. Each of these models is constructed using 15,999 data items from the data set and each model is then used to classify the data item not used in the model construction. Since 9,312 out of 16,000 models succeed in classifying the unseen data item correctly, one may assume that this would happen in the future as well. However, simply stating the classification performance of 58.2% is not too meaningful. It has to be compared with the performance obtainable by a "default" classification procedure that always guesses the class of the data item to be the class of the majority (class "0" in this case). This simple method would yield the performance rate of 50.5%.
3. Classification performance and its reliability by groups. The overall result of 58.2% is just an average performance rate. Suppose our model classifies a

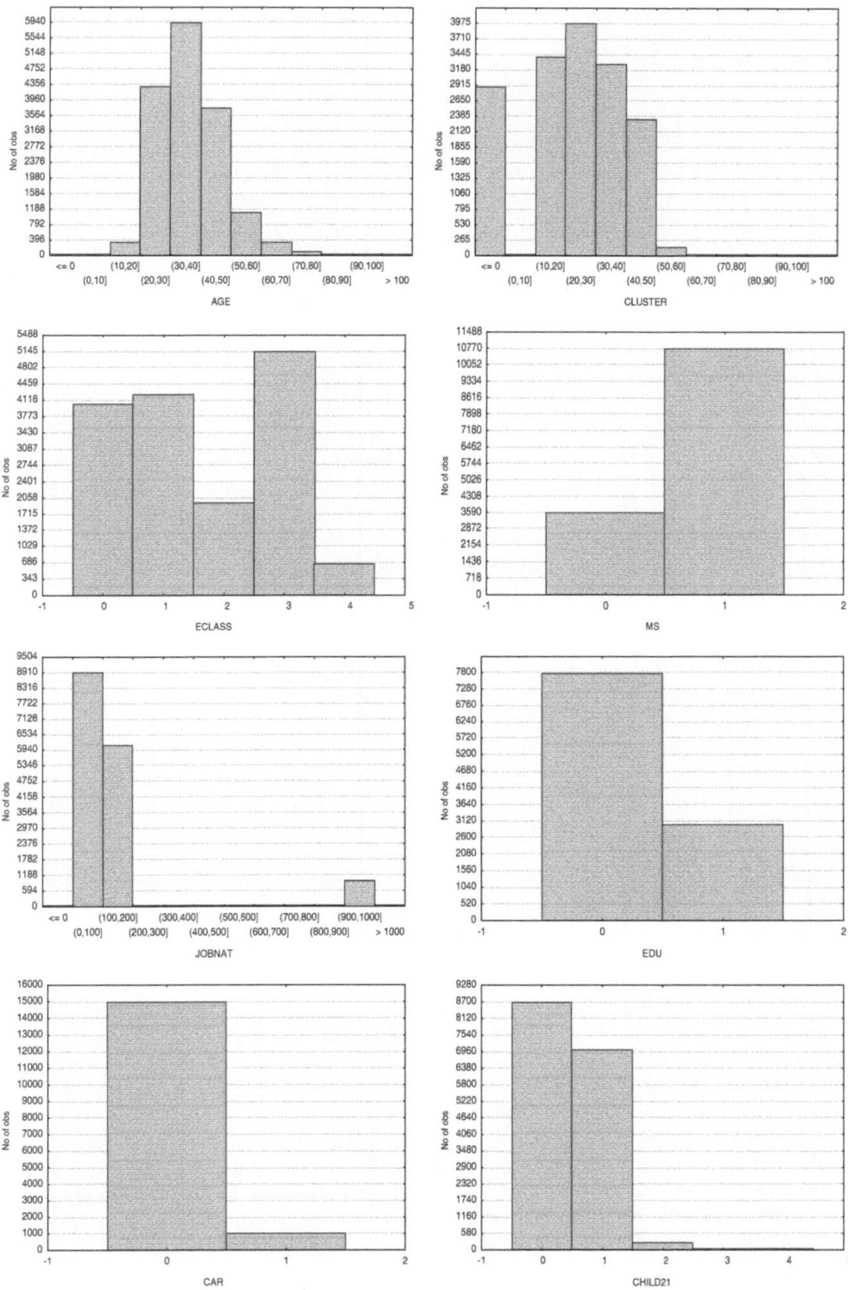

Fig. 3.2 (Continued)

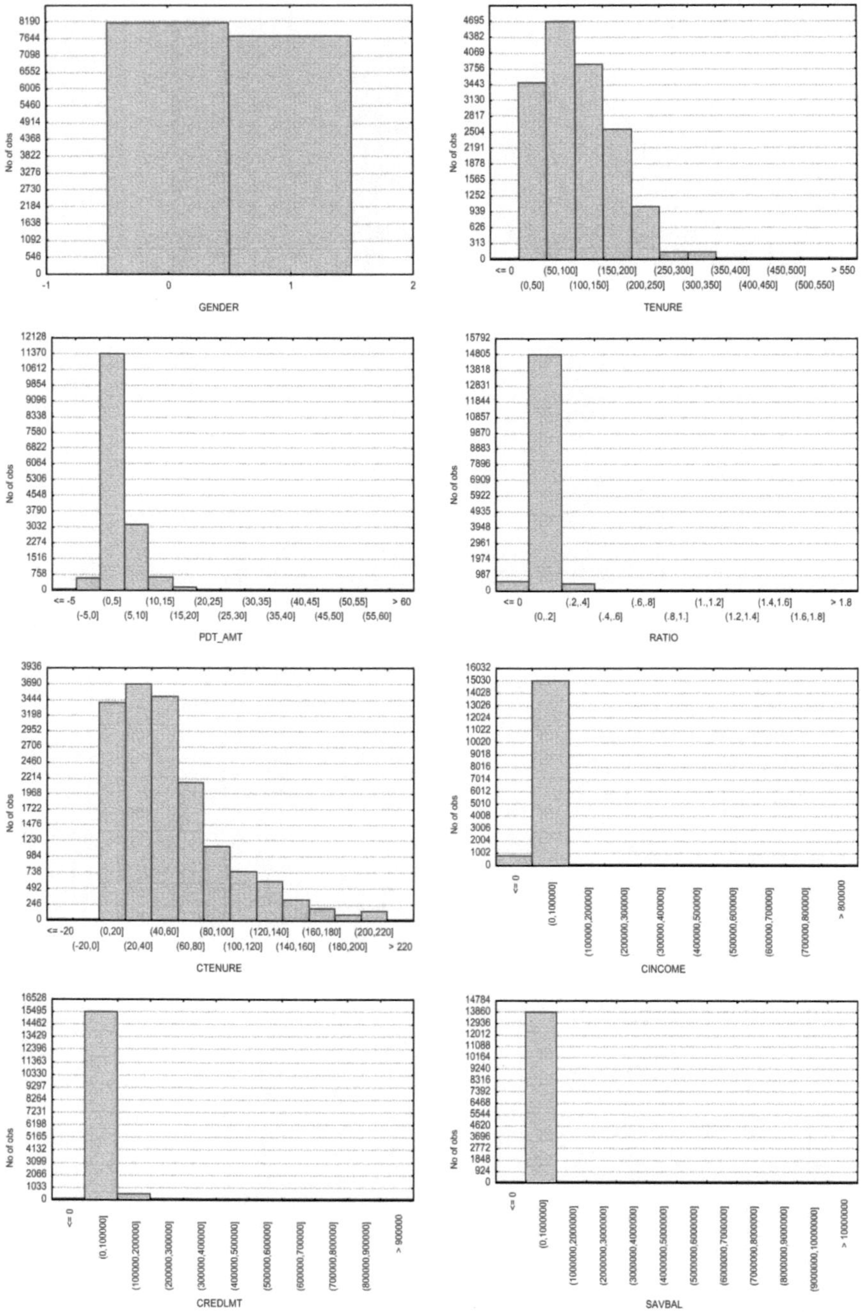

Fig. 3.2 (Continued)

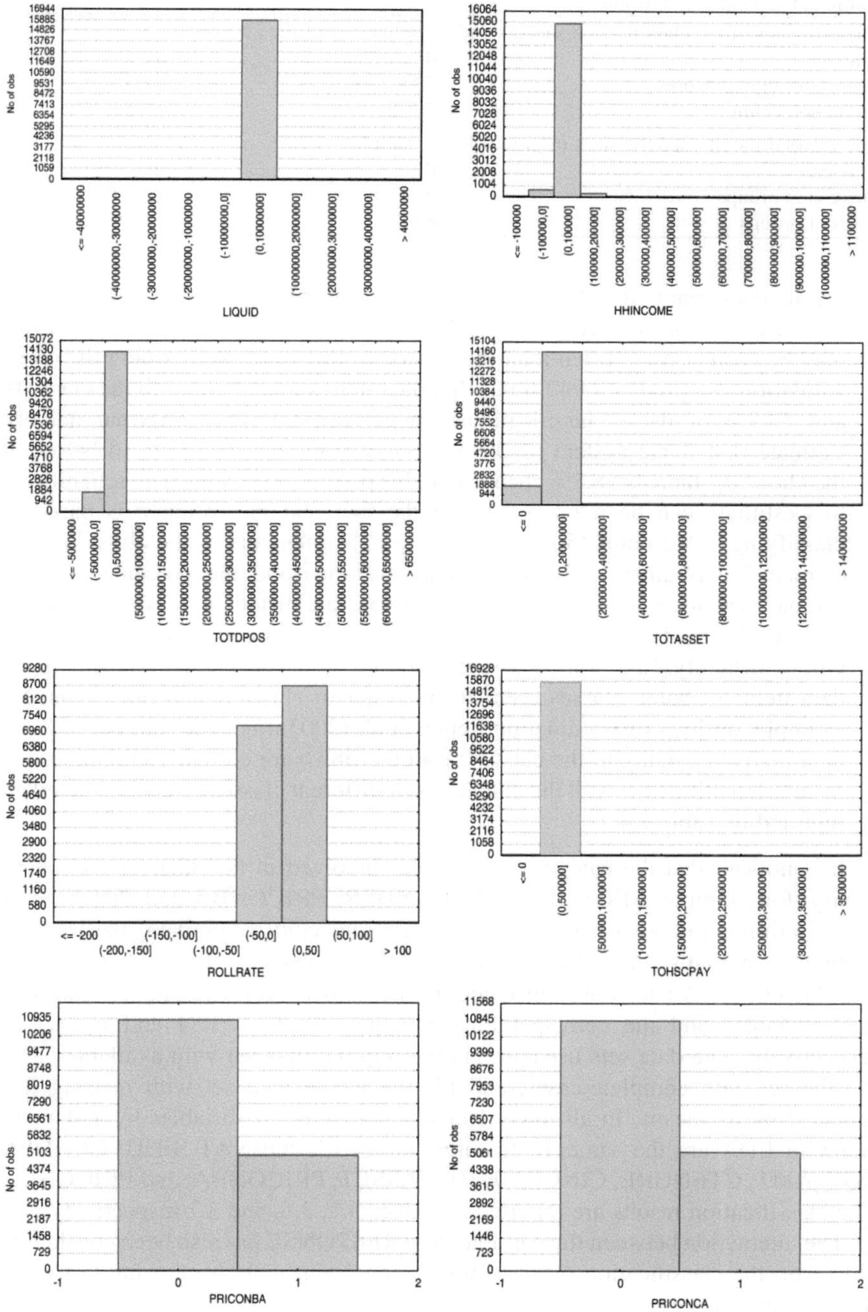

Fig. 3.2 Histograms for the selected variables

Table 3.4 Classification results obtained by Naive Bayes

The estimates of the classification success and the reliability of the estimates		
Classes	0	1
Success of classification	59%	58%
Prediction count	7,942	8,058
The group difficulty and the sizes of groups in the sample		
Classes	0	1
Success in groups	58%	59%
Sizes of groups	8,074	7,936

certain data item to the class "0." Does this mean that there is 58.2% chance that this classification is correct? It is not necessarily true, since some classifications may be correct more often than the others. In this case, while doing the cross-validation, we predict 7,942 times that data item should belong to the class "0" and 58.7% of these classifications are correct. So we (somewhat naively) estimate that if the system predicts previously unseen data item to belong to the class "0," there is 58.7% chance that this prediction is right. The reliability of this estimate can be rated by stating the fact that the estimate is based on classifying 7,942 items (50% of the sample) as members of the class "0." The estimated correctness of different classifications and the percentage of the sample size used to calculate each estimate are listed in Table 3.4. If this estimate is based on very few predictions, then it is of course not very reliable.

4. Group difficulty. Like some classifications are more reliable than the others, the data items of some classes seem to be easier to classify than the others. For example, during cross-validation Leung et al. (2003a) noticed that out of 8,074 data items belonging to the class "0," 4,664 (58%) are correctly classified. The results depicting how well the data items of different classes can be predicted are reported in Table 3.4.

It is noticed that the categorical variables involved in the data set are mainly binary, for example, MS, EDU, CAR, GENDER, PRICONBA and PRICONCA. For discrimination with a mixture of binary and continuous data, the simplest method is perhaps the LDA (Vlachonikolis and Marriott 1982). Leung et al. (2003a) choose Mahalanobis distance as the statistic for entering or removing new variables and the corresponding F values are $F_1 = 3.84$ and $F_2 = 2.71$, respectively. The data sets used are (1) the original data set with available-cases, (2) the set with complete-cases, and (3) the whole data set with missing data replaced by its means. In all three data sets, the selected variables by a standard stepwise LDA are the same 10 features, i.e. AGE, JOBNAT, EDU, GENDER, PDT_AMT, CTENURE, CINCOME, TOTASSET, PRICONBA, and PRICONCA. The classification results are reported in Tables 3.5, 3.6, and 3.7 respectively.

The interaction between the variables and RESPONSE has also been considered. However, the classification result is almost consistent with those using LDA, and the difference is very small.

From Tables 3.5, 3.6 and 3.7, Leung et al. (2003a) conclude that replacing missing values by the means is not a good way to deal with it, at least in their

Table 3.5 Classification results obtained by LDA with available-cases

Response			Predicted group membership		Total	Percent of cases correctly classified
			0.00	1.00		
Original	Count	0.00	3,161	2,185	5,346	59.1%
		1.00	2,232	3,212	5,444	
	%	0.00	59.1	40.9	100.0	
		1.00	41.0	59.0	100.0	
Cross-validated[a]	Count	0.00	3,152	2,194	5,346	58.9%
		1.00	2,245	3,199	5,444	
	%	0.00	59.0	41.0	100.0	
		1.00	41.2	58.5	100.0	

[a]Cross validation is done only for those cases in the analysis. In cross validation, each case is classified by the functions derived from all cases other than that case

Table 3.6 Classification results obtained by LDA with complete-cases

Response			Predicted group membership		Total	Percent of cases correctly classified
			0.00	1.00		
Original	Count	0.00	2,349	1,726	4,075	59.6%
		1.00	1,674	2,660	4,334	
	%	0.00	57.6	42.4	100.0	
		1.00	38.6	61.4	100.0	
Cross-validated	Count	0.00	2,341	1,734	4,075	59.3%
		1.00	1,685	2,649	4,334	
	%	0.00	57.4	42.6	100.0	
		1.00	38.9	61.1	100.0	

Table 3.7 Classification results obtained by LDA for the whole data set with missing data replaced by the means

Response			Predicted group membership		Total	Percent of cases correctly classified
			0.00	1.00		
Original	Count	0.00	4,103	3,897	8,000	57.5%
		1.00	2,907	5,093	8,000	
	%	0.00	51.3	48.7	100.0	
		1.00	36.3	63.7	100.0	
Cross-validated	Count	0.00	4,087	3,913	8,000	57.3%
		1.00	2,924	5,076	8,000	
	%	0.00	51.1	48.9	100.0	
		1.00	36.6	63.5	100.0	

data set. It is also noticed that the feature subsets selected by Naive Bayes and LDA are almost the same. It means that the subsets do not shed too much light on RESPONSE due to missing data. Based on these feature variables, more precise classification models can be utilized.

Remark 3.1. Although recent empirical evaluations have found the Bayesian classifier surprisingly accurate (Kohavi 1996), its learning is not efficient when the data set is not complete. In this case, the exact estimate of each conditional probability distribution is a mixture of the estimates that can be computed in each data set generated by the combination of all possible values of the missing entry. For lower computational cost, it may be necessary to develop other methods for classification. In addition, the randomness assumption made on the missing data is difficult to check. The development of the robust Bayesian estimator (RBE) (Ramoni and Sebastiani 1996), which does not rely on any assumption about the pattern of missing data, is to overcome such problem.

3.2.4 Robust Bayesian Classification Model

The RBE produces probability estimates that are robust with respect to the pattern of missing data by providing probability intervals that contain the estimates learned from all possible completions of the incomplete data set. Based on this estimate, a robust Bayesian classifier (RoC) is developed in Ramoni and Sebastiani (1999). RoC is trained on an incomplete data set using RBE and performs classification by propagating probability intervals.

Since the RoC handles discrete attribute variables only, continuous variables need to be discretized first. For simplicity, the discretization method chosen in Leung et al. (2003a) splits the range between minimum and maximum into four intervals containing the same number of cases. Then the conditional probability $p(x_i = a_{ij}|y = k)$ of each attribute and the prior probability $p(y = k)$ of each class may be taken as uniform. The prior precision in the RoC is set at 10 and the whole data set is divided into four equal-size data subsets. Applying the RoC to each subset five times and using two different assignment criteria (the stochastic dominance and weak dominance), the cross-validation classification results are obtained in Table 3.8.

Although the 62% accuracy can be achieved by the stochastic dominance criteria, the percent of cases that can be classified is only 29%, while the weak dominance criterion gives only 56% accuracy. If the RoC is trained with the weak dominance for the whole data set, the accuracy is 56.54%. Clearly, RoC is not ideal in the data mining task. It seems to imply that the MAR pattern of missing data is

Table 3.8 Cross validation results of using two assignment criteria

	Stochastic dominance	Weak dominance
Correct	14,388	44,943
Incorrect	8,815	35,057
Not classified	56,797	0
Accuracy	62.0092229453088%	56.17875%
Coverage	29.00375%	100.0%

correct and informative. Without considering the pattern, RoC sacrifices this useful information for robustness so that its performance is not too satisfactory. Thus the Naïve Bayes method appears to be more reliable.

3.3 Mixture Discriminant Analysis

3.3.1 A Brief Statement About Mixture Discriminant Analysis

Though LDA enjoys a number of favorable properties, such as reasonable robustness and non-normality, its performance is significantly hampered when problems involve a large number of highly correlated features, and the class boundaries in the predictor space are complex and nonlinear. In finance, particularly banking business, features are correlated (Thomas et al. 1992) and in consumer behavior, consumption may exhibit non-linear dynamics (Fitzpatrick 1976). Therefore, in terms of discriminant analysis, we need to consider a more precise classification model.

LDA can be derived as the ML method for normal populations with different means and common covariance matrix. In practice the normal assumption for each population class is rarely met, and a class might even have multiple modes and arbitrary distribution. Fortunately, a finite mixture of normal distributions can provide an approximation to arbitrary distribution. In other words, finite mixture approximation is theoretically suitable to almost all populations. By assuming that each observed class is a mixture of unobserved normally distributed subclasses, LDA can be generalized into the mixture discriminant analysis (MDA) which has many advantages over traditional classification methods (Hastie and Tibshirani 1996).

MDA models the class densities of the predictors $p(\mathbf{x}|y)$ by normal mixture models. That is,

$$p(\mathbf{x}|y=j) = |2\pi\mathbf{\Sigma}|^{-1/2}\sum_{r=1}^{R_j}\pi_{jr}\exp[-M(\mathbf{x},\mu_{jr})/2], \tag{3.6}$$

where R_j is the number of subclasses in class j, π_{jr} is the mixing probability for the rth subclass, Σ is the common covariance matrix, and $M(\mathbf{x},\mu)$ is the Mahalanobis distance between $\mathbf{x}$ and μ. Thus, the posterior class probabilities are

$$p(y=j|\mathbf{x}) \propto p(y=j)p(\mathbf{x}|y=j) \propto p(y=j)\sum_{r=1}^{R_j}\pi_{jr}\exp[-M(\mathbf{x},\mu_{jr})/2]. \tag{3.7}$$

The classification rule chooses j to maximize $p(y=j|\mathbf{x})$. It should be noted that this does not have the same form as the LDA, and it is likely to be non-linear. From the discriminant analysis point of view, the decision boundary provided by MDA is non-linear while that by LDA is linear. It is expected that the performance of classification using MDA is more successful to the bank data set.

In LDA with K classes, we can choose a subspace of rank $r < K$ that maximally separates the class centroids. This is useful for descriptive purposes mainly but is also a form of regularization. It often improves classification performance. Similar to LDA, MDA also has such a property.

3.3.2 Mixture Discriminant Analysis by Optimal Scoring

To achieve significant computational advantages, both the LDA and the MDA are fitted by using "optimal scoring," i.e. multiple linear regression followed by an eigenvector analysis. The algorithm is based on the well-known fact that LDA is equivalent to canonical correlation analysis. As for the estimation of mixture model parameters, the EM algorithm is suggested in Hastie and Tibshirani (1996). The implementation is as follows:

Initialization

Start with a set of R_j subclasses c_{jr} for each class, and he associated subclass probabilities

$$\hat{p}(c_{jr}|\mathbf{x}_i, j) = \text{Prob } (\mathbf{x} \text{ belongs to } r \text{ th subclass of class } j|\mathbf{x}, j)$$
$$= \frac{\pi_{jr} \exp[-M(\mathbf{x}, \mu_{jr})/2]}{\sum_{k=1}^{R_j} \pi_{jk} \exp[-M(\mathbf{x}, \mu_{jk})/2]} \tag{3.8}$$

obtained from the preprocessing of data. Let $R = \sum_j R_j$ be the total number of all subclasses.

Iteration

Step 1. *Compute the blurred response* $Z_{n \times R}$: define the blurred response matrix as follows: If $y_i = j$, then fill the jth block of R_j entries in the ith row with the values $\hat{p}(c_{jr}|\mathbf{x}_i, j)$, $r = 1, \ldots, R_j$, and the settle with 0s.

Step 2. *Perform multivariate linear regression*: fit a multi-response, linear regression of Z on X. Let $\hat{Z}$ be the fitted values and $\eta(\mathbf{x})$ be the vector of fitted regression functions.

Step 3. *Obtain optimal scoring*: let Θ be the largest r non-trivial eigenvectors of $Z^T\hat{Z}$, with normalization $\Theta^T D_p \Theta = I_r$. Here D_r is a diagonal $R \times R$ matrix of weights, with the kth entry being the sum of the elements of the kth column of Z (the total weight for subclass k).

Step 4. *Update*: the fitted model from step 2 using the optimal scores $\eta(\mathbf{x}) \leftarrow \Theta^T \eta(\mathbf{x})$.

Step 5. *Update*: $\hat{p}(c_{jr}|\mathbf{x}_i, j)$ and $\hat{\pi}_{jr}$ using (3.8) and $\hat{\pi}_{jr} \propto \sum_{y_i = j} p(c_{jr}|\mathbf{x}_i, j)$.

Classification
The final optimally scaled regression fit is the (K-1) vector function $\eta(\mathbf{x}) = \Theta^T\mathbf{x}$. Assign an observation $\mathbf{x}$ to the class j that minimizes

$$\delta(\mathbf{x},j) = ||D(\eta(\mathbf{x}) - \bar{\eta}^j)||^2 \tag{3.9}$$

where $D = diag(d_1, ..., d_r)$, $d_k^2 = 1/[\alpha_k^2(1 - \alpha_k^2)]$, α_k^2 is the kth largest eigenvalues computed in step 3, $\bar{\eta}^j = \sum_{y_i=j} \eta(\mathbf{x}_i)/n_j$ is the fitted centroid of the jth class in this space of canonical variates.

3.3.3 Analysis Results and Interpretations

Leung et al. (2003a) apply the above MDA procedure to the bank data set without any missing value. The result obtained is $r = 5$, i.e., the classification problem can be solved in 5-dimensional canonical variate space by (3.9). The coefficients of the function $\eta(\mathbf{x}) = \Theta^T\mathbf{x}$ are tabulated in Table 3.9. As a comparison, the three MDA results are given in Tables 3.10, 3.11 and 3.12 respectively. Each uses 10 groups of data subsets and each subset contains 800 complete-cases.

From Table 3.13, it can be observed that the MDA results with feature variables selected by NB are on the average almost the same as those obtained by LDA.

Table 3.9 Coefficients obtained by MDA

	[1]	[2]	[3]	[4]	[5]
Intercept	−0.0850116	1.2215510	0.0899665	−0.8217654	1.9675210
AGE	−0.0084463	−0.0054678	−0.0075755	0.0176246	−0.0228753
CLUSTER	−0.0006324	0.0095413	0.0021439	0.0033997	0.0080512
ECLASS	−0.0533932	−0.0934749	−0.1200723	−0.0320507	−0.0043241
MS	0.1247881	0.3437687	−0.3714430	−0.1676564	0.1037491
JOBNAT	0.0001674	0.0004556	−0.0006812	−0.0000419	−0.0014111
EDU	0.1305043	−0.0784225	−0.0942460	0.0968627	−1.0757600
CAR	−0.1207183	1.2580590	0.7001054	0.1339236	0.0166667
CHILD21	0.0537040	−0.1816621	0.1114166	0.1720327	−0.3083423
GENDER	0.1046692	−0.0332733	0.0177266	−0.4373076	−0.6267653
TENURE	0.0003726	0.0027991	0.0023134	0.0026482	−0.0040459
PDT.AMT	−0.0861770	−0.1641653	0.0461901	0.1485395	0.0577677
RATIO	0.3125694	−2.6678610	6.3422000	7.2232060	0.0242849
CTENURE	−0.0004971	−0.0037301	0.0002703	−0.0075157	0.0068805
CINCOME	0.0000081	−0.0000105	−0.0000035	−0.0000111	−0.0000091
CREDLMT	−0.0000111	−0.0000152	0.0000121	−0.0000127	0.0000062
SAVBAL	−0.0000032	−0.0000035	0.0000044	−0.0000041	0.0000004
LIQUID	−0.0000025	−0.0000019	−0.0000086	0.0000023	0.0000002
HHINCOME	−0.0000035	0.0000053	0.0000013	0.0000028	0.0000008
TOTDPOS	0.0000107	−0.0000020	0.0000059	−0.0000002	0.0000011
TOTASSET	−0.0000013	0.0000033	−0.0000017	−0.0000009	−0.0000010
ROLLRATE	0.0094289	0.0052612	−0.0010156	0.0090758	0.0942741
TOHSCPAY	0.0000023	0.0000057	−0.0000033	−0.0000011	−0.0000002
PRICONBA	0.1614056	0.2667631	−0.1810600	0.0449305	−1.1230800
PRICONCA	−0.0657292	−0.1973182	−0.2215508	−0.3177543	−0.6636434

Table 3.10 Results obtained by MDA with feature variables selected by LDA

	Percent between-group variance explained					Correct	Deviance
	V1	V2	V3	V4	V5		
1	94.7	99.16	99.62	99.85	100	0.58	1286.936
2	96.29	98.84	99.7	99.91	100	0.6175	1096.195
3	97.68	98.85	99.54	99.85	100	0.60375	1092.998
4	98.57	99.16	99.62	99.83	100	0.61125	1108.54
5	98.56	99.42	99.8	99.94	100	0.5775	1118.636
6	99.24	99.89	99.96	99.99	100	0.61625	1128.123
7	97.56	98.77	99.7	99.94	100	0.6275	1096.77
8	96.1	98.43	99.47	99.84	100	0.58875	1107.493
9	95.84	98.52	99.64	99.9	100	0.60625	1205.035
10	97.44	98.86	99.68	99.89	100	0.60875	1052.611
Mean	97.198	98.99	99.673	99.894		0.60375	1129.334
Median	97.5	98.855	99.66	99.895		0.6075	1108.017
Std. Dev.	1.431524	0.433923	0.135651	0.052324		0.01658	67.46167

Table 3.11 Results obtained by MDA with feature variables selected by NB

	Percent between-group variance explained					Correct	Deviance
	V1	V2	V3	V4	V5		
1	91.65	98.48	99.26	99.89	100	0.58	1128.888
2	98.4	99.45	99.87	99.98	100	0.6125	1089.119
3	97.88	98.95	99.68	99.92	100	0.6175	1015.46
4	98.91	99.31	99.65	99.89	100	0.59.75	1121.656
5	80.46	96.34	97.79	99.13	100	0.58125	1164.255
6	91.53	97.37	98.84	99.58	100	0.5975	1094.978
7	86.79	97.54	99	99.9	100	0.62125	1093.166
8	95.22	97.85	99.02	99.62	100	0.575	1107.391
9	95.4	97.54	99.03	99.8	100	0.61375	1033.472
10	97.63	98.77	99.31	99.79	100	0.59125	1169.284
Mean	93.387	98.16	99.145	99.75		0.59667	1101.77
Median	95.31	98.165	99.145	99.845		0.59438	1101.18
Std.Dev.	5.94726	0.99368	0.58559	0.25338		0.19013	49.4629

Table 3.12 Results obtained by MDA with all feature variables

	Percent between-group variance explained					Correct	Deviance
	V1	V2	V3	V4	V5		
1	92.21	97.02	99.22	99.63	100	0.58375	1189.344
2	88.71	94.54	98.6	99.4	100	0.63	1024.858
3	94.54	97.3	98.38	99.33	100	0.6375	1015.025
4	87.17	97.92	99.16	99.6	100	0.64625	1064.057
5	56.47	94.18	99.03	99.56	100	0.61375	1059.393
6	95.28	99.58	99.81	99.94	100	0.60625	1078.014
7	92.61	96.13	98.1	99.47	100	0.63125	1078.552
8	90.52	95.57	98.86	99.5	100	0.595	1048.562
9	90.39	94.56	97.87	99.42	100	0.615	1102.324
10	72.51	95.3	98.79	99.56	100	0.62125	1076.817
Mean	86.041	96.21	98.782	99.541		0.618	1073.695
Median	90.455	95.85	98.825	99.53		0.61813	1070.437
Std.Dev.	12.21886	1.732102	0.572593	0.169013		0.01937	48.30761

Table 3.13 Comparison of results obtained by MDA with LDA, NB and All

Groups	Percent of correctly classified cases		
	NB	LDA	ALL
1	0.58	0.58	0.58375
2	0.6125	0.6175	0.63
3	0.6175	0.60375	0.6375
4	0.59.75	0.61125	0.64625
5	0.58125	0.5775	0.61375
6	0.5975	0.61625	0.60625
7	0.62125	0.6275	0.63125
8	0.575	0.58875	0.595
9	0.61375	0.60625	0.615
10	0.59125	0.60875	0.62125
Mean	0.59667	0.60375	0.618
Median	0.59438	0.6075	0.61813
Std.Dev.	0.19013	0.01658	0.01937

However, if all feature variables are used, the MDA results are more precise for the apparent reason that it uses more information. Nevertheless, it pays the cost for a more complex model, even not feasible in practice sometimes.

Remark 3.2. Several parametric statistical classifiers have been introduced and applied to client segmentation in the promotion of credit cards. These models have their own advantages and limitations. With reference to classification performance, MDA is more accurate but requires more feature variables. NB and LDA both have almost the same classification performance and need almost the same feature variables. NB and MDA can both provide the class posterior probability of each case so that the clients can be segmented into different groups with different class posterior probabilities. The promotion strategy can thus zero in on the clients with larger class probability. Due to the very poor quality of the data, particularly with a large number of missing values, the classification accuracies tend to be low and are more or less similar among these methods. So, the results do not imply that the methods are not appropriate. With better data quality, it is expected that the level of accuracy is much higher, particularly MDA. We just use the examples to demonstrate what can possibly be done under a very bad situation.

3.4 The Logistic Model for Data Classification

3.4.1 A Brief Note About Using Logistic Regression as a Classifier

Some classifiers for the informative approach to data classifications have been discussed in Sects 3.2 and 3.3. The logistic regression model is employed in this section to illustrate how the discriminative approach can be employed to classify data with a different perspective.

The discriminative approach models the class posteriors $p(y|\mathbf{x})$ and makes inference directly. Logistic regression model: $p(y|\mathbf{x}) = \exp(\beta^T\mathbf{x})/[1+\exp(\beta^T\mathbf{x})]$ is a typical example (Menard 1995). The discriminative approach is more flexible with regard to the class densities. Parameters are estimated by maximizing the conditional log-likelihood function

$$\hat{\theta}_{MCL} = \arg\max_{\theta} \sum_{i=1}^{n} \log p_\theta(y_i|x_i). \tag{3.10}$$

However, this approach ignores part of the information from the data, namely, the marginal distribution $p(\mathbf{x})$.

The technique has been used quite successfully in many finance and marketing studies (Guadagni and Little 1983; Allenby and Rossi 1994). Using the client segmentation for product marketing as an example, let $\mathbf{x} = (x_1, x_2, \cdots, x_m)$ be the vector of the predictors, where $x_i (i = 1, 2, \cdots, m)$ can be either categorical or continuous variables, and $p(\mathbf{x})$ be the probability of a client (characterized by $\mathbf{x}$) having positive response to the credit-card promotion. Specifically, the logistic regression model has the form

$$\log\left(\frac{p(\mathbf{x})}{1-p(\mathbf{x})}\right) = \beta_0 + \beta_1 x_1 + \cdots + \beta_m x_m, \tag{3.11}$$

or,

$$p(\mathbf{x}) = \frac{\exp(\beta_0 + \beta_1 x_1 + \cdots + \beta_m x_m)}{1+\exp(\beta_0 + \beta_1 x_1 + \cdots + \beta_m x_m)}, \tag{3.12}$$

where $\beta_0, \beta_1, \cdots, \beta_m$ are the model parameters that need to be estimated. By building an appropriate logistic regression model, we can use it to predict the probability of a client having response in the future campaigns or to find out the categories that include those people who are more likely to have positive response. Strategies for promoting the credit card in the future can then be provided.

3.4.2 Data Manipulation for Client Segmentation

In their study, Leung et al. (2003b) first drop from the data set those variables that have more than 8,000 missing values because these variables cannot provide sufficient information for the analysis. To explore the significance of the predictors in details, they then fit a logistic regression model with the remaining predictors one by one to check whether or not each predictor is having a significant effect on the probability of a client having positive response. It is found that most of the continuous variables are, in their original status, not significant at 0.05 level of significance. Since the values of some variables are estimated or enriched with auxiliary information in a certain way, then they may introduce extra error to the analysis, especially for the estimated values of the continuous variables such as the financial variables. To lessen such error, it may then be better to transform the continuous variables into categorical variables to make the values granular.

After preprocessing the original variables, a total of 19 binary variables and a three-level variable "education level" (where the value "UNKNOWN" is treated as one of the levels) are selected. These newly-formed variables are used as the set of predictors for building the logistic regression models.

Three types of logistic regression models are built for different purposes. The first model uses all newly-formed binary variables to fit a logistic regression model in which the variables are selected by the stepwise procedure at 0.001 level of significance. In the second model, interactions are examined between each pair of the selected variables at the same significance level. The final model includes nine newly-formed binary variables and three interaction terms which may be used for prediction in practice.

To facilitate our discussion, Table 3.14 first list all of the variables that will appear in the above three types of logistic regression models with their names, descriptions, codes of the levels for the newly-formed binary or three-level variables and definitions of the codes (for convenience, the same names of the original variables are used to name the corresponding newly-formed variables).

Generally, observations with missing values in the selected variables are deleted for the fitting of a logistic regression model. However, it is noted that "UNKNOWN" is still treated as one of the levels in the variable EDU, i.e. the education levels of clients. For this variable, there is, on one hand, a significant difference between the response rates at its two levels of "COLLEGE" and "NON-COLLEGE" and, on the other hand, there are 5,203 out of 16,000 observations with the levels of education being recorded as "UNKNOWN". Considering the importance of this variable and to avoid too many observations being deleted when fitting a model, this variable is transformed into a three-level variable with "UNKNOWN" being its second level. By such a categorization, this newly-formed three-level variable is very significant when fitting the logistic regression model. After all, the information for this variable is not complete and the values of this variable are not easy to obtain in practice either. So, this variable is only employed to fit the model for developing new clients.

3.4.3 Logistic Regression Models and Strategies for Credit Card Promotion

3.4.3.1 The Model for Prediction

According to the methodology stated in the previous section, Leung et al. (2003b) first fit a logistic regression model with the 19 newly-formed categorical variables, and employ the stepwise procedure to select the final variables to remain in the

Table 3.14 Variable list for the credit card promotion problem

Variable name	Description	Level code	Code definition
GENDER	Gender of clients	0	GENDER = MALE
		1	GENDER = FEMALE
AGE	Age of clients	0	AGE < 25 or AGE > 45
		1	25 ≤ AGE ≤ 45
		0	EDU = COLLEGE
EDU	Education levels	1	EDU = UNKNOWN
		2	EDU = NON-COLLEGE
JOBNAT	Job nature of clients	0	Businessman; Restaurant; Service provider; Blue collar; Other working population except for those in level 1; Non-working population
JOBNAT	Job nature of clients	1	Import, export, wholesale, retail and manufacturing; Education; Medical services; Real estate and construction; White collar; Government and public organizations; Professional
PRICONBA	Price consciousness for BIA/BSA	0	PRICONBA = Yes
		1	PRICONBA = No
CTENURE	Number of the months from the time that the first credit card account was opened to the time stamped on Aug. 31, 1999	0	CTENURE ≤ 36
		1	CTENURE > 36
ROLLRATE	Rollover rate of credit card	0	ROLLRATE ≥ 0
		1	ROLLRATE < 0
PDT_AMT	Number of active products held by each client in the latest month	0	PDT_AMT ≤ 2
		1	PDT_AMT > 2
RATIO	Ratio of number of active products to the number of months from the time that the first account was opened in the bank to the time stamped on Aug. 31, 1999	0	RATIO < 0.02
		1	RATIO ≥ 0.02
CINCOME	Estimated income of clients	0	CINCOME < 7,000 or CINCOME > 30,000
		1	7,000 ≤ CINCOME ≤ 30,000
RESPONSE	Whether or not a client having at least one response in the campaigns of the credit card promotion	0	RESPONSE = Yes
		1	RESPONSE = No

model at 0.001 level of significance. Then interactions are examined between each pair of the selected variables. An interaction term is retained in the model only if its p-value of the Wald χ^2 test (see for example, SAS Institute Inc. 1995) is equal to or less than 0.001 and the corresponding two variables are still significant at this level. When the interaction terms already in the model are made insignificant after another significant interaction term has been added, they are deleted from the model.

The rationale for using a relatively small significance level of 0.001 for selecting the variables and interaction terms in the analysis is based on the following considerations:

1. In the process of analyzing the data set, there is no significant improvement on the model performance, such as prediction and classification abilities, by adding a variable with the significance level set for around 0.01 or larger
2. By considering the practical aspect and the high cost in obtaining the observations of the variables through survey, it is more pragmatic to keep as few a number of variables in the model as possible.

With the aforementioned rules for building the model, a logistic regression model (Model-1) which includes nine newly-formed binary variables and three interaction terms is obtained as follows:

$$\begin{aligned}\log\left(\frac{p(\mathbf{x})}{1-p(\mathbf{x})}\right) &= -1.8282 + 0.4422\ \text{GENDER} + 0.3804\ \text{PRICONBA}\\ &\quad + 0.2633\ \text{AGE} + 1.1038\ \text{CTENURE} + 0.6830\ \text{ROLLRATE}\\ &\quad + 0.3726\ \text{CINCOME} + 0.2927\ \text{PDT\_AMT} + 0.1549\ \text{RATIO}\\ &\quad + 0.1664\ \text{JOBNAT} - 0.2534\text{PRICONBA} * \text{CTENURE}\\ &\quad - 0.4434\text{ROLLRATE} * \text{CTENURE} - 0.3565\text{CINCOME}\\ &\quad * \text{CTENURE}.\end{aligned}$$

To further understand the importance of each variable in the model, we can examine in Table 3.15 part of the results.

From this table, we can observe that the newly-formed binary variables GENDER, CTENURE, and ROLLRATE play an important role in the model since their observed values of the Wald X^2 statistic are, respectively, 174.02, 175.29 and 156.40 which are much larger than those of the other variables in the model.

Model-1 contains many variables and has better prediction ability. Based on this model, procedure for promoting the credit card can be made as follows:

1. For each plausible potential client targeted for promotion of the credit card, determine the coded value (that is, 0 or 1 by the definition listed in Table 3.14) of each variable in Model-1 according to the client's information recorded in the bank or obtained by some method of estimation. It should be observed that the exact values of the continuous variables are not needed in this case.
2. For a potential client characterized by $\mathbf{x}$, Substitute the corresponding coded value of each variable into Model-1 and calculate his/her response probability

Table 3.15 Partial output by SAS logistic procedure for Model-1

Variable	Estimated Parameter	Standard Error	Wald χ^2	p-value
INTERCEPT	−1.8282	0.0863	448.94	< 0.0001
GENDER	0.4422	0.0335	174.02	< 0.0001
PRINCONBA	0.3804	0.0618	37.85	< 0.0001
AGE	0.2633	0.0365	52.17	< 0.0001
CTENURE	1.1038	0.0834	175.29	< 0.0001
ROLLRATE	0.6830	0.0546	156.40	< 0.0001
CINCOME	0.3726	0.0592	39.62	< 0.0001
PDT_AMT	0.2927	0.0444	43.46	< 0.0001
RATIO	0.1549	0.0450	11.88	0.0006
JOBNAT	0.1664	0.0335	24.66	< 0.0001
PRICONBA*CTENURE	−0.2534	0.0732	11.97	0.0005
ROLLRATE*CTENURE	−0.4433	0.0731	36.80	< 0.0001
CINCOME*CTENURE	−0.3565	0.0766	21.69	< 0.0001

$p(\mathbf{x})$. Choose an appropriate cut-off point p_0 (for example, $p_0 = 0.5$) according to the size of the population of potential clients and the number of users of the credit card the bank expects. Then send the credit card promotion to this candidate only if $p(\mathbf{x}) \geq p_0$. Under such a strategy, the larger is the cut-off point p_0, the smaller the size of the mail is needed and the larger is the probability of having a positive response.

3.4.3.2 The More Practical Model

Although Model-1 can provide a useful guide for promotion of the credit cards in future campaigns, it contains too many variables and some of them, even their coded values, such as CINCOME and JOBNAT, may be too expensive to obtain in practice. And, for a larger cut-off pointp_0, there may only be a few people who are eligible for the mail promotion. In order to overcome the insufficiency of Model-1, only a few variables in Model-1 which are most significant to the response probability and relatively easy to obtain their information for determining the coded values are used to fit a logistic regression model. Based on this model, the population of the candidates can be divided into several groups in descending order of probable responsiveness. Choosing potential clients from the first several groups as main targets for credit card promotion will then have a higher response rate.

Along this line of reasoning, Leung et al. (2003b) start with the model obtained by the stepwise procedure and delete the variables one by one according to their degree of significance until the observed values of the Wald χ^2 statistic of the variables in the model are all larger than 30. As a result, only five newly-formed binary variables (i.e., GENDER, AGE, CTENURE, ROLLRATE and PDT_AMT) are retained in the model. Then interactions among each pair of these five variables are checked and only the interaction terms whose observed values of the Wald χ^2

statistic are larger than 30 are kept in the model. In this way, the final model (Model-2) is obtained as follows:

$$\log\left(\frac{p(\mathbf{x})}{1-p(\mathbf{x})}\right) = -\,1.1101 + 0.4712 \text{ GENDER} + 0.2895 \text{ AGE}$$
$$+\,0.6705 \text{ CTENURE} + 0.6852 \text{ ROLLRATE}$$
$$+\,0.2412 \text{ PDT\_AMT} - 0.4531 \text{ ROLLRATE} * \text{CTENURE}.$$

The other information related to Model-2 is given in Table 3.16.

Since all of the variables in Model-2 are binary, the population of the candidates can then be divided into 32 groups according to the predicted probabilities computed from the model. Table 3.17 lists all of the groups with their predicted probabilities in descending order. The codes of the variables in each group (explained in Table 3.14), taken together, portray the characteristics of the candidates in the respective groups. For example, G1, actually constitutes a classification rule which can be stated as:

"The group with the largest predicted response probability includes those women whose age are between 25 and 45 and their rollover rate of the credit card is negative and who have opened the first credit card account for more than 36 months and had more than two times of active products in the latest month."

In future promotions of the credit card, we can divide all potential clients into different groups according to the characteristics listed in Table 3.17. The first several groups of people, for example, can be chosen as the main targets for promoting the credit card because they have higher response rate. The number of groups to be selected will then be on the discretion of the bank.

Remark 3.3. It should be pointed out that the predicted response probability of each group listed in the last column of Table 3.17 is not the estimate of the actual response rate of this group because the sample on which the analysis is based is not randomly drawn from the target population. In practice, we may not expect such high response rate for each group as listed in Table 3.17. The predicted probability should perhaps be evaluated by using 0.5 as the base point since in the sample 50% registered positive response to previously carried out credit-card promotion.

Table 3.16 Partial output by SAS logistic procedure for Model-2

Variable	Estimated Parameter	Standard Error	Wald χ^2	p-value
INTERCEPT	−1.1101	0.0479	537.14	<0.0001
GENDER	0.4712	0.0331	203.06	<0.0001
AGE	0.2895	0.0356	66.30	<0.0001
CTENURE	0.6705	0.0410	266.85	<0.0001
ROLLRATE	0.6852	0.0540	160.95	<0.0001
PDT_AMT	0.2412	0.0339	50.78	<0.0001
ROLLRATE*CTENURE	−0.4531	0.0723	39.29	<0.0001

Table 3.17 Target groups of potential clients derived from Model-2

Group	Code of the variables					Predicted
	GENDER	AGE	CTENURE	ROLLRATE	PDT_AMT	Probability
G1	1	1	1	1	1	0.6888
G2	1	1	0	1	1	0.6404
G3	1	1	1	0	1	0.6370
G4	1	1	1	1	0	0.6349
G5	1	0	1	1	1	0.6236
G6	1	1	0	1	0	0.5832
G7	0	1	1	1	1	0.5801
G8	1	1	1	0	0	0.5796
G9	1	0	0	1	1	0.5714
G10	1	0	1	0	1	0.5678
G11	1	0	1	1	0	0.5655
G12	0	1	0	1	1	0.5264
G13	0	1	1	0	1	0.5228
G14	0	1	1	1	0	0.5205
G15	1	0	0	1	0	0.5116
G16	0	0	1	1	1	0.5084
G17	1	0	1	0	0	0.5079
G18	1	1	0	0	1	0.4730
G19	0	1	0	1	0	0.4662
G20	0	1	1	0	0	0.4626
G21	0	0	0	1	1	0.4542
G22	0	0	1	0	1	0.4506
G23	0	0	1	1	0	0.4483
G24	1	1	0	0	0	0.4135
G25	1	0	0	0	1	0.4019
G26	0	0	0	1	0	0.3953
G27	0	0	1	0	0	0.3918
G28	0	1	0	0	1	0.3591
G29	1	0	0	0	0	0.3455
G30	0	1	0	0	0	0.3056
G31	0	0	0	0	1	0.2955
G32	0	0	0	0	0	0.2479

3.4.3.3 The Model for Developing New Users of the Credit Card

It is in general necessary for a bank to develop new users of a credit card outside the population of its original clients. Since the values (or the codes) of the variables such as CTENURE, ROLLRATE and PDT_AMT are not available for those who are not the clients of the bank, then Model-1 and Model-2 as well as the related strategies for credit card promotion cannot be used in practice. For a person with no records in the bank, however, the most obtainable information may be his/her socio-economic demographics such as job nature and income level whose values (or codes) may be relatively easy or inexpensive to obtain from some existing organizational records or surveys. Based on this consideration, nine variables of this type are chosen from a total of 20 newly-formed categorical variables in Sect. 3.4 for fitting of the logistic regression model.

Table 3.18 Partial output by SAS logistic procedure for Model-3

Variable	Estimated Parameter	Standard Error	Wald χ^2	p-value
INTERCEPT	−0.7114	0.0442	258.86	<0.0001
GENDER	0.3780	0.0326	134.54	<0.0001
EDU	0.1574	0.0211	55.59	<0.0001
AGE	0.3378	0.0353	91.32	<0.0001
JOBNAT	0.1857	0.0326	32.50	<0.0001

In this case, EDU, education level of a candidate, is one of the variables chosen because of its importance and its 5,203 observations with "UNKNOWN" value are treated as one of the three levels which is coded as 1 (see Table 3.14). Once again, the stepwise procedure with the 0.001 significance level is used to select the variables. Interactions are checked between each pair of the selected variables and no significant interaction terms are found even at the level of 0.05. The final model (Model-3) is obtained as

$$\log\left(\frac{p(\mathbf{x})}{1-p(\mathbf{x})}\right) = -0.7114 + 0.3780\ \text{GENDER} + 0.1574\ \text{EDU} + 0.3378\ \text{AGE} + 0.1857\ \text{JOBNAT}.$$

The other estimation information obtained from the SAS logistic procedure is shown in Table 3.18.

The three binary variables (GENDER, AGE and JOBNAT) and the three-level variable (EDU) divide all potential clients into 24 groups with varying response probabilities arranged in descending order (Table 3.19).

When promoting the credit card, the bank can choose those people with the characteristics stipulated in the first several groups. Higher predicted probability means people have higher response rate. These predicted probabilities, similar to that in Table 3.17, again should be evaluated by using 0.5 as the base point.

3.4.4 Model Comparisons and Validations

In the previous subsection, three kinds of models are constructed and the corresponding strategies for promoting the credit card are provided. However, how do these models perform? Are they valid to the data set? What is the gain in model performance from categorizing values of the variables? We shall, to a certain extent, provide answers to these questions.

First of all, categorizing values of the original variables, especially the continuous variables, not only makes value estimation, as mentioned in Sect. 3.4, easier to carry out, but also improves the statistical significance in model fitting. To further demonstrate the advantages of categorizing values of the variables, we fit a logistic regression model with the same 19 variables that have been chosen in Sect. 3.4 (not including EDU) in their original form (that is, without categorizing their values)

Table 3.19 Target groups of potential new clients derived from Model-3

Group	Code of the variables				Predicted
	GENDER	EDU	AGE	JOBNAT	Probability
G1	1	2	1	1	0.6236
G2	1	1	1	1	0.5860
G3	1	2	1	0	0.5791
G4	1	0	1	1	0.5474
G5	1	2	0	1	0.5417
G6	1	1	1	0	0.5404
G7	0	2	1	1	0.5317
G8	1	1	0	1	0.5024
G9	1	0	1	0	0.5011
G10	1	2	0	0	0.4954
G11	0	1	1	1	0.4924
G12	0	2	1	0	0.4853
G13	1	0	0	1	0.4632
G14	1	1	0	0	0.4561
G15	0	0	1	1	0.4531
G16	0	2	0	1	0.4475
G17	0	1	1	0	0.4461
G18	1	0	0	0	0.4174
G19	0	1	0	1	0.4089
G20	0	0	1	0	0.4077
G21	0	2	0	0	0.4021
G22	0	0	0	1	0.3715
G23	0	1	0	0	0.3649
G24	0	0	0	0	0.3293

except for the variable JOBNAT (this categorical variable in its original form has too many levels and is not significant at 0.05 level of significance when fitting a logistic regression model with only this variable). The procedures and significance level are same as those in Model-1 in the selection of variables and examination of interaction terms. The final model (Model-4) is as follows:

$$\begin{aligned}\log\left(\frac{p(\mathbf{x})}{1-p(\mathbf{x})}\right) = &-0.8296 + 0.4308\ \text{GENDER} + 0.2858\ \text{PRICONBA} \\ &+ 0.3356\ \text{PRICONCA} + 0.1220\ \text{CHILD21} - 0.0104\ \text{AGE} \\ &+ 0.00193\ \text{CTENURE} + 0.0780\ \text{PDT\_AMT} \\ &+ 0.3825\ \text{JOBNAT} - 0.0411\ \text{PDT\_AMT} * \text{JOBNAT}.\end{aligned}$$

where the variables PRICONCA and CHILD21, which are not included in Table 3.14, represent, respectively, price consciousness for the credit card (with two levels being yes and no) and number of children in a family (with three levels being 0, 1 and 2). Other results are partially tabulated in Table 3.20.

Compared with Table 3.15, we find that three continuous variables (ROLL-RATE, CINCOME and RATIO) become significant after their values are categorized. In contrast, the categorical variables are almost the same in both Model-1 and Model-4 except for PRICONCA and CHILD21. Besides, the model performance,

Table 3.20 Partial output by SAS logistic procedure for Model-4

Variable	Estimated Parameter	Standard Error	Wald X^2	p-value
INTERCEPT	−0.8296	0.0826	100.99	<0.0001
GENDER	0.4308	0.0334	166.85	<0.0001
PRICONBA	0.2858	0.0465	37.74	<0.0001
PRICONCA	0.3356	0.0432	60.37	<0.0001
CHILD21	0.1220	0.0306	15.90	<0.0001
AGE	−0.0104	0.00176	34.94	<0.0001
CTENURE	0.00193	0.000533	13.11	0.0003
PDT_AMT	0.0780	0.00864	81.36	<0.0001
JOBNAT	0.3825	0.0517	54.78	<0.0001
PDT_AMT*JOBNAT	−0.0411	0.00961	18.27	<0.0001

such as the goodness of fit and prediction ability, has been improved by categorizing values of the variables. For example, we can observe in the outputs that the variables and the interaction terms in Model-1 reduce the -2LOG L (here LOG L represents the log-likelihood) by 873.7 with 12° of freedom and the variables as well as the interaction term in Model-4 reduce it by 527.1 with 9° of freedom. Also, the measures used for describing the association of the predicted probabilities and observed responses such as Somers' D, Gamma, Tau-a and C (see SAS (1995) for explanations of these measures) for Model-1 are all larger than those for Model-4, which are, respectively, 0.266, 0.269, 0.133 and 0.633 for Model-1 and 0.212, 0.213, 0.106 and 0.606 for Model-4. Although only a few very significant variables and interaction term are included in Model-2, its performance is still comparable to that of Model-1. For example, the five binary variables and one interaction term in Model-2 reduce -2LOG L by 753.7 with 6° of freedom and Somers' D, Gamma, Tau-a and C of Model-2 are, respectively, 0.245, 0.256, 0.122 and 0.622. Model-3 performs less satisfactorily because of the limited information used for fitting the model. The four variables in Model-3 reduce -2LOG L by 367.1 with 4° of freedom; Somers' D, Gamma, Tau-a and C of Model-3 are 0.176, 0.188, 0.088 and 0.588, respectively.

On the other hand, the validity of Model-2 and Model-3 can be observed, to some extent, by comparing the predicted probability with the observed response rate in each group. The predicted probability for each group is calculated by substituting the coded values of the variables for that group into Model-2 or Model-3 and solving for $p(X)$, and the corresponding observed response rate is the ratio of the number of responses to total number of observations in that group. The results are shown in Tables 3.21 and 3.22 for Model-2 and Model-3, respectively.

From the results, we can observe that the predicted probability and the corresponding observed response rate of each group are in general comparable for both Model-2 and Model-3.

To compare prediction ability of the models that we have built, we use the first 10,000 observations of the data set to fit the four logistic regression models that, respectively, include the same variables and interaction terms as those in Model-1,

Table 3.21 Comparison of the predicted probabilities and the observed response rate for each group based on Model-2

Group	Number of observations	Number of responses	Response rate	Predicted probability
G1	690	473	0.6855	0.6888
G2	567	332	0.5855	0.6404
G3	1,561	992	0.6355	0.6370
G4	255	154	0.6039	0.6349
G5	195	119	0.6103	0.6236
G6	391	271	0.6931	0.5832
G7	636	375	0.5896	0.5801
G8	837	475	0.5675	0.5796
G9	202	108	0.5347	0.5714
G10	518	300	0.5792	0.5678
G11	71	46	0.6479	0.5655
G12	338	171	0.5059	0.5264
G13	1,539	821	0.5335	0.5228
G14	244	132	0.5410	0.5205
G15	207	105	0.5072	0.5116
G16	283	141	0.4982	0.5084
G17	273	131	0.4799	0.5079
G18	899	420	0.4672	0.4730
G19	217	110	0.5069	0.4662
G20	960	427	0.4448	0.4626
G21	151	69	0.4570	0.4542
G22	770	366	0.4753	0.4506
G23	121	53	0.4380	0.4483
G24	625	278	0.4448	0.4135
G25	375	166	0.4427	0.4019
G26	149	53	0.3557	0.3953
G27	442	169	0.3824	0.3918
G28	701	243	0.3466	0.3591
G29	443	141	0.3183	0.3455
G30	426	123	0.2887	0.3056
G31	290	92	0.3172	0.2955
G32	359	81	0.2256	0.2478

Model-2, Model-3 and Model-4 (for convenience, we still write the fitted model with the first 10,000 observations as Model-1, Model-2, Model-3 and Model-4). Then we use the fitted models to calculate the predicted probability for each of the last 6,000 observations. Classifying these 6,000 observations by the predicted probabilities with the cut-off point being 0.5, we have obtained the correct classification rate for each of the four models in Table 3.23 (the observations with missing values are, as usual, deleted):

Of the four models, Model-1 performs best in classification. Compared with the classification result of Model-4, it once again shows that some gain can be achieved by categorizing values of the variables. Given that only five variables are included in Model-2, its classification ability is quite satisfactory compared with that of Model-1. It is understandable for Model-3 to have relatively low correct classification rate because only limited information is used for model fitting. Besides, if we

Table 3.22 Comparison of the predicted probabilities and the observed response rate for each group based on Model-3

Group	Number of observations	Number of responses	Response rate	Predicted probability
G1	1,902	1,164	0.6120	0.6236
G2	781	484	0.6197	0.5860
G3	1,291	751	0.5817	0.5791
G4	596	310	0.5201	0.5474
G5	649	347	0.5347	0.5417
G6	866	472	0.5450	0.5404
G7	1,006	531	0.5278	0.5317
G8	176	99	0.5625	0.5024
G9	365	200	0.5479	0.5011
G10	629	306	0.4865	0.4954
G11	641	336	0.5242	0.4924
G12	1,140	543	0.4763	0.4853
G13	107	49	0.4579	0.4632
G14	338	166	0.4911	0.4561
G15	663	271	0.4087	0.4531
G16	496	225	0.4536	0.4475
G17	1,161	523	0.4505	0.4461
G18	360	126	0.3500	0.4174
G19	397	173	0.4358	0.4089
G20	455	191	0.4198	0.4077
G21	628	247	0.3933	0.4021
G22	157	64	0.4076	0.3715
G23	554	204	0.3682	0.3649
G24	292	93	0.3185	0.3293

Table 3.23 The correct classification rates of the last 6,000 observations by the respective models fitted with the first 10,000 observations

Model	Model-1	Model-2	Model-3	Model-4
Correct classification rate	0.6023	0.5950	0.5655	0.5712

use models fitted with the whole data set to reclassify all of the observations with 0.5 as the cut-off point, the correct classification rates of the four models are 0.5968, 0.5900, 0.5661 and 0.5793 respectively. There is not much of a difference between these classification results and those listed in Table 3.23.

To recapitulate, three logistic regression models have been built for different purposes and the related strategies for credit card promotion have been discussed in Leung et al. (2003b). Model comparisons and validations have shown that if more information on each client of the bank is available, Model-1 is the first choice in practice. Given the possibly high expenses and the difficulty in obtaining the values of some variables in Model-1, Model-2 is more practical and easier to use. Furthermore, prediction ability of Model-2 is comparable to that of Model-1. Model-3 can provide useful guides for the bank to develop new credit-card clients. It has been demonstrated in this study that categorizing the values of the variables is helpful not only in making the model constructed easier to use but also in improving the statistical significance of the variables.

Nevertheless, it should be noted that the given data set is extremely noisy. Also, some of the variables have been enriched by some methods which may introduce potential errors. According to our experience in analyzing such data set, it seems to be difficult to further improve the performance of the models in the context of logistic regression.

Remark 3.4. In general, when the population is Gaussian, informative classification is more efficient then discriminative, i.e. fewer training observations are required, or, for a fixed number of training observations, better classification can be obtained. Even when the class densities are not Gaussian there are circumstances, such as when the classes are well separated, in which informative training does about as well as discriminative. For example, it turns out that the advantage of LDA and logistic regression are dataset specific and each of them has no absolute advantages over the other (Efron 1975)

Rubinstein and Hastie (1997) suggest that an informative approach should be used if confidence in the model correctness is high. This suggests a promising way of combining the two approaches: partition the feature space into two. Train an informative model on those dimensions for which it seems to be correct, and a discriminative model on the others. Even when the goal is discrimination between classes, it pays to investigate the performance of the corresponding informative model which borrows strength from the marginal density.

3.5 Support Vector Machine for Spatial Classification

3.5.1 *Support Vector Machine as a Classifier*

In recent pattern recognition research, a powerful method for small-sample learning problems, called Support Vector Machine (SVM), has been developed on the basis of statistical learning theory (Vapnik 1995, 1998, 1999). The basic idea of SVM is to map the input vectors into a high-dimensional feature space in which an optimal hyperplane separating two classes with maximal margin is constructed. It has subsequently been extended to contain polynomial classifiers, neural networks, radial basis function (RBF) networks and other architectures so that special and complicated classes can be nonlinearly separated. It has been demonstrated that SVMs deliver good results in pattern classification, regression estimation, and operator inversion for ill-posed problems, even though they do not capitalize on domain specific knowledge (Burges 1998; Cristianni and Shawe-Taylor 2000).

Based on the principle of structural risk minimization and the capacity with pure linear combinatorial definitions, the quality and complexity of the SVM solutions do not depend directly on the dimensionality of the input space. The optimal decision surface of a SVM is constructed from its finite support vectors, a subset of training examples. Its parameters are conventionally determined by solving a quadratic programming (QP) problem. Therefore, SVM differs from conventional statistical approach (which uses distribution function) and neural network approach (which uses connection weights) in that it uses the minimal number of support vectors for the construction of decision function for pattern recognition. For SVM, the dimension of the feature space is not limited because it provides much more significant measurements of complexity independent of the dimension of the feature space. Based on finite support vectors, the linear decision boundaries are constructed in the feature space of higher dimension, which corresponds to the input space of lower dimension. Conceptually, SVM provides better and more flexible generalization of approximation functions in high-dimensional space by linear estimates using multiple basis functions. Computationally, the QP optimization problem can be solved by the dual kernel of linear functions in high-dimensional feature space. One distinct characteristic of SVM is that it aims to find the optimal hyperplane from a set of training samples such that the expected recognition error for the test samples is minimized. Therefore, SVM has recently attracted much attention owing to its rigorous theoretical derivations from statistical learning theory and good empirical results in some classification tasks. While the success of SVM still awaits a lot more applications, it appears that a wide application can be made in feature extraction and classification of spatial data. As a novel approach, SVM models, albeit simple, have been applied to perform automatic "pure pixel" selection and classification from remote-sensing data based on linear spectral mixture model (Hermes et al. 1999; Brown et al. 1999, 2000).

I present in here a SVM-based spatial classification model for remote sensing data (Leung et al. 2002b). The proposed model is a supervised classification procedure based on multi-dimensional training vectors selected by a window template of a certain size rotated in multiple angles. Two SVM-based algorithms, SVM1 and SVM2, are formulated. SVM1 dose not use any preprocessing procedure and takes inputs as the original multi-dimensional vectors. For more efficient training and effective classification, SVM2 employs a preprocessing procedure to reduce the dimension of the input vectors while keeping the maximum information. A comparison with BP-MLP and ARTMAP is also made.

3.5.2 Basics of Support Vector Machine

3.5.2.1 Two-Class Problem

The fundamental learning principle of SVM is essentially the construction of a hyperplane to linearly separate a data set into two classes in the feature space. Since the multiple-class problem can be reduced to a set of independent two-class problems, we only need to concentrate our discussion on the two-class problem.

Let the data set be training samples of n pairs: $(x_1, x_2), \cdots (x_n, y_n), \mathbf{x} \in \boldsymbol{R}^d$, and let $y \in \{+1, -1\}$ be the class label set. Suppose that there exists a hyperplane which can separate the positive examples from the negative examples. The decision function of the hyperplane is then defined as:

$$D(\mathbf{x}) = (\mathbf{w} \cdot \mathbf{x}) + w_0, \ w \in \boldsymbol{R}^N, w_0 \in \boldsymbol{R} \tag{3.13}$$

where $\mathbf{w}$and $w_0$are suitable coefficients, and $\mathbf{w}$ is a normal vector to the hyperplane.

Given a training data set, the hyperplane for linearly separable data should satisfy the following constraints:

$$\begin{cases} (\mathbf{w} \cdot \mathbf{x}_i) + w_0 \geq +1, & y_i = +1, \quad i = 1, \cdots, n, \\ (\mathbf{w} \cdot \mathbf{x}_i) + w_0 \leq -1, & y_i = -1, \quad i = 1, \cdots, n. \end{cases} \tag{3.14}$$

These can be combined into one set of inequalities:

$$y_i[(\mathbf{w} \cdot \mathbf{x}_i) + w_0] \geq 1, \ i = 1, \cdots, n. \tag{3.15}$$

Let the "margin" of a separating hyperplane be the shortest distance from the separating hyperplane to the closest positive or negative point, which is represented by τ. The margin is directly relevant to the generalization ability of the separating hyperplane. The larger is the margin, the more separable two classes become. Therefore, for the linearly separable case, the SVM algorithm simply looks for the hyperplane that separates the data with maximal margin, i.e. the optional separating hyperplane. The support vectors are the points located at the edge of the margin, or equivalently they satisfy:

$$y_i[(w \cdot \mathbf{x}_i) + w_0] = 1, \tag{3.16}$$

Though it is very difficult to separate a space by a single point, we can parametrically combine these points to determine the decision surface of the optimal hyperplane. The training algorithm of SVM is thus the search for the support vectors and their combination coefficients.

Let the perpendicular distance of point $\mathbf{x}'$ to the separating hyperplane be $|D(\mathbf{x}')|/||\mathbf{w}||$. If the margin really exists, then all training samples should satisfy the following inequality:

$$\frac{y_k D(\mathbf{x}_k)}{||\mathbf{w}||} \geq \tau, \ y_k \in \{-1, 1\}, \ k = 1, \cdots, n, \tag{3.17}$$

where $||\mathbf{w}||$ is the Euclidean norm of $\mathbf{w}$. The target of finding the optimal hyperplane is equivalent to the estimation of $\mathbf{w}$ in the maximum margin. Subject to the above constraints, we can find the pair of hyperplanes which gives the maximum margin by minimizing $\|\mathbf{w}\|^2$.

To construct this optimal hyperplane, we solve the following primal problem:

$$\begin{cases} \min & h(\mathbf{w}) = \frac{1}{2}\|\mathbf{w}\|^2 \\ s.t. & y_i[\mathbf{w}\cdot\mathbf{x}_i + w_0] \leq 1, \quad i = 1, \cdots, n. \end{cases} \tag{3.18}$$

This constrained optimization problem can be solved via the Lagrange function:

$$Q(\mathbf{w}, w_0, \alpha) = \frac{1}{2}(\mathbf{w}\cdot\mathbf{w}) - \sum_{i=1}^{n} \alpha_i\{y_i[(\mathbf{w}\cdot\mathbf{x}_i) + w_0] - 1\}, \tag{3.19}$$

where α_i is the Lagrange multiplier. The Lagrange function Q has to be minimized with respect to the primal variable $\mathbf{w}$ and $\mathbf{w}_0$ and maximized with respect to the dual variables α_i.

According to Karush–Kuhn–Tucker complementarity's conditions in optimization theory, the primal variable $\mathbf{w}$ and w_0 can be represented by the Lagrange multiplier α_i. The optimization of Q is then converted into a dual maximization problem in which only the Lagrange multiplier α_i is relevant. It should be noted that for all constraints that are not precisely satisfied as equalities, the corresponding $\alpha_i$must be 0: this is a solved value of α_i that maximizes Q. At the saddle point, the derivatives of Q with respect to the primal variables must vanish. By solving the derivatives, the solutions, including vectors $\mathbf{w}^*, w_0^*, \alpha^*$ of Q, can be obtained. The vectors possess the following characteristics:

1. The Lagrange multipliers $\alpha_i^*(i = 1, \cdots, n)$ satisfy:

$$\sum_{i=1}^{n} \alpha_i^* y_i = 0, \alpha_i^* \geq 0, \quad i = 1, \cdots, n. \tag{3.20}$$

2. The vector $\mathbf{w}^*$ is a linear combination of the training patterns:

$$\mathbf{w}^* = \sum_{i=1}^{n} \alpha_1^* y_i \mathbf{x}_i, \alpha_i^* \geq 0, \quad i = 1, \cdots, n. \tag{3.21}$$

3. The solution vector has an expansion in terms of a subset of the training patterns, namely those patterns whose α_i is non-zero, called Support Vectors. By the Karush–Kuhn–Tucker complementarity's conditions

$$\alpha_i^*[y_i(\mathbf{w}^*\cdot\mathbf{x}_i + w_0^*) - 1] = 0, \quad i = 1, \cdots, n. \tag{3.22}$$

That means the support vector lies on the margin. All remaining examples of the training set are irrelevant: their constraint does not play a role in the optimization, and they do not appear in the expansion. This nicely captures our intuition of the problem: as the hyperplane is completely determined by the patterns closest to it, the solution should not depend on the other examples.

By substituting the above three characteristics into Q, one eliminates the primal variables and arrives at the Wolfe dual of the primal problem:

Find multipliers α_i by solving:

$$\begin{cases} \max \quad Q(\alpha) = \sum_{i=1}^{n} \alpha_i - \frac{1}{2} \sum_{i,j=1}^{n} \alpha_i \alpha_j y_i y_j (\mathbf{x}_i \cdot \mathbf{x}_j) \\ s.t. \quad \sum_{i=1}^{n} y_i \alpha_i = 0, \quad \alpha_i \geq 0, \quad i = 1, \cdots, n. \end{cases} \tag{3.23}$$

The hyperplane decision function can thus be expressed as:

$$D(\mathbf{x}) = \sum_{i=1}^{n} \alpha_i^* y_i (\mathbf{x} \cdot \mathbf{x}_i) + w_0^*. \tag{3.24}$$

It should be noted that non-zero α_i^* is the multiplier of the support vector $\mathbf{x}_j$ which can be represented by the dot product $(\mathbf{x} \cdot \mathbf{x}')$ with the input vectors. The structure of the optimization problem closely resembles the Lagrange's formulations in mechanics. The hyperplane is an ideal decision function for the problem of separating points in the feature space because its complexity deviates from the dimension of the input space and can be carefully controlled in order to obtain good generalization.

3.5.2.2 Kernel-Function-Based Nonlinear SVM

Extending on the idea of optimal margin classifier, which however needs linearly separable classes, SVM can have nonlinear kernel functions and soft margins with slack variables. Such extension leads to a rather successful employment of support vector machines to solve practical classification problems. It provides a way to overcome the "curse of dimensionality." The generalization ability of SVM is based on the factors described in the theory for controlling the generalization of the learning processes. SVM permits the use of samples to directly describe the hyperplane by which the classification problem can be directly solved and the probabilistic density estimate is no longer necessary.

The hyperplane is defined as a linear decision function, and the optimal hyperplane can be obtained by the input space. To allow for more general decision surfaces, one can first nonlinearly transform a set of input vectors into a high-dimensional space by a mapping procedure and then perform a linear separation there. Thus, SVM can be extended to realize nonlinear mapping function by constructing high-dimensional basis functions, and the corresponding mapping space is called the feature space (Scholkopf et al. 1997).

Let $g_j(\mathbf{x}), j = 1, \cdots, m$, be a nonlinear transformation function that maps the input vector $\mathbf{x}$ into a point in m-dimensional feature space. Then by the linear SVM, the hyperplane is constructed in this feature space, and the linear decision function can be mapped into nonlinear decision function in the input space. If the feature is produced by a nonlinear transformation function $g_j(\mathbf{x})$, the decision function can be written as:

$$D(\mathbf{x}) = \sum_{j=1}^{m} w_j g_j(\mathbf{x}), \tag{3.25}$$

where the sum only depends on the dimension of the feature space. Here the zero rank of the threshold item w_0 is removed for it can be represented by adding a constant basis function ($g(\mathbf{x}) = 1$) into the feature space.

Compared to linear separation by hyperplane in the feature space, nonlinear SVM needs to compute the dot product between vectors in order to map the input space into the feature space. Maximizing and evaluating then require the computation of dot products in the feature space. Let $g_j(\mathbf{x}), j = 1, \cdots, m$, be the large set of basis function. Then the key step is to determine the dot product of basis functions. In its dual form, the decision function is transformed into:

$$D(\mathbf{x}) = \sum_{i=1}^{n} \alpha_i y_i H(\mathbf{x}_i, \mathbf{x}). \tag{3.26}$$

The dot product kernel H is a representative form of the basis function $g_j(\mathbf{x})$. For certain basis function $g_j(\mathbf{x})$, H can be determined as:

$$H(\mathbf{x}, \mathbf{x}') = \sum_{j=1}^{m} g_j(\mathbf{x}) g_j(\mathbf{x}'), \tag{3.27}$$

where m is the dimension.

In other words, one constructs nonlinear decision functions in the input space that are equivalent to the linear decision functions in the feature space. In the feature space, we form the convolutions of the inner products between the support vector and the input vector.

Using different expressions for inner products, one can construct different learning machines with different types of nonlinear decision surfaces in the input space. In 1909, Mercer proved a theorem which defines the general form of inner products in the Hilbert space. Therefore, any function satisfying Mercer's condition can be used as a construction rule which is equivalent to constructing an optimal separating hyperplane in some feature space.

For example, in order to specify polynomials of any fixed order q in a SVM, we can use the following polynomial kernel:

$$H(\mathbf{x}, \mathbf{x}') = [(\mathbf{x} \cdot \mathbf{x}') + 1]^q \tag{3.28}$$

to construct a polynomial learning machine.

On the other hand, a radial basis function machine can be implemented by using the exponential kernel:

$$H(\mathbf{x}, \mathbf{x}') = \exp\left(-\frac{|\mathbf{x} - \mathbf{x}'|^2}{\sigma^2}\right). \tag{3.29}$$

In this case, the SVM will find both the center **x**' and the corresponding spread width σ of the kernel function. Figure 3.3a is the classification result of a 2-class space with simulated samples based on the radial basis function kernel. There are eight support vectors (solid bright dots) determining the overall separation.

Similarly, a two-layer neural network can be realized by introducing the neural network kernel $H(\mathbf{x}, \mathbf{x}') = S(\rho(\mathbf{x}, y) + \theta)$ where$S$is a sigmoid function; $\rho$and $\theta$are constants satisfying certain conditions.

Remark 3.5. The extension from the two-class problem to the multiple-class problem is an important step in the SVM approach (Scholkopf et al. 1999; Angulo and Catala 2000). Although SVM is initially proposed for the two-class problem, it can easily be extended to solve multiple-class problems. In the standard approach, the two-class decision function is extended to K classes by constructing a two-valued decision function for all k classes as follows:

$$f_k : \boldsymbol{R}^N \to \{\pm 1\} \begin{cases} +1, & \textit{for all samples in class } k, \\ -1, & \textit{otherwise}. \end{cases} \tag{3.30}$$

It means samples in class k are regarded as one class, and other samples that are not attributed to class k are regarded as another class. Therefore, in a k (k > 2) class problem, k groups of decision functions are represented by k groups of support vectors which can realize the separation in the input space. Figure 3.3b is the classification result of a 6-class space based on simulated samples. This is, in fact, the solution to the problem depicted in Fig. 1.3 in Sect. 1.5.

3.5.3 *Experiments on Feature Extraction and Classification by SVM*

3.5.3.1 The SVM-Based Procedure

Extracting and classifying spatial features from remote sensing images is a significant and challenging task in remote sensing research. Though classification can be pixel-based, it is more reliable if multiple pixels can be used. That is, spatial features can be captured in template windows. The number of pixels contained in a template naturally becomes the dimension of the input vector for a classifier. However, if the size of a template is too large, the dimension of the input vector

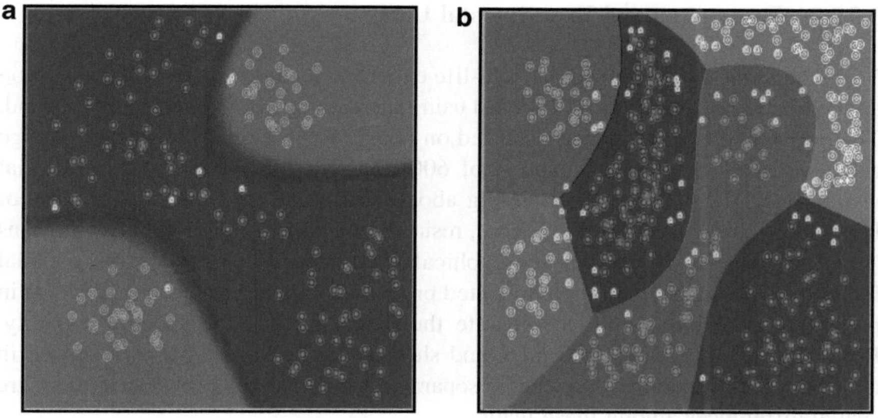

Fig. 3.3 Experimental separation results with SVM classification. (**a**) A two-class problem. The solid bright dots represent the support vectors. (**b**) A multiple-class problem. The solid bright dots represent the support vectors

increases accordingly. It is often difficult to analyze and process large templates containing complicated geographical and spectral information by conventional approaches. SVM, on the other hand, is a good alternative to solve such problems. As a supervised classification algorithm, the SVM-based procedure consists of the following steps:

Step 1. Selection of training samples. Training samples labeled with each classification category are visually chosen with a moving template window.

Step 2. Preprocessing of training samples. In order to speed up the training and classification phases, especially for very large template, it is essential to lower the dimension of the input vectors while keeping relevant spatial information. The karhunen-Loeve (K-L) transformation is thus employed to lower the dimension of the input vector.

Step 3. Constructing the kernel function for the nonlinear SVM. According to the inner product kernel function, the vector of spatial feature in the input space is mapped into the corresponding vector in high-dimensional feature space. Generally, the Gaussian radial basis function is utilized as the kernel function.

Step 4. Training Phase of SVM. In order to separate a feature with other patterns, the nonlinear decision function of the separating hyperplane is computed. The hyperplane in the feature space is determined by a set of support vectors and the corresponding multipliers α_i.

Step 5. Extracting or classifying phase. Finally, the unknown spatial pattern read from the primary image is inserted into the decision function of each category of the spatial features with inner product mapping into the feature space. Then by the winner-takes-all scheme, the input vector is classified.

3.5.3.2 Study Area and Experimental Data

To evaluate the SVM algorithms, real-life experiments on land cover classification are conducted by Leung et al. (2002b) using the data of SPOT panchromatic band. The primary SPOT image was acquired on Feb 3, 1999. The size of the sub-image cut down from the whole image is of 600 rows by 700 columns with a spatial resolution of 10 m by 10 m, covering about 42 km^2 of the central urban area of Hong Kong along the Victoria Harbor, resided by a population of over 2.5 millions (Fig. 3.4). Due to its special geographical location and complex terrain, spatial features in the image exhibit complicated properties. On the panchromatic image in particular, it is impossible to separate the pixels by just the spectral property. Reflection of water body, hill area, and shadows of buildings are very similar in appearance. Fortunately, they can be separated by human vision because there are spatial differences among these features.

Basing on the knowledge acquired from related materials and practical surveying, together with the visual interpretation of the corresponding remote sensing data, five main types of spatial features covering the area are identified for classification:

C1 — Water Body; *C2* — Hilly Area; *C3* — Barren Area;
C4 — Concrete Land; *C5* — Built up Area

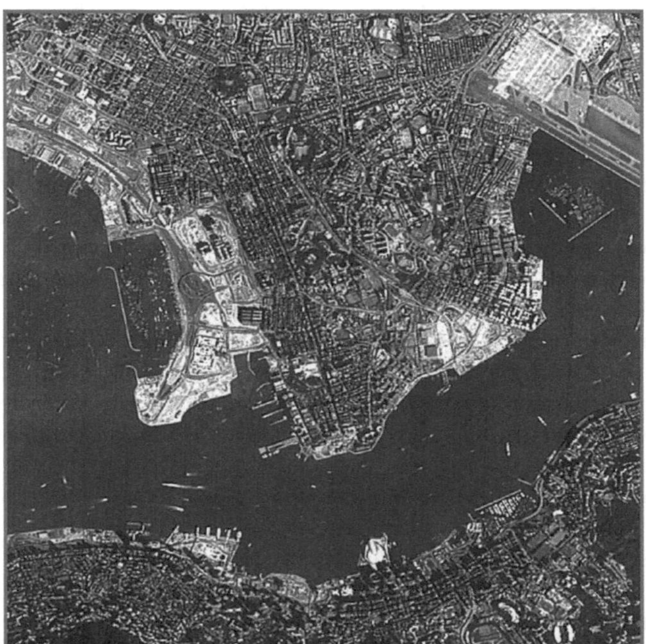

Fig. 3.4 Original SPOT panchromatic image covering central urban area in Hong Kong

The task of the experiment is to separate apart these five land covers. Based on the SVM, the experiments of extracting and classifying urban features in the SPOT panchromatic image were carried out and the results are cross compared with that of the BP-MLP and ARTMAP.

3.5.3.3 Analysis Results and Interpretations

In these experiments, template window of four different sizes, respectively, including 2×2, 3×3, 4×4, 5×5, are used to represent the basic unit of the spatial features. By visual interpretation with reference to urban maps and intrinsic contextual knowledge, about 3,200 training samples (400 locations with 8 directions) and 500 test samples (100 samples in each class) are chosen for the classification task (Table 3.24). Then, four classifiers (including BP-MLP, ARTMAP, and SVM

Table 3.24 Comparisons of parameters of the classifiers for land cover classification

Classifiers template		2×2	3×3	4×4	5×5
Number of training samples		3,360	3,000	3,300	3,320
Number of test samples		500	500	500	500
1. BP-MLP	Training time (second)	600	No convergence	No convergence	No convergence
	Classification time (second)	8	No convergence	No convergence	No convergence
	Test accuracy	59.00%	No convergence	No convergence	No convergence
2. ARTMAP	Training time (second)	148	122	230	280
	Classification time (second)	115	102	166	246
	Test accuracy	75.80%	82.00%	75.00%	82.60%
3. SVM-1 (No K-L)	Input dimension	4	9	16	25
	Training time (second)	8	12	15	18
	Classification time (second)	290	308	519	720
	Test accuracy	82.60%	87.60%	87.60%	92.00%
4. SVM-2 (with K-L)	Input dimension	3	4	8	9
	Training time (second)	6	7	8	9
	Classification time (second)	250	260	308	320
	Test accuracy	84.00%	87.60%	87.80%	92.00%

(SVM1 and SVM2)) are applied and their results are compared by using these training and test samples.

1. BP-MLP: Multilayer perceptron with back-propagation (BP) algorithm.
2. ARTMAP (Carpenter et al., 1991): A neural network evolved from the biological theory of cognitive information processing with self-associative memory and incremental learning mechanism.
3. SVM1: SVM trained with original training samples.
4. SVM2: SVM trained with training samples of K-L-transformation-reduced dimension.

(In both SVM1 and SVM2, the Gaussian radial basis function (RBF) is employed as the nonlinear mapping kernel function. After normalizing the input vectors, the uniform parameter of the RBF is determined, where the width σ ranges from 0.1 to 0.3.)

The experimental results are tabulated in Table 3.24. The following observations can be made:

1. Comparison of SVM1 and SVM2. SVM is more efficient in training and more accurate in classification.
2. Comparison of SVM (SVM1 and SVM2) and BP-MLP. It can be observed that the SVMs give more accurate classification, shorter training time, and guaranteed convergence. On the other hand, if the dimension of the input vector is very high or the number of training samples is very large, the corresponding structure of the BP-MLP becomes very complicated with a very slow training time, or the inability to converge due to sharp oscillation.
3. Comparison of SVM (SVM1 and SVM2) and ARTMAP. It is apparent that the SVMs are more effective and efficient. Since SVM has smaller correlation with the dimension of the input vector, it is lower in complexity and computational cost in the training and classification procedure. Furthermore, the separating procedure of SVM can be parametrically represented by hyperplane decision functions.
4. The principle of selecting the window template. As discussed above, the size of the template exerts great impact on the classification result. Too small a template will not give sufficient information and too large a template will make contained information too varying. Therefore, the size of the window template should be determined in accordance with the contextual information. In this study, the final size of the window template is 5×5, resulting in the error matrix shown in Table 3.25 and the classification result depicted in Fig. 3.5.
5. Over-Fitting: In any training procedure, over-fitting is a common phenomenon. In SVM, when the obtained set of support vectors is too large, the decision function becomes very complicated and its effectiveness will be reduced, albeit good approximation can be made from the training samples. In order to avoid over-fitting, human knowledge can be integrated into the training phase so that the best separating result can be obtained.

Table 3.25 The error matrix resulting from the 5 × 5 window (Accuracy = 92.00%, kappa = 0.900)

	C1	C2	C3	C4	C5	SUM
C1	91	3	0	5	0	99
C2	4	92	0	0	3	99
C3	0	1	99	2	8	110
C4	2	0	1	91	2	96
C5	3	4	0	2	87	96
SUM	100	100	100	100	100	500

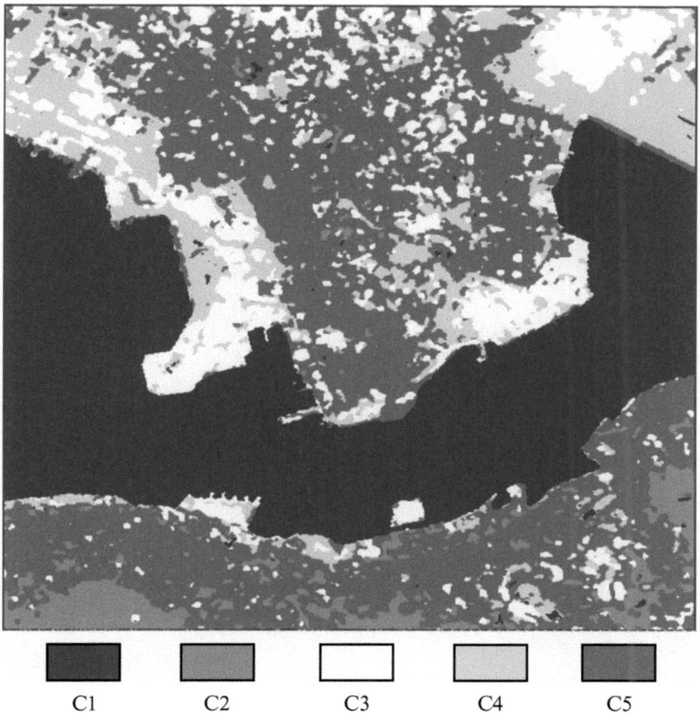

Fig. 3.5 The result of urban land cover classification with 5×5 windows

In further studies, SVM models for remote-sensing feature extraction and classification can be further improved along several directions (Burges and Scholkopf 1997; Hearst et al. 1998; Keerthi et al. 2000). (1) After a large set of support vectors is acquired in the training phase, the decision function of the hyperplane in the feature space should be simplified in order to alleviate the computational overhead. (2) Multiple sources of spatial data should be integrated into the SVM decision function so that a variety of spatial data types can be considered in the decision-making process. (3) It is essential to formulate SVM-based spatial knowledge processing system (including the processing of shapes, shadows, networks, and relationships) and to construct a serial processing system, including the training, memory, extraction and classification phases. (4) It is useful to study contextual information extraction and detection by SVM, such as the extraction of regular spatial contextual features from high resolution satellite or aerial images.

Chapter 4
Algorithmic Approach to the Identification of Classification Rules or Separation Surface for Spatial Data

4.1 A Brief Background About Algorithmic Classification

As discussed in Chap. 3, naïve Bayes, LDA, logistic regression, and support vector machine are statistical or statistics related models developed for the classification of data. Breaking away from the statistical tradition is a number of classifiers which are algorithmic in nature. Instead of assuming a data model which is essential to the conventional statistical methods, these algorithmic classifiers attempt to work directly on the data without making any assumption about them. It has been regarded by many, particularly in the pattern recognition and artificial intelligence communities, as a more flexible approach to discover how data should be classified. Decision trees (or classification trees in the context of classification), neural networks, genetic algorithms, fuzzy sets, rough sets are typical paradigms. They are in general algorithmic in nature. In place of searching for a separation surface, like the statistical classifiers, some of these methods attempt to discover classification rules that can appropriately partition the feature space with reference to pre-specified classes. A decision tree is a segmentation of a training data set (Quinlan 1986; Friedman 1977). It is built by considering all objects as a single group, with the top node serving as the root of the tree. Training examples are then passed down the tree by splitting each intermediate node with respect to a variable. A decision tree is constructed when a certain stopping criterion is met. Each leaf, terminal, node of the tree contains a decision label, e.g., a class label. The decision tree partitions the feature space into sub-spaces corresponding to the leaves. Specifically, a decision tree that handles classification is known as a classification tree and a decision tree that solves regression problems is called a regression tree (Breiman et al. 1984). A decision tree that deals with both the classification and regression problems is referred to as a classification and regression tree (Breiman et al. 1984). Decision tree algorithms differ mainly in terms of their splitting and pruning strategies. They usually aim at the optimal partitioning of the feature space by minimizing the generalization error. The advantages of the decision tree approach are that it does not need any assumptions about the underlying distribution of the data, and it can

Y. Leung, *Knowledge Discovery in Spatial Data*, Advances in Spatial Science,
DOI 10.1007/978-3-642-02664-5_4, © Springer-Verlag Berlin Heidelberg 2010

handle both discrete and continuous variables. Furthermore, decision trees are easy to construct and interpret if they are of reasonable size and complexity. Their disadvantages are that splitting and pruning rules can be rather subjective. The theory is not as rigorous in terms of the statistical tradition. They also suffer from combinatorial explosion if the number of variables and their value labels are not appropriately controlled. Typical decision tree methods are ID3 (Quinlan 1986), C4.5 (Quinlan 1993), CART (Breiman et al. 1984), CHAID (Kass 1980), QUEST and newer versions, and FACT (Loh and Vanichsetakul 1988).

Treating a classifier as a massively connected network, neural network models such as the perceptron (Rosenblatt 1958), multilayer feedforward neural network with back propagation (Rumelhart and McClelland 1986), and radial basis function neural network (Girosi 1994; Sundararajan et al. 1999) attempt to approximate the separation surface via more or less a black-box approach (Bishop 1995; Ripley 1996). Learning algorithms in these networks are distribution-free and can actually be treated as hypersurface reconstruction that tries to estimate the hypersurface partitioning nonlinearly separable classes from training data/examples. Though they can be effective for a classification task, they are plagued by local minima, high computational overhead, training unpredictability, and poor generalization. Their interpretations are by no means straightforward. Instead of being unidirectional, recurrent neural networks construct classifiers as fully connected networks that partition the feature space into attraction basins (Hopfield 1982). The self-organization map (Carpenter and Grossberg 1988; Kohonen 1988) and various types of associative memories (Xu et al. 1994)) are typical examples. These models, among other things, need to deal with the problems of storage capacity, convergence, and error correction capability. As a classifier, their interpretations are again non-trivial. Neural networks are actually large sample algorithms for large sample problems. They might give unsatisfactory results in real-life classification problems where small samples are common.

To obtain an optimal set of rules that can partition the feature space, genetic algorithms employ an encoding scheme for a rule set and evolve it through selection, crossover and mutation under the survival of the fittest principle (Goldberg 1989). Upon convergence, the Darwinian approach generates an optimal rule set that separates pre-specified classes in an appropriate way. Its results tend to be more interpretable compared to that generated by the neural network methods. However, the ways in which the crossover and mutation operators work are sometimes confusing. Differing from neural networks that do not need a pre-specified model, genetic algorithms are essentially model-based.

By allowing imprecision in rules, the fuzzy sets approach attempts to construct a fuzzy partition of the feature space. In place of well-defined rules, fuzzy classification rules do not yield an all-or-nothing separation surface. That is, partial and multiple class memberships are allowed in a classification. The building block of the approach consists of the membership functions defining fuzzy sets and operators that work on them. Its results are often easy to interpret. The fuzzy sets approach itself does not directly deal with the discovery of classification rules unless it is integrated with methods such as genetic algorithms (Leung et al. 2001b) and neural networks (Kosko 1992). The closest fuzzy sets method is perhaps the projection of fuzzy clusters in high dimensional space onto the feature and class label dimensions to obtain fuzzy classification rules. To cater for granular and incomplete information and to allow data to speak for themselves, the rough sets approach even works more directly on data (Pawlak 1991). Unlike the statistical approach which often depends on some kinds of assumption about the probabilistic distribution of data, and the neural network approach which relies on certain network topology, and the fuzzy sets approach which needs pre-specification of membership functions, the rough set methods discover classification rules without any prerequisites. Rules are unraveled by an information deduction mechanism. Though it is developed for data mining in qualitative data, recent developments have freed it from this restriction (Leung et al. 2007, 2008a).

In this chapter, I will discuss with illustrations the pros and cons of some major algorithmic approaches to the discovery of separation surfaces or classifications rules for spatial data. The examination is of course not exhaustive. Breaking away from the conventional statistical tradition, the classification tree approach is investigated in Sect. 4.2. Computational paradigms accounting for local effect and nonlinearity in classification are then introduced in Sects. 4.3 and 4.4, where neural networks and genetic algorithms are discussed, respectively. Freeing us from the restrictions of data distributions and model assumptions, the rough sets approach for the discovery of classification rules is introduced in Sect. 4.5. To make our classification method less mechanical and closer to human perception, a vision-based method which directly treats noise and scale is highlighted in Sect. 4.6. A remark on the choice of classifiers is made in Sect. 4.7.

4.2 The Classification Tree Approach to the Discovery of Classification Rules in Data

4.2.1 A Brief Description of Classification and Regression tree (CART)

Classification and regression tree (CART) has been an important data mining methodology for the analysis of large data sets via a binary partitioning procedure (Breiman et al. 1984). It consists of a recursive division of N cases on which a response variable and a set of predictors are observed. Such a partitioning procedure is known as a regression tree when the response variable is continuously valued and as a classification tree when the response variable is categorical. A classification tree procedure provides not only a classification rule for new cases of unknown class, but also an analysis of the dependence structure in large data sets.

Figure 4.1 depicts a simple tree structure with three layers of nodes. The top level node is the root node. The second layer consists of an internal node which needs to be further partitioned and a terminal node where partitioning is no longer

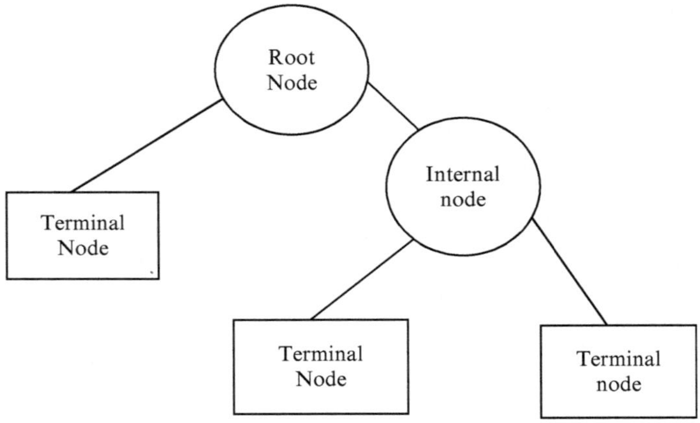

Fig. 4.1 A simple tree structure

required. Finally, two terminal nodes are obtained in the last layer. It should be noted that the root and the internal nodes are both marked with circles and are connected to two nodes in the next layer, called the left and right offspring nodes. The root node contains the entire learning sample, and the other nodes correspond to subgroups of the learning sample. The two subgroups in the left and right offspring nodes are disjoint, and their union comprises the subgroups for the parent node. A critical step of the tree-based technique is to determine the split from one parent node to two offspring nodes.

Let (Y, **X**) be a multivariate random variable where **X** is the predictor vector $(X_1, \cdots, X_m, \cdots, X_M)$, where $X_1, \cdots, X_m, \cdots, X_M$ can be a mixture of ordered and categorical variables; and Y is the criterion variable taking values in the set of prior classes $\Gamma = \{1, \cdots, j, \cdots, J\}$.

Four elements are needed in the classification tree growing procedure:

1. A set of binary questions of the form$\{ is\, \mathbf{X} \in A?\}$.
2. A goodness of split criterion $\Delta i(s|t)$ that can be evaluated for any split s of any node t.
3. A splitting termination rule.
4. A rule for assigning every terminal node to a class.

For each ordered variable X_m, all questions in a set of binary questions are of the form$\{ is\ \mathrm{X_m} \leq c?\}$ for all c ranging over $(-\infty, \infty)$.

If X_m is categorical taking values, say, in$\{ b_{1,} b_2, \cdots, b_u\}$, then all questions in a set of binary questions are questions of the form $\{ is\ \mathrm{X_m} \in s?\}$, as s ranges over all nontrivial subsets of $\{ b_{1,} b_2, \cdots, b_u\}$. For example, the variable GENDER of the bank data set has two values {MALE, FAMALE}. So there are two nontrivial subsets {MALE} and {FAMALE}, and the set of binary questions are, {Is the client male?} and {Is the client female?}.

The set of binary questions generates a set Q of splits s of every node t. For those cases in t answering "yes" to a question will go to the left descendant node $\{t_l\}$ and those answering "no" will go to the right descendant node $\{t_l\}$.

There are many impurity functions one can use to define splitting rules, such as the Gini index of heterogeneity:

$$G(t) = 1 - \sum_j P(j|t)^2, \tag{4.1}$$

and the entropy index:

$$H(t) = -\sum_j P(j|t) \log P(j|t), \tag{4.2}$$

where $P(j|t)$ is the proportion of cases with class j at node t. From these impurity functions, we can see that impurity of a node is largest when all classes in it are evenly mixed, and smallest when a node contains only one class. Based on Leung et al. (2003c), no apparent advantage is gained from using a specific index. This echoes the empirical results reported in the literature. Since the Gini index is simple, so it is used in the analysis. The CART splitting criterion is as follows:

At each intermediate node t, the split s selected is the one which maximizes

$$\Delta i\,(s|t) = i\,(t) - (p_l i\,(t_l) + p_r i\,(t_r)), \tag{4.3}$$

where $i\,(.)$ is the impurity function, p_l and p_r are the proportions of cases at the left node and the right node, respectively.

Let $h_Y(t)$ be the impurity of the categorical variable Y at node t, $h_{y|i_m}(t)$ be the impurity of the conditional distribution of Y given the modality i_m of predictor X_m at node t. The proportion reduction in the impurity of variable Y due to the information of X_m is given by the following general statistical index:

$$\gamma_{Y|X_m}(t) = \frac{h_Y(t) - \sum_{i_m=1}^{I_m} P(i_m|t) h_{Y|i_m}(t)}{h_Y(t)}, \tag{4.4}$$

where $p(i_m|t)$ is the proportion of cases having modality i_m of X_m at node t. Equation (4.3) takes values in [0,1]. It gives the degree of dependency of Y on predictor X_m when it is globally considered. Special cases of (4.3) are the predictability index τ of Goodman and Kruskal (1979):

$$\tau_{Y|X_m} = \frac{\sum_{i_m} \sum_j p^2(i_m, j|t)/p(i_m|t) - \sum_j p(j|t)^2}{1 - \sum_j p(j|t)^2}. \tag{4.5}$$

In Leung et al. (2003c), a stopping rule is formulated on the basis of the following CATANOVA statistic:

$$c_{Y|X_m^*}(t) = (G-1)(j(t)-1)\tau_{Y|X_m^*}(t), \tag{4.6}$$

where $\tau_{Y|X_m^*}(t) = \max_m (\tau_{Y|X_m})$, $j(t)$ is the number of categories of response variable Y at node t, G is the number of categories of X_m^* at node t. The distribution of $c_{Y|X_m^*}(t)$ is approximated by a χ^2-distribution with $(j(t)-1)(G-1))$ degrees of freedom under the null hypothesis that Y and X_m^* are independent at node t. Thus, when the null hypothesis is accepted at a significance level α, the node t is called the "terminal node." Splitting also stops at a node when it is pure or contains less than a pre-specified number of cases.

The class j assigned to terminal nodes t is done using the plurality rule: "j equals the class for which $p(j|t)$ is largest." In Leung et al. (2003c), the variable RESPONSE has two possible values (Yes, No). According to the assignment rule, if the rate RESPONSE equals Yes is larger than 50%, then the terminal node t is assigned "Y," otherwise it is assigned "N."

In CART, all possible splits are inspected to find the best split at each node. For efficient computation, Leung et al. (2003c) introduce the following fast algorithm basing on a property of the index τ (see Mola and Siciliano (1997) for details).

At each node t , a split s divides the I Categories of X into two subgroups, i.e., the value i of X in the left node t_l or the value i of X in the right node t_r, which in turn defines a splitting variable X_s with two categories denoted by l and r. For a split s induced by the splitting variable X_s, (4.4) becomes

$$\tau(Y|X_s) = \frac{\sum_j p_{t_l}^2(j|l)p_{t_l} + \sum_j p_{t_r}^2(j|r)P_{t_r} - \sum_j p_t^2(j)}{1-\sum_j p_t^2(j)}. \tag{4.7}$$

It can be proved that $\tau(Y|X_m) \geq \tau(Y|X_s)$.

The fast splitting algorithm for finding the best split s^* at node t consists of the following major steps:

Step 1. Calculate the value of $\tau(Y|X_m)$ for each predictor variable X_m and order the predictors with respect to the value of the index $\tau(Y|X_m)$. Denote the ordered predictors by $X_{(1)}, \cdots, X_{(m)}, \cdots, X_M$, so that $\tau(Y|X_{(m)})$ is the m-th higher value.

Step 2. Define the set $s_{(k)}$ of all possible splits of the categories $X_{(k)}$. Find the best split s_k^* of the predictor $X_{(k)}$ such that $\tau(Y|s_{(k)}^*) = \max_{s \in s_{(k)}} \tau(Y|s)$.

Step 3. If $\max_k \tau(Y|s_{(k)}^*) \geq \tau(Y|X_{(k+1)})$, and $\tau(Y|s^*) = \max_k \tau(Y|s_{(k)}^*)$, then s^* is the best split.

4.2.2 *Client Segmentation by CART*

4.2.2.1 Preprocessing of Variables

The same data set (Table 3.1) is employed in Leung et al. (2003c) to discover classification rules by CART. Again variables of the data set have to be preprocessed. For the predictors, those variables that have more than 3,200 missing values are dropped because these variables cannot provide sufficient information for the analysis. A total of 17 variables are selected: RESPONSE, AGE, JOBNAT, CAR, CHILD21, GENDER, TENURE, PDT_AMT, RATIO, CTENURE, CINCOME, CREDLMT, LIQUID, HHINCOME, ROLLRATE, PRICONBA and PRICONCA, where GENDER with 81 missing data, TENURE with 62 missing data, RATIO with 62 missing data, ROLLRATE with 72 missing data.

Since values of some of the variables are estimated or enriched in a certain way, they may introduce extra error to the analysis, especially to the estimated values of the continuous variables such as the financial variables. To decrease error and for practical purpose, the values of each of the remaining variables are categorized as follows:

For an original binary variable such as GENDER (male or female), PRICONBA (Yes, No), it is kept unchanged. For a categorical variable with several levels such as JOBNAT, basing on the contingency table of the target variable (that is, RESPONSE) versus this variable, levels with very high response rate and very low response rate are combined, respectively, into two new levels, and those with medium response rates are combined into another new level. So JOBNAT becomes a 3-ary variable. Similarly, the variable CHILD21 is transformed into a binary variable by the method of contingency table.

For continuous variables, they are categorized into binary variables or 3-ary variables by finding appropriate cut-off point(s) such that (a) a distinctive distribution of response rate is achieved in the contingency table formed by the new binary variable or 3-ary variable versus the target binary variable "RESPONSE," and (b) the numbers of observations that fall into the newly-formed two or three categories are as comparable as possible.

After this preprocessing, the relevant variables are either categorical or categorized numerical variables that are easier to use and explain in practice. Furthermore, for the original variables whose values have to be estimated in some way, it is obviously much easier and less erroneous to estimate their values in categorical terms.

All of the selected variables and the associated codes (for convenience, the newly-formed variables use the same names of the corresponding variables) are listed in Table 4.1.

4.2.2.2 Client Segmentation: Tree Structure of the data set

Figure 4.2 shows the final binary tree with 46 nodes and 24 terminal nodes at the significance level $\alpha = 0.01$. It indicates the predictor and split at each non-terminal node and the assigned classes for the response variable RESPONSE at each terminal node. The value of the stopping statistics and the distributions of RESPONSE at the terminal nodes are tabulated in Table 4.2. As an example, at non-terminal node 5 we can observe in Fig. 4.2 that the best predictor is JOBNAT and the best split sends cases having category 1 to the left-node and categories (0, 2) to the right node. From Fig. 4.2 and Table 4.2 we may conclude that clients having the characteristics corresponding to terminal nodes 24, 25, 31, 35, 38, 41, and 46 should be chosen as the main targets for promoting the credit card because they

Table 4.1 Variables used in the CART

Variable name	Description	Level code	Code definition
GENDER	Gender of clients	0	GENDER = FEMALE
		1	GENDER = MALE
AGE	Age of clients	0	AGE< 25
		1	25 ≤AGE≤ 45
		2	AGE≥ 45
JOBNAT	Job nature of clients	0	010, 033, 041, 042, 051, 070, 072, 073, 082, 091, 092, 101, 120, 121, 123
		1	011, 021, 022, 023,030, 050, 061, 063, 080, 081, 083, 110, 910, 920, 990
		2	012, 013, 020, 031, 032, 040, 043, 052, 053, 060, 062, 071, 090, 093, 102, 103, 111, 112, 113, 122
PRICONBA	Price consciousness for BIA/BSA	0	PRICONBA = YES
		1	PRICONBA = NO
PRICONCA	Price conscious sense for CARD: PRICONCA = "y"; else PRICONCA="N."	0	PRICONCA = "Y"
		1	PRICONCA = "N"
CTENURE	Number of the months from the time that the first credit card account was opened to the time stamped on Aug. 31, 1999	0	CTENURE≤ 36
		1	CTENURE> 36
ROLLRATE	Rollover rate of credit card	0	ROLLRATE≥ 0
		1	ROLLRATE< 0
PDT_AMT	Number of active products held by each client in the latest month	0	PDT_AMT≤ 2
		1	PDT_AMT> 2
RATIO	Ratio of number of active products to the number of months from the time that the first account was opened in the bank to the time stamped on Aug. 31, 1999	0	RATIO< 0.02
		1	RATIO≥ 0.02
CINCOME	Estimated income of clients	0	CINCOME< 7,000 or CINCOME>30,000
		1	7,000 ≤CINCOME≤ 30,000
RESPONSE	Whether or not a client having at least one response in the campaigns of the credit card promotion	0	RESPONSE="Y"
		1	RESPONSE="N"
TENURE	No. of months ago (time stamp on Aug.31, 1999) for the first account opened in H.S.	0	TENURE≤ 55
		1	55 <TENURE< 151
		2	TENURE≥ 151
CREDLMT	Total Credit card limit for customer	0	CREDLMT< 15,000
		1	CREDLMT≥ 40,000
		2	15,000 ≤CREDLMT< 40,000

(*continued*)

Table 4.1 (continued)

Variable name	Description	Level code	Code definition
LIQUID	Liquidity = saving a/c balance (SAVBAL) + Check a/c (CHKBAL) + Unused OD {OD Limit (UODLMT) − OD balance (ODBAL)} + Credit Card limit (CREDLMT)	0	LIQUID≤ 21,000
		1	21,000<LIQUID<100058.14
		2	LIQUID≥ 100058.14
HHINCOME	Estimated household income = est. client's income + est. spouse's income	0	HHINCOME≤ 13098.65
		1	13098.65<HHINCOME<42818.375
		2	HHINCOME≥ 42818.375
CHILD21	Enriched NO. of Children Aged< 21	0	CHILD21≤ 1
		1	CHILD21≥ 2
CAR	Enriched Car Ownership: OWNER, NOT-OWNER	0	CAR = OWNER
		1	CAR = NOT-OWNER

have higher response rate(larger than 55%). For example, terminal node 24 represents clients having characteristics CTENURE=1, GENDER=0, RATIO=1, JOBNAT =2 with response rate 0.68. Therefore, if a woman with one of the jobs "011," "021," "022," "023," "030," "050," "061," "063," "080," "081," "083," "110," "910," "920," or "990," has opened the first credit card account for more than 36 months and with the ratio of number of Active Products exceeding 0.02, then she should be chosen as a target for credit card promotion. For clients with characteristics corresponding to terminal nodes 14,19,20,29,30,32,34,39,43,44, and 47, we should not choose them as targets for promotion since they have lower response rates(less than 45%). On the other hand, there is not much difference between the response and non-response rates at terminal nodes 12, 28, 33, 36, 42, and 45. We need other information to analyze clients in these nodes. These are actually classification rules unraveled by the CART for client segmentation.

By increasing the significance level α to 0.05, Fig. 4.3 shows the final binary tree with 113 nodes and 58 terminal nodes. Since terminal node 75 has only one case, according to the stopping rule we should amalgamate it with terminal 74. Terminal node 98 also has one case, we should also amalgamate it with terminal node 99. After amalgamation, the final tree have 56 terminal nodes. The distribution of "Y" and "N" of the variable RESPONSE at the terminal nodes and the number of cases involved are given in Table 4.3.

Comparing the two trees, we can observe that by increasing the significance level, final trees with more terminal nodes can be obtained. Define re-substitution estimate $r(T)$ for correct classification rate of the classification tree as:

$$r(T) = \sum_{t \in \tilde{T}} (\max_j P(j|t))P(t), \tag{4.8}$$

where $P(j|t)$ is the proportion of cases with class j at terminal node t, $P(t)$ is the proportion of cases which fall into the terminal node, and $\tilde{T}$ is the set of all terminal nodes.

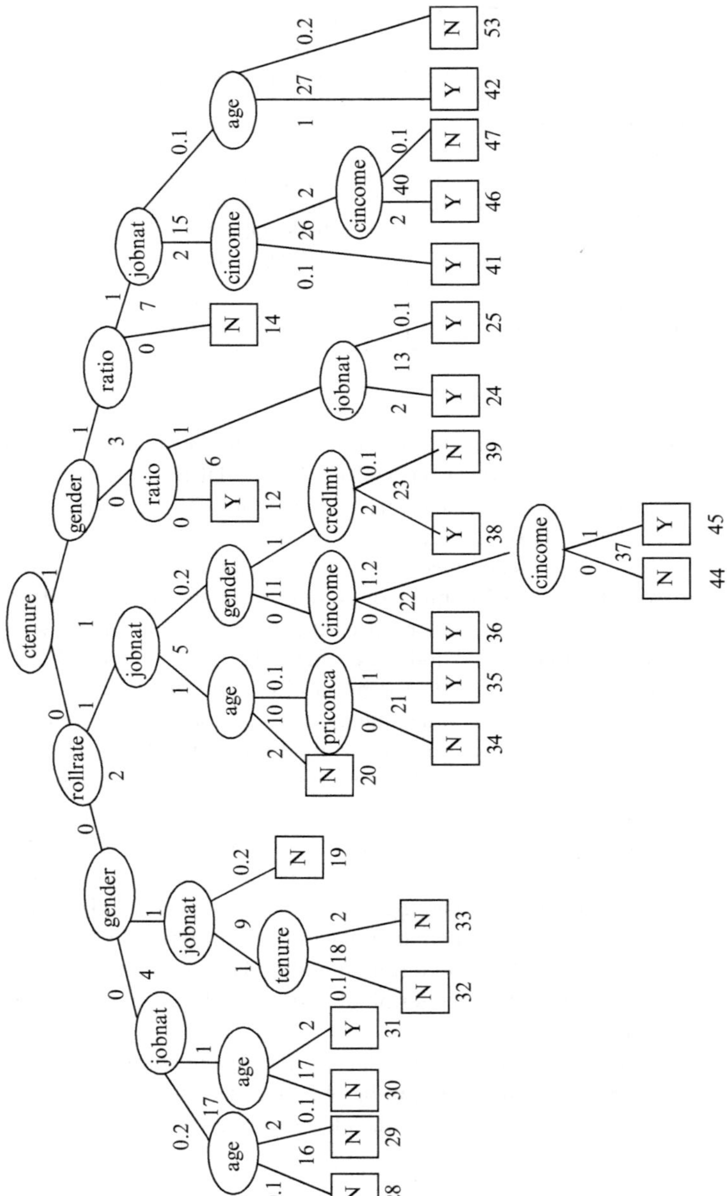

Fig. 4.2 Final binary tree with 46 nodes and 24 terminal nodes at $\alpha = 0.01$

Table 4.2 Terminal nodes information for $\alpha = 0.01$

Node t	N(t)	The value of stop statistic	Degree of freedom	Distribution of terminal node (Y, N)	Assigned classes
12	1,158	8.9675	2	(0.54, 0.46)	Y
14	1,747	7.9108	2	(0.44, 0.56)	N
19	1,488	8.2292	2	(0.32, 0.68)	N
20	308	7.4783	2	(0.35, 0.65)	N
24	1,253	2.8700	1	(0.68, 0.32)	Y
25	1,957	6.2495	2	(0.62, 0.38)	Y
28	1,854	7.7872	2	(0.46, 0.54)	N
29	249	4.3800	1	(0.33, 0.67)	N
30	261	7.2670	2	(0.25, 0.75)	N
31	34	4.2200	1	(0.68, 0.32)	Y
32	306	5.5378	2	(0.19, 0.81)	N
33	26	4.1808	2	(0.46, 0.54)	N
34	21	1.9139	2	(0.10, 0.90)	N
35	11	6.8600	1	(0.55, 0.45)	Y
36	174	2.2403	2	(0.51, 0.49)	Y
38	391	6.0001	2	(0.55, 0.45)	Y
39	320	5.3243	2	(0.42, 0.58)	N
41	698	3.6300	2	(0.62, 0.38)	Y
42	1,652	5.7403	2	(0.53, 0.47)	Y
43	786	8.6452	2	(0.46, 0.64)	N
44	54	3.9797	2	(0.44, 0.56)	N
45	120	4.7589	2	(0.53, 0.47)	Y
46	28	2.9098	2	(0.75, 0.25)	Y
47	114	2.5400	1	(0.39, 0.61)	N

N(t) is the number of cases in node t

The estimate of correct classification rate of the tree corresponding to Fig. 4.2 is 0.59702, while the estimate of correct classification rate of the tree corresponding to Fig. 4.3 is 0.6040. Even though the tree in Fig. 4.3 is larger than that in Fig. 4.2, there is not much of an improvement in classification rate.

To sum up, 17 variables extracted from the bank data set have been chosen to grow two trees with significance level of 0.01 and 0.05 respectively. The tree with significance level of 0.01 has the correct classification rate of 0.59702 and the tree with significance level 0.05 has the correct classification rate of 0.6040. Since there is not much of a difference between the two classification rates, we recommend adopting the smaller trees for simplicity sake. To improve the correct classification rate, variables with a large number of missing values may be used. To handle missing data, the Surrogate Splits may be used. The idea is to define a measure of similarity between any two splits$s$ and $s^{'}$ of a node t. If the best split of node t is the split s on X_m, then find the split s' on the variables other than X_m that is most similar to s and call s' the best surrogate for s. Similarly, define the second best surrogate, the third best, and so on. If a case has X_m missing in its measurement, decide whether it goes to t_L or t_R by using the best surrogate split. If it is missing the variable containing the best surrogate split, use the second best, and so on (see Breiman et al. (1984), pp.140–143 for details).

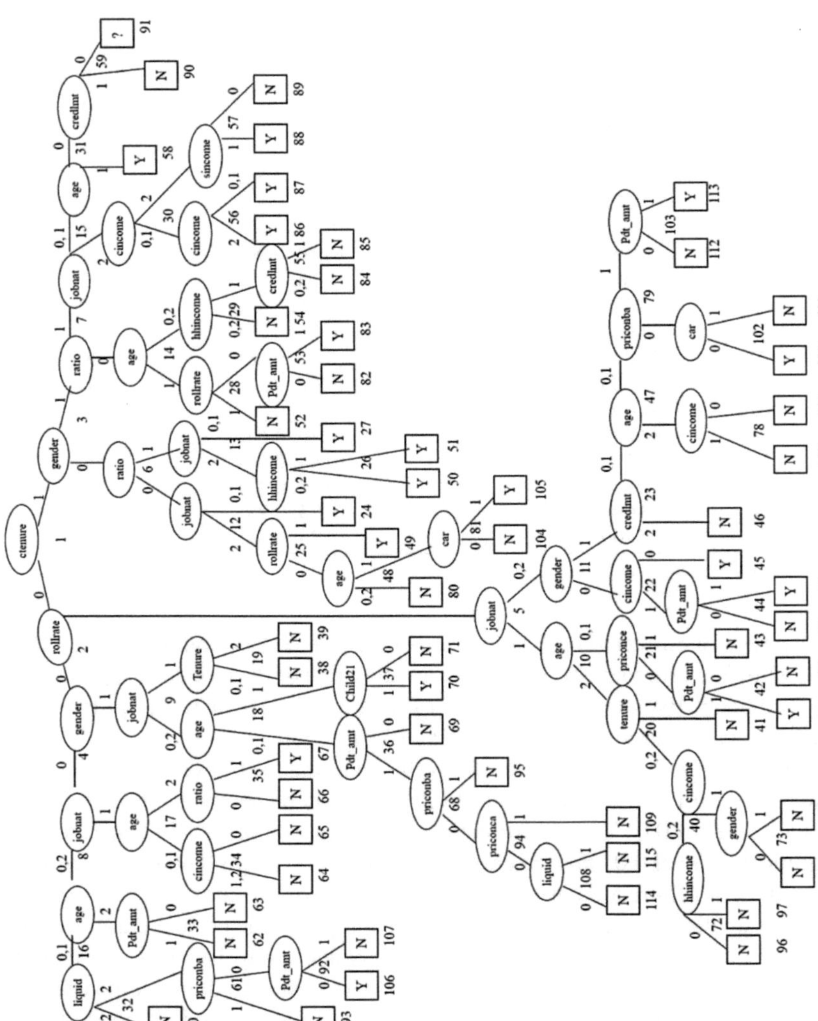

Fig. 4.3 Final binary tree with 113 nodes and 58 terminal nodes at $\alpha = 0.05$

Table 4.3 Terminal nodes information for $\alpha = 0.05$

Node t	N(t)	The distribution of terminal nodes ('Y,' 'N')	Assigned classes
24	320	(0.61, 0.39)	Y
27	1,253	(0.68, 0.32)	Y
38	26	(0.46, 0.54)	N
39	306	(0.19, 0.81)	N
41	21	(0.10, 0.90)	N
43	97	(0.46, 0.54)	N
45	174	(0.51, 0.49)	Y
46	320	(0.42, 0.58)	N
49	199	(0.58, 0.42)	Y
50	1,004	(0.59, 0.41)	Y
51	953	(0.64, 0.36)	Y
52	240	(0.53, 0.47)	Y
54	368	(0.35, 0.65)	N
58	1,652	(0.53, 0.47)	N
60	671	(0.42, 0.58)	N
62	167	(0.38, 0.62)	N
63	82	(0.24, 0.76)	N
64	89	(0.35, 0.65)	N
65	172	(0.20, 0.80)	N
66	8	(0.38, 0.63)	N
67	26	(0.77, 0.23)	Y
69	498	(0.30, 0.70)	N
70	6	(0.67, 0.33)	Y
71	233	(0.23, 0.77)	N
74	7	(0.86, 0.14)	Y
75	1	(0.00, 1.00)	N
76	54	(0.44, 0.56)	N
77	120	(0.53, 0.47)	Y
80	178	(0.41, 0.59)	N
82	659	(0.42, 0.58)	N
83	171	(0.51, 0.49)	Y
84	150	(0.38, 0.62)	N
85	159	(0.52, 0.48)	Y
86	635	(0.61, 0.39)	Y
87	63	(0.73, 0.27)	Y
88	28	(0.75, 0.25)	Y
89	114	(0.39, 0.61)	N
90	290	(0.39, 0.61)	N
91	496	(0.50, 0.50)	?
93	525	(0.45, 0.55)	N
95	255	(0.47, 0.53)	N
96	155	(0.26, 0.74)	N
97	46	(0.46, 0.54)	N
98	1	(1.00, 0.00)	Y
99	9	(0.11, 0.89)	N
100	56	(0.48, 0.52)	N
101	11	(0.09, 0.91)	N
104	10	(0.20, 0.80)	N
105	451	(0.52, 0.48)	Y
106	298	(0.57, 0.43)	Y

(continued)

Table 4.3 (continued)

Node t	N(t)	The distribution of terminal nodes ('Y,' 'N')	Assigned classes
107	360	(0.47, 0.53)	N
109	119	(0.41, 0.59)	N
110	13	(0.77, 0.23)	Y
111	108	(0.44, 0.56)	N
112	169	(0.47, 0.53)	N
113	101	(0.70, 0.30)	Y
114	209	(0.27, 0.73)	N
115	83	(0.22, 0.78)	N

N(t) is the number of cases at node t

The correct classification rates in Leung et al. (2003c) are given by the re-substitution estimate. The following method is also recommended: choose the significance level $\alpha_1 < \alpha_2 < \cdots < \alpha_K$, and denote the tree corresponding to α_k by T_k. The Cross-Validation method can then be used to estimate the correct classification rates of T_k, $k = 1, \cdots, K$. The data set A is randomly divided into v subsets of nearly equal size. Denote the subsets by $A_1, A_2, \cdots, A_V$. For every set $A - A_v$, assume that the procedure for constructing the tree T_k can be applied to it, then we can obtain a re-substitution estimate of correct classification rate $\hat{\gamma}_{vk}$ of the tree T_k. Following this, an estimate of correct classification rate can be given by $\hat{\gamma}_k = \frac{1}{V}\sum_{v=1}^{V}\hat{\gamma}_{vk}$. Choose the tree with the largest correct classification rate $\hat{\gamma}_k$ as the final tree for decision making.

Similar to other data mining methods, CART is vulnerable to overfitting. Recursive partitioning is a technique to lessen the overfitting problem in CART, and random forests (Breiman 2001) is an approach built on this idea. However, more research is required to overcome the overfitting problem.

4.3 The Neural Network Approach to the Classification of Spatial Data

4.3.1 *On the Use of Neural Networks in Spatial Classification*

Over the years, three major paradigms have been developed for the classification of spatial data in general and remote sensing data in particular. They are statistical classifiers based on probabilistic decision functions discussed in the previous sections, artificial neural network (ANN) classifiers (Benediktsson et al. 1990; Wilkinson et al. 1995; Leung 2001), and symbolic knowledge-based reasoning classifiers (Leung and Leung 1993a,b; Richards and Jia 1998; Mather 1999). Artificial neural networks have been extensively applied to classify geo-referenced and remotely sensed data (Pao 1989; Kulkarni 1994; Atkinson and Tatnall 1997;

Fisher and Getis 1997; Fischer and Leung 2001). The emergence of ANN is biologically inspired. Artificial neural networks have a parallel distributed architecture with a large number of units and connections. The connection from one unit to another is associated with a numeric weight. Although a single unit in an ANN can perform only simple computation functions, all units assembled within the network constitute a highly non-linear dynamic computation system by which complicated computing functions can be realized.

One may wonder why neural networks are necessary while existing statistical methods appear to be handling rather adequately the task of high dimensional data classification. The answer to the question is essentially threefold (Leung 2001):

First, classification problems are in general highly nonlinear. Mathematically speaking, data classification is basically the search of separation surface in high dimensional space. Our mission is to find hypersurfaces to separate data into classes. For linearly separable problems, statistical methods are usually effective and efficient. The methods, however, fail when nonlinearity exists in spatial classification. Spatial data are ordinarily not linearly separable. While linear separability implies spherical separability and spherical separability implies quadratic separability, the reverse is not true. Thus, to be versatile, classification methods should also be able to handle nonlinearity. Neural networks, especially multilayer feedforward neural networks, are appropriate models for such a task. Learning algorithms in most neural networks can be viewed as a problem of hypersurface reconstruction which tries to approximate hypersurfaces partitioning nonlinearly separable classes from training examples.

Second, most statistical methods assume certain types of probability distributions, e.g., Gaussian distribution in the maximum-likelihood Gaussian classifier. Nevertheless, spatial data may be non-Gaussian. In the classification of remotely sensed images, for example, data are multi-source. Gaussian distribution is generally an inadequate assumption when spectral (color, tone) and spatial (shape, size, shadow, pattern, texture, and temporal association) data are simultaneously employed as a basis of classification (see for example Benediktsson et al. 1990). Neural network models, on the other hand, can deal with such restriction.

Third, on-line and real-time computing is a desirable feature of a classifier. Most statistical methods are not no-line and real-time. Neural network algorithms, on the other hand, usually strive for on-line or real-time computation. This is especially important when we are classifying a large volume of data and we do not wish to reclassify the data with, for example, a new datum, but still would like to train the classifier with that additional piece of information through a single computation. The adaptive mechanism of some neural-network learning algorithms can perform on-line or real-time computation.

A great variety of neural network models have been proposed in the past few decades (Arbib 1995). The multilayer perceptron (MLP) with back-propagation (BP) algorithm (widely known as the multilayer feedforward neural network with BP) is one of the most widely used models for the classification of spatial, particularly remote sensing, data (Bischof et al. 1992; Civco 1993; Paola and Schowengerdt 1995; Zhou 1999). Compared to conventional statistical classifiers, the BP

Neural Network (BP-MLP) is distribution-free and non-parametric, and is more robust, especially when the distributions of features are strongly non-Gaussian. However, BP-MLP exhibits some serious drawbacks such as slow convergence of the learning algorithm, potential convergence to local minimum, common chaotic behavior of non-linear systems, and inability to detect over-fitting. Though improvements of the BP learning algorithm, such as the self-adaptation of the learning rate and momentum (Heermann and Khazenie 1992; Kanellopoulous and Wilkinson 1997), and the optimization of connection weights and network topology by evolutionary computations (Fischer and Leung 1998; Yao 1999) have been made in recent years, the above fundamental problems still remain.

Apparently, as a universal classifier, BP-MLP is not adequate to account for localized effect in a classification task. Parallel to the statistical classifiers, kernel or basis expansion methods might need to be employed in the neural network formulations. Radial basis function (RBF) networks (Powell 1987; Moody and Darken 1989) are the kind of multilayer network which can handle localized effects but is very different from the BP-MLP in its training algorithm. A RBF network consists of an input, a kernel (hidden), and an output layer. Its output units form a linear combination of the basis functions in the kernel layer, and the basis functions produce a localized response to the input. The basis function can be viewed as an activation function in the kernel layer in which each unit has a localized receptive field to the input vector. RBF networks can overcome some of the above limitations of BP-MLP by relying on a rapid training phase, avoiding chaotic behavior, having simpler architecture while keeping the complicated mapping capability. Such characteristics and the intrinsic simplicity of the RBF networks make them an interesting alternative to pattern recognition in general and classification of remotely sensed images in particular (Chen and Chen 1995; Bishop 1995; Bruzzone and Prieto 1999).

One of the problems of the neural classification models is that they could only simulate low level cognitive functions of human vision and neural system resulting in a low degree of understanding of images (Fu 1994; Medsker 1994). They are not effective in reasoning with deep knowledge that is generally represented in a symbolic way. In addition to spectral information, recognition and classification of remotely sensed images usually require domain specific knowledge such as DEM and its derivatives. To achieve more accurate classification, neural networks and symbolic knowledge should be integrated into a single system (Benediktsson et al. 1990; Foody 1995a,b; Peddle 1995; Gong 1996; Gong et al. 1996; Murai and Omatu 1997; Leung 1997). Integrating geographical knowledge on top of a geographical information system has become an approach to increase the accuracy and effectiveness in the classification of remotely sensed images. For example, topographical knowledge describing the relationships between land covers and topographical characteristics can be represented in semantic form, such as rules. Such knowledge can be employed to fine tune neural network classification based on spectral information.

To account for the local effects and to make use of domain specific knowledge, Leung et al. (2002a) have constructed a knowledge-integrated radial basis function model for the classification of remotely sensed images. The integrated model, called the RBF model, employs a RBF network to classify images with spectral information. Geographical knowledge represented as rules is used in parallel to classify the images with topographical information. Classification results obtained from both methods are then combined by an evidence combination method to derive the ultimate classification of the images. The model is detailed in the following discussion.

4.3.2 The Knowledge-Integrated Radial Basis Function (RBF) Model for Spatial Classification

4.3.2.1 The Architecture of the Knowledge-Integrated RBF Model

There are basically four major components in the knowledge-integrated RBF model (Leung et. al. 2002a) for the classification of remotely sensed images (Fig. 4.4): (1) data source, (2) the RBF network, (3) rule-based inference, and (4) evidence combination. The first component is data source management that processes and prepares remotely sensed data for the neural network classification, and geographical information (from GIS) for the rule-based inference. The neural network component is essentially a RBF network that performs land-cover classification by hypersurface reconstruction in high dimensional (multispectral bands) space. Embedded in the RBF network is the ART network which facilitates the learning phase by performing efficient clustering in the kernel layer. Parallel to the neural

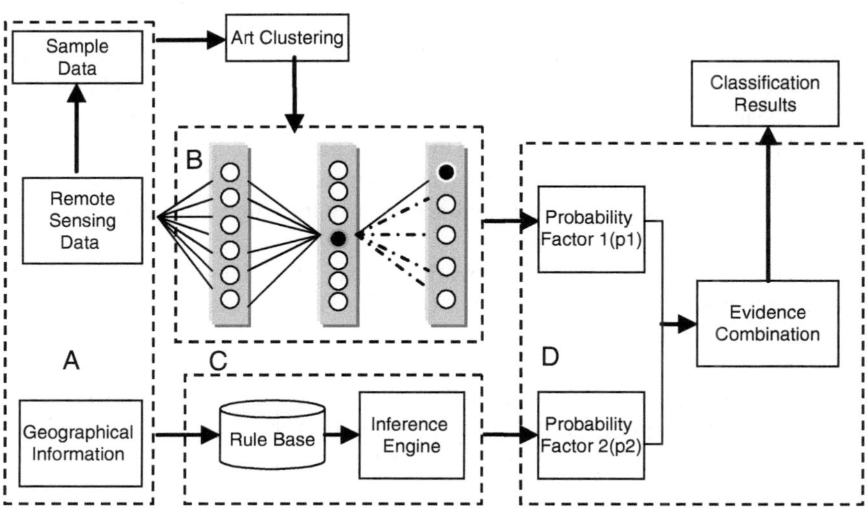

Fig. 4.4 The general architecture of the knowledge-integrated RBF model. (**a**) Data source; (**b**) RBF network; (**c**) Rule-base inference (**d**) Evidence combination

network component is the rule-based inference engine which classifies land covers by topographical knowledge built on top of a vector-based geographical information system. Classification results of the RBF network and the rule-based inference are then integrated via evidence combination to produce the final classification of land covers.

Specifically, the knowledge-integrated RBF model operates according to the following procedures:

Step 1. Data sets, including satellite spectral data, topographical data and its derivatives, are registered geometrically.
Step 2. Train the RBF network using spectral samples.
Step 3. Output vector (p_1) that is computed by the trained RBF network after an unknown pattern A is fed into the network. Vector p_1 can be regarded as the first probability factor that assigns the pixel to each of the categories.
Step 4. Another probability factor (p_2) can be obtained by feeding the corresponding vector containing auxiliary topographical data, such as elevation and slope, into the inference engine and firing the relevant rules in the rule base.
Step 5. According to Dempster-Shafer rule for evidence combination, the final vector of probability factor (p_3) is determined accordingly. The class at which the maximum value of p_3 is achieved is then the class to which the pixel belongs.

In what follows, the components of the RBF model are analyzed in details.

4.3.2.2 The RBF Network

RBF networks can be regarded as a special neural network with multilayer feedforward structure in which the parametric statistical distribution model and non-parametric linear perceptron algorithm are combined together in sequence (Lippmann 1994; Girosi 1994). Based on the kernel theory for pattern recognition (Serpico et al. 1996; Scholkopf et al. 1999), the problem of segmentation of data sets located in low-dimensional space when projected nonlinearly into high-dimensional space is more likely to be linearly separable.

The basic structure of a RBF network consists of an input layer (I), a kernel (hidden) layer (K), and an output layer (O) (Fig. 4.5). In the context of a neural network, the units in the kernel layer provide a set of kernel basis functions (called radial basis functions) that constitute the "basis" for the input vectors when they are expanded into the kernel unit space. The basis functions can be viewed as the activation functions in the kernel layer. The output of the RBF network is a linear combination of the radial basis (kernel) functions computed by the kernel units. Each kernel unit has a localized receptive field. The basis functions of the kernel layer are provided with the cluster centers which have statistical significance (Lippmann 1994). Though basis function can be chosen with respect to practical

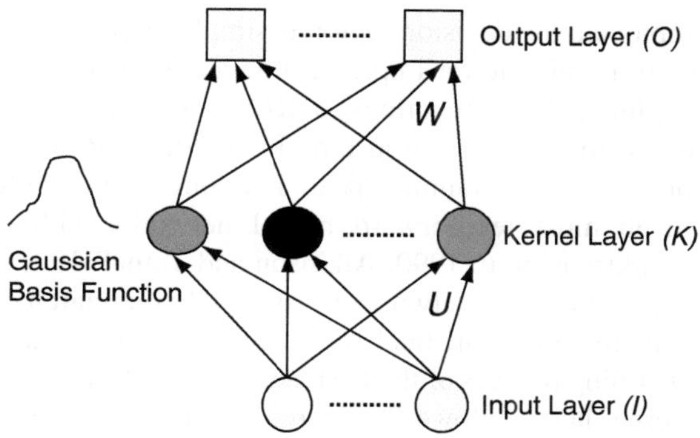

Fig. 4.5 The basic architecture of a RBF network

needs, the most widely used basis function is the simple Gaussian function in which the activation, O_j, of kernel unit j is obtained as

$$O_j(\mathbf{x}) = e^{-\frac{(\mathbf{x}-\boldsymbol{\mu}_j)\cdot(\mathbf{x}-\boldsymbol{\mu}_j)}{2\sigma^2{}_j}}, \tag{4.9}$$

where $\mathbf{x}$ is the input vector, $\boldsymbol{\mu}_j$ is the vector determining the center of the basis function associated with kernel unit j, and σ^2 is the normalization factor. The output value in the kernel unit lies between 0 and 1. The closer is an input to the center of the Gaussian function, the larger the response of the unit becomes. The normalization factor σ^2 represents a measure of the spread of the data around the cluster center associated with the kernel unit. It is usually determined by the average distance between the cluster center and each training instance (point) around the center. Therefore, for kernel unit j the factor σ^2 is obtained as:

$$\sigma_j^2 = \frac{1}{m}\sum_{i=1}^{m}(\mathbf{x}_i - \boldsymbol{\mu}_j)(\mathbf{x}_i - \boldsymbol{\mu}_j), \tag{4.10}$$

where x_i is training point i around the cluster center, $\boldsymbol{\mu}_j$ is the cluster center associated with kernel unit j, and m is the number of the training points associated with that center. The activation level O_j of unit j in the output layer is determined by the linear combination: $O_j = \sum_{i=1}^{n} w_{ji}o_i$, where w_{ji} is the weight from kernel unit i to output unit j. In the output layer, the value of a unit is obtained through a linear combination of the nonlinear outputs from the kernel layer. Thus, the overall network essentially performs nonlinear transformation from inputs to outputs.

Therefore, RBF networks can be regarded as a bottom up approach to data classification by treating the design of a neural network as an approximation (curve-fitting) problem in high-dimensional space. For the problem of classification, RBF networks can determine how closely a given input is to the center of a kernel by the response of the corresponding kernel unit. If only a single kernel unit is employed, the decision region is simply circular. From this perspective, the RBF network is suitable for implementing an efficient classification model. Using a set of nonlinear basis functions, a RBF network is capable of approximating very arbitrary mapping relationship. In addition, RBF networks can overcome the problems of slow training speed and convergence to local minimum, which often occur in most feedforward neural networks with back-propagation algorithm (Benediktsson et al. 1990; Atkinson and Tatnall 1997).

Leaning in RBF networks is essentially the search of a surface that provides the best fit (in some statistical sense) to the training data in multidimensional space. The learning process of the RBF network can be divided into two stages: learning in the kernel layer followed by learning in the output layer. Typically, the process of learning in the kernel layer is to determine the status of the units in the kernel layer, and it is usually performed by unsupervised clustering method. The supervised methods like Least Mean Square (LMS) algorithm are used for learning from the kernel layer to the output layer. The two learning algorithms implemented in Leung et al. (2002) are discussed as follows:

4.3.2.3 ART Clustering Algorithm in the Kernel Layer

The ART model is developed from the biological theory for processing human cognitive information (Grossberg 1976), and it is essentially a cluster discovery model useful for pattern recognition (Carpenter and Grossberg 1988). The research in ART has led to an evolving series of real-time models for unsupervised and supervised category learning and pattern recognition, including the early ART model (an unsupervised learning system to categorize binary input patterns), fuzzy ART (to categorize analog input patterns), and ARTMAP (supervised network architecture) (Carpenter et al. 1991). The ART provides a solution to the stability-plasticity dilemma during the design process of a learning system, and it has two useful properties: real-time learning and self-organization. It has been demonstrated that ART is more sensitive to noise than other conventional clustering methods such as the K-means algorithm and ISODATA. The complicated connective structure between ARTa and ARTb in ARTMAP, which have been applied to supervised classification for remotely sensed images (Gopal and Fischer 1997; Carpenter et al. 1997; Mannan et al. 1998), on the other hand, makes its effectiveness in classification of large numbers of patterns inevitably low. Thus, only the ART model is utilized to discover clusters in the kernel layer for the RBF network.

The basic mechanism of the ART network is the use of resonance of a pattern in the output layer with a pattern in the input layer to establish good hetero-associative pattern matching. The ART network has three main layers (*F0, F1, F2*) (Fig. 4.6). *F0* is the input layer for encoding the input patterns, and *F1* receives and holds the input patterns. The *F2* layer responds with a pattern classification or association to an input pattern and verifies it by sending a return pattern to the *F1* layer. The statues of neurons in *F1* and *F2* form short-term memory (STM), and the connective weights (top-to-bottom weights and bottom-to-top weights, w_{ij} and w_{ji},

Fig. 4.6 Fuzzy ART model for clustering

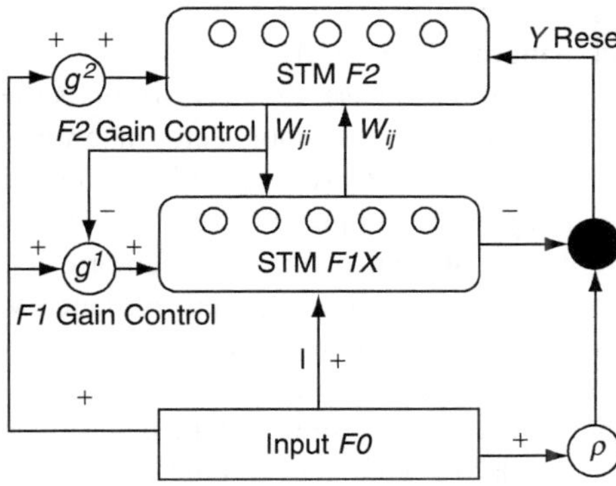

respectively) between *F1* and *F2* form the long-term memory (LTM) of the system. In the learning phase, the network performs the following vigilance (ρ) test to judge if the input pattern is the familiar pattern stored in the trained memory:

$$\frac{\sum_i w_{ji} I_i}{\sum_i I_i} > \rho \tag{4.11}$$

where I_i is the activation level of input unit i, and w_{ji} is the top-down weight from output unit j to unit i in *F1*, and ρ is a vigilance parameter ($0 < \rho < 1$). The vigilance threshold is usually a fraction indicating how close an input pattern must be to a stored cluster to provide a desirable match. A value close to one indicates that a close match is desirable while a smaller value indicates that a poorer match is permitted. When ρ is close to 0, all patterns in the feature space will be clustered into one category. When ρ is close to 1, each pattern in the feature space is treated as a category.

Both *F1* and *F2* are associated with a gain control, called Gain-1 and Gain-2 respectively. Both gain units give an output 1 if at least one component of the input vector is 1. Each neuron in *F1* receives inputs from the input vector, Gain-1, and the feedback signal from *F2*. At least two of these three inputs must be 1 for a neuron to give 1 as output. This is the two-thirds rule. In the comparison phase, if any component of the top-down vector is 1, Gain-1 is forced to 0. At this point, a neuron will fire only if its top-down signal matches its input signal according to the rule. The connective weights between *F1* and *F2* are adjusted according to the rule:

$$w_j(t+1) = \beta(I \cap w_j(t)) + (1-\beta)w_j(t) \tag{4.12}$$

where I is the input vector of *F1*, and $\beta 0 < \beta < 1$is the learning rate. When β is close to 1, it represents a fast learning. As a self-organizing neural network, ART adopts competitive learning to realize the network organization. The mutual connective structure between *F1* and *F2* can realize the feedback of signals. The signal pattern will unceasingly vibrate between *F1* and *F2* until all input patterns are matched with feedback patterns stored in the system.

In the next recognition phase, the network finds the output neuron in *F2* whose bottom-up weight is close to the input vector in terms of their scalar product. This is essentially the winner-takes-all strategy. When a familiar pattern is entered into *F1*, the system will adjust the connective weights between *F1* and *F2* to stabilize the coding of the whole familiar patterns in memory. If a novel input pattern cannot match all stored patterns within a given vigilance value, a new pattern will be formed and stored in the system by creating a newly coded neuron in *F2* and its connective weights between *F1* and *F2*.

The number of clustering centers can be determined by the vigilance value. It is often the case that a lower vigilance value favors the formation of fewer clusters, while a higher value leads to more clusters. From the physiological point of view, the ART network is a simple simulation of the human vision which can obtain different degrees of attention to objects when adjusting the vision scale. Compared to some conventional methods, ART clustering algorithm possesses a more natural property for partitioning a feature space. Standing alone, it can actually be employed to discover natural clusters.

4.3.2.4 Learning Algorithm from the Kernel Layer to the Output Layer

The connective structure of the single-layer perceptron is adopted for learning between the kernel layer and the output layer in the RBF network. If there is a connection from unit *i* in the kernel layer to unit *j* in the output layer, the weight w_{ij} is assigned to this connection. In the learning phase of the RBF network, w_{ij} can be iteratively adjusted by the following procedure until it attains convergence:

$$w_{ij}(t+1) = w_{ij}(t) + \Delta w_{ij} \tag{4.13}$$

where $w_{ji}(t)$ is the weight from unit *i* in the kernel layer to unit *j* in the output layer at time t (or the tth iteration) and Δw_{ij} is the weight adjustment at the current step which may be computed by the delta rule:

$$\Delta w_{ij} = \eta \delta_j o_j \tag{4.14}$$

where η is a trial-independent learning rate ($0 < \eta < 1$) and δ_j is the error at unit j in the output layer, i.e.,

$$\delta_j = T_j - O_j \tag{4.15}$$

where T_j is the desired activation level of output unit *j*, and O_j is the actual activation level of output unit *j* in the output layer.

4.3.2.5 The Rule-Based Inference and Evidence Combination

Neural computation model is a simulation of human vision with relatively low level of intelligence. An intelligent pattern recognition system should be able to process higher level of knowledge which is often symbolic in nature. A suitable integration of both will enhance the understanding of remotely sensed imagery. Based on spectral information, for example, neural networks such as the RBF network can be employed as an efficient means to perform the low level image classification. Higher level knowledge can then be used to cross check, sharpen, or modify patterns recognized on the basis of spectral values.

Integration of ancillary data or knowledge in image classification has been shown to be effective in enhancing discrimination and classification accuracy (Benediktsson et al. 1990; Eiumnoh and Shrestha 2000). The RBF classification model in Leung et al. (2002a) employs geographical data or knowledge as ancillary information to improve classification. Terrain features and their derived elements, such as slope and aspect, are integrated with spectral information to determine the final pattern of distribution. There are several ways in which geographical data and expert knowledge can be captured and represented by a knowledge-based system (Wang 1991; Leung 1997). Leung et al. (2002a) employ the simplest and perhaps the most common approach to symbolic knowledge representation, namely the production rules, to take into account the geographical knowledge of land covers. The captured rules are mostly fuzzy in meaning and uncertain in belief, i.e., consisting of two major types of uncertainty, imprecision and randomness in knowledge representation and inference (Leung 1997). The uncertainty of rule is represented by a notion of probability taking the following format:

IF (condition), THEN (conclusion), PF (Probability Factor)

where *PF* reflects the degree of uncertainty about a rule, and $PF \in [0, 1]$. When PF is equal to 0, the rule is absolutely uncertain. When *PF* is equal to 1, the conclusion of the rule is absolutely certain. When $0 < \mathrm{PF} < 1$, the conclusion is certain to a degree. To allow for imprecision in probabilistic statement, linguistic hedges such as "close-to" and "very-close-to" can be used to modify PF. For example:

IF *((dem > 0.0) or (slope > 0.0))*
THEN *id is WATER* PF *is close-to 0.01*
IF *(dem <= 1.0)*
THEN *id is URBAN* PF *is very_close_to 0.06……*

If we treat the classification results obtained from both the RBF network and the rule-based inference as evidence leading to the final classification of land covers, then we need some methods of evidence combination to integrate the initial results in order to derive the final classification. Among different methods for evidence combination (Leung 1997), the Dempster-Shafer theory is adopted in this study.

According to the Dempster-Shafer theory (Shafer 1976) of evidence combination, the final vector of probability (p_3) can be determined by the technique of orthogonal summation (denoted by $\oplus$) of p_1 (Vector of probability obtained from the RBF network) and p_2 (Vector of probability obtained from the rule-based inference) as follows:

Let p_1 = {0.2,0.3,0.3,0.1,0.1}, and p_2 = {0.1,0.3,0.4,0.1,0.1}. Then the final PF (p_3) can be calculated as:

$$p_3(G) = p_1 \oplus p_2(G) = \frac{\sum_{x \cap y} p_1(x) \bullet p_2(y)}{1 - \sum_{x \cap y = \phi} p_1(x) \bullet p_2(y)} \tag{4.16}$$

where G is a subset which represents the category, ϕ is the empty set, and $p_3(G)$ is the partial PF attributed to G from the final *PF*. For example, the pattern should be attributed to category 3 if the final PF vector is obtained as p_3 = {0.08, 0.36,0.48,0.04,0.04}.

4.3.2.6 An Application in Land Cover Classification

As an evaluation, the RBF model is applied to classify land covers from TM image. In the application, experiments are conducted using LANDSAT TM image with six non-thermal bands (1–7). The resolution is 30×30 m^2. The image was acquired on March 3, 1996 by LANDSAT10 satellite. The study area covers the Yuen Long region, northwest of Hong Kong (Fig. 4.7). The size of the sub image cut out from the whole image is 600 rows by 600 columns, covering about 3,200 km^2. According to the survey of the study area and with the vision interpretation of the corresponding data, there are 12 main types of land covers:

C1 – Sea water	C2 – Beach	C3 – Inland water
C4 – Wet land	C5 – Mangrove	C6 – Urban area
C7 – Concrete land	C8 – Barren land	C9 – Green land
C10 – Forest	C11 – Hill grass	C12 – Rock grass

where water (C1, C3), build area (C6, C7) , vegetation area (C9,C10,C11), water (C1, C3) and shadow of building (C6, C7) cannot be easily separated because of their similarity in spectral characteristics. The knowledge-integrated RBF model is thus used for the task.

The four dimensional input vector for the RBF network is A = (PCA1, CH4, CH5, CH7), where PCA1 is the principle component of PCA transformation of visible bands (CH1-Blue, CH2-Green, CH3-Red) for dimension reduction. PCA1 contains the majority of information from the Red, Green, Blue bands. CH4 is the near-infrared band, and CH5 and CH7 are middle infrared bands whose spectrums are 1.55–1.75 um and 2.08–2.35 um, respectively. A total of 1,700 training samples are selected through visual interpretation of the scenes by comparing with a land-use map. In the training phase, the data sets include 1,700 training sample data and 800 test sample data.

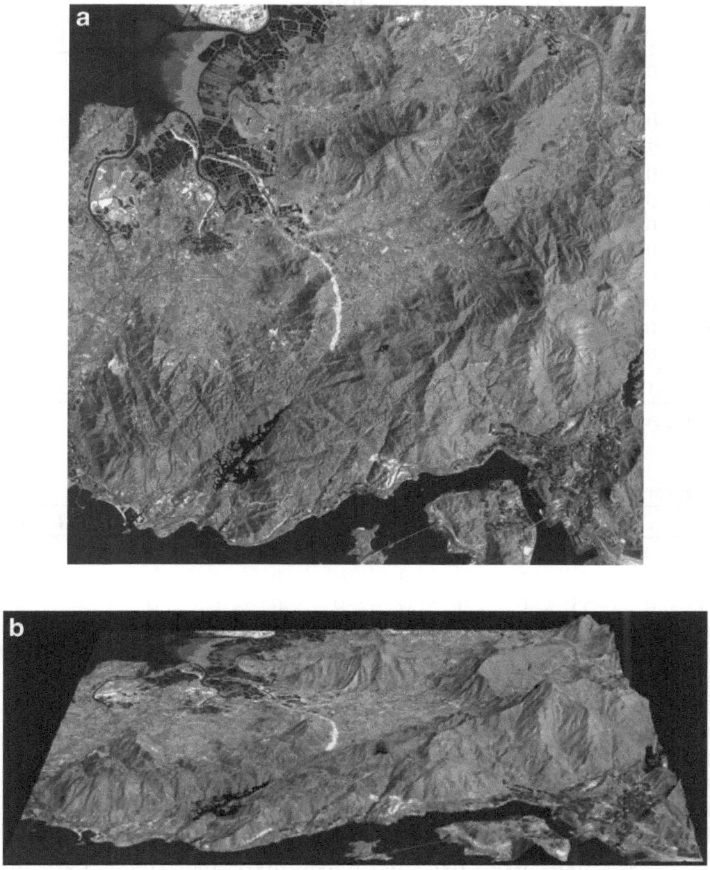

Fig. 4.7 The TM image of the study area. (**a**) The TM image covering the experimental area. (**b**) The three-dimensional display of the same image showing the topographical situation of the area

The RBF network with a kernel layer of size 120 and learning rate η of 0.01 is first trained by the training sample data. The test error matrix (Table 4.4) is then obtained. The training time of the RBF network is about 50 s, and the overall accuracy is 90.17%. The maximum likelihood classifier (MLC) and multilayer perceptron with back-propagation algorithm (BP-MLP) are also applied to the same data sets. The obtained structure of the BP-MLP is of three layers with four input nodes, 24 hidden nodes and 12 output nodes. The test error matrices are, respectively, listed in Tables 4.5 and 4.6. We can observe that the overall accuracy of the MLC is 85.25%, and that of BP-MLP is 89.92%. However, the learning time of the BP-MLP is about 1,200 s after about 650,000 iterations. Comparing the three classifiers, the following observations can be made:

Table 4.4 Error matrix of classification by the RBF network

	C1	C2	C3	C4	C5	C6	C7	C8	C9	C10	C11	C12	Total
C1	92	0	9	0	0	0	0	0	0	0	0	0	101
C2	0	98	0	11	0	6	0	0	0	0	0	0	115
C3	8	0	78	1	0	11	0	0	0	0	0	0	98
C4	0	2	1	77	0	5	1	0	0	0	0	0	86
C5	3	0	0	0	99	0	0	0	0	0	0	0	99
C6	0	0	12	4	0	74	7	0	0	0	0	0	97
C7	0	0	0	1	0	4	92	6	5	0	0	0	108
C8	0	0	0	0	0	0	0	93	2	0	0	0	95
C9	0	0	0	5	0	0	0	1	91	0	0	0	97
C10	0	0	0	1	1	2	0	0	1	97	3	1	104
C11	0	0	0	0	0	0	0	0	1	3	95	2	101
C12	0	0	0	0	0	0	0	0	0	0	2	97	99
Total	100	100	100	100	100	100	100	100	100	100	100	100	1200

(Time = 50 s, Accuracy = 90.17%, Kappa = 0.893)

Table 4.5 Error matrix of classification by the MLC

	C1	C2	C3	C4	C5	C6	C7	C8	C9	C10	C11	C12	Total
C1	83	0	2	0	0	0	0	0	0	0	0	0	85
C2	0	95	0	2	0	1	0	0	0	0	0	0	98
C3	17	0	55	1	0	7	0	0	0	0	0	0	80
C4	0	5	3	88	0	7	2	1	8	5	4	0	123
C5	0	0	0	0	99	0	0	0	0	0	0	0	99
C6	0	0	40	2	0	75	12	0	0	0	0	0	129
C7	0	0	0	5	0	10	83	7	6	0	0	0	108
C8	0	0	0	0	0	0	3	91	3	0	2	1	100
C9	0	0	0	2	0	0	0	1	81	0	0	1	87
C10	0	0	0	0	1	0	0	0	1	95	1	3	101
C11	0	0	0	0	0	0	0	0	0	0	92	5	97
C12	0	0	0	0	0	0	0	0	1	0	1	90	92
Total	100	100	100	100	100	100	100	100	100	100	100	100	1200

(Accuracy = 85.25%, Kappa = 0.839)

Table 4.6 Error matrix of classification by the BP-MLP

	C1	C2	C3	C4	C5	C6	C7	C8	C9	C10	C11	C12	Total
C1	99	0	6	0	0	0	0	0	0	0	0	0	105
C2	0	97	3	10	0	1	0	1	0	0	0	0	112
C3	1	0	83	2	0	12	0	0	0	0	0	0	98
C4	0	3	1	80	0	6	2	0	0	0	0	1	93
C5	0	0	0	1	99	0	0	0	1	0	0	0	101
C6	0	0	7	1	0	75	5	0	0	0	0	0	88
C7	0	0	0	1	0	6	90	16	6	0	0	0	119
C8	0	0	0	0	0	0	0	80	0	0	0	0	80
C9	0	0	0	3	0	0	0	2	90	1	1	0	97
C10	0	0	0	2	1	0	0	0	2	99	4	3	111
C11	0	0	0	0	0	0	0	0	1	0	93	2	96
C12	0	0	0	0	0	0	3	1	0	0	2	94	100
Total	100	100	100	100	100	100	100	100	100	100	100	100	1200

(Training Time = 1,200 s, Accuracy = 89.92%, Kappa = 0.890)

1. Training time of the RBF network is less than that of the BP-MLP and the former attains higher accuracy of classification.
2. Neural network classifiers are distribution-free and have more capability than conventional parametric statistical classifiers to separate the categories of mixture distribution in the feature space, like urban area C5 and land water C6. Therefore, the RBF network yields the most effective classification both in the learning phase and the test phase.

To select the reasonable number of units in the kernel layer of the RBF network is a key to the success of the classification. As shown in Table 4.7, different number of units in the kernel layer, including 30, 40, 50, 60, 90, 125, 160, 200, and 250, are, respectively, selected. In other words, the patterns in the feature space are partitioned into different areas by the clustering method. Figure 4.8 shows the overall accuracy achieved with different size of the kernel layer. The results indicate that the overall accuracy can be improved by increasing the size of the kernel layer, but the computational overhead also increases. However, the overall classification accuracy levels off when the kernel layer is increased to a certain size.

As discussed above, better classification may be achieved if suitable geographical knowledge can be integrated into the RBF classification model. As an evaluation, knowledge established in terms of DEM and slope is used. For examples, it is impossible to have sea water distributed in places with DEM being higher than 0; mangrove fields usually distribute within the DEM range from 0 to 2 m; urban area is generally built on places with DEM between 5 and 100 m; hill grass generally distributes on hill tops with DEM higher than 100 m; the green forest land, such as urban park, should only distribute on level plains around urban areas; the angle of slope in land water should only be very close to 0°, and etc. The knowledge described above is represented as rules and is employed to classify land covers in the study area by the knowledge-integrated RBF model depicted in Fig. 4.9.

The same test samples are used to assess the overall accuracy of the knowledge-integrated RBF classification model. As shown in Table 4.8 and Fig. 4.9, the results of the test indicate that the overall accuracy of the knowledge-integrated RBF model evidently increases to 93.17% in comparison with the 90.17% achieved by the RBF network. Furthermore, the results obtained by the knowledge-integrated RBF model are visually more natural, especially the distribution of land covers such as wetland, rock grass and inland water. It should be noted that only very limited coarse domain specific knowledge is used in the integrated model. Better performance is expected if finer and deeper knowledge could be integrated into the classification process.

Remark 4.1. Though relatively successful in classifying remotely sensed data, there are limitations of the RBF method. It is well known that the classification error made by the RBF networks depends on the selection of the centers and widths of the kernel functions constituting the hidden layer (Bruzzone and Prieto 1999;

Table 4.7 Relationship between accuracy and size of the kernel layer

		BPNN	Size of kernel layer in RBFNN								
			30	40	50	60	90	125	160	200	240
Learning speed (seconds)		900	8	10	12	15	25	35	45	60	75
ACCURACY (%)											
	C1	99.00	57.00	67.00	75.00	88.00	83.00	94.00	87.00	93.00	94.00
	C2	97.00	30.00	44.00	90.00	96.00	94.00	99.00	98.00	97.00	98.00
	C3	83.00	75.00	74.00	46.00	66.00	65.00	67.00	82.00	82.00	73.00
	C4	80.00	8.00	58.00	67.00	70.00	74.00	81.00	69.00	72.00	78.00
	C5	99.00	86.00	94.00	99.00	99.00	97.00	96.0	100.0	99.00	97.00
	C6	75.00	88.00	81.00	76.00	74.00	76.00	66.00	72.00	77.00	73.00
	C7	90.00	95.00	95.00	93.00	96.00	93.00	91.00	93.00	91.00	93.00
	C8	80.00	93.00	93.00	92.00	93.00	96.00	93.00	95.00	93.00	94.00
	C9	90.00	78.00	86.00	89.00	91.00	92.00	93.00	94.00	93.00	94.00
	C10	99.00	98.00	96.00	97.00	96.00	98.00	99.00	98.00	97.00	97.00
	C11	93.00	93.00	86.00	83.00	94.00	94.00	94.00	94.00	95.00	94.00
	C12	94.00	47.00	58.00	94.00	85.00	94.00	96.00	97.00	96.00	94.00
	Average	89.92	70.67	77.67	83.42	87.33	88.00	89.08	89.92	90.42	89.92

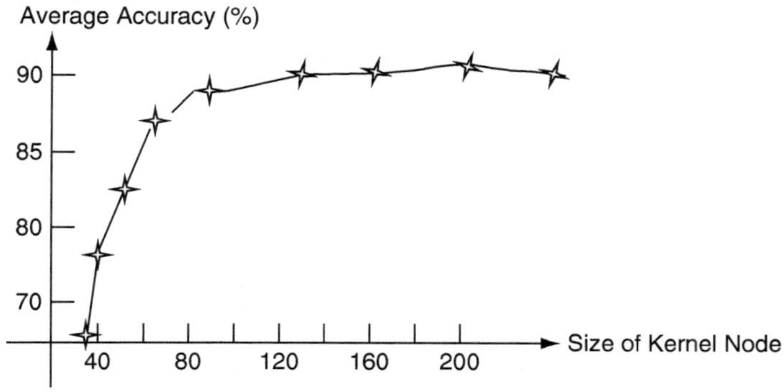

Fig. 4.8 The relationship between average accuracy and the number of kernel unit

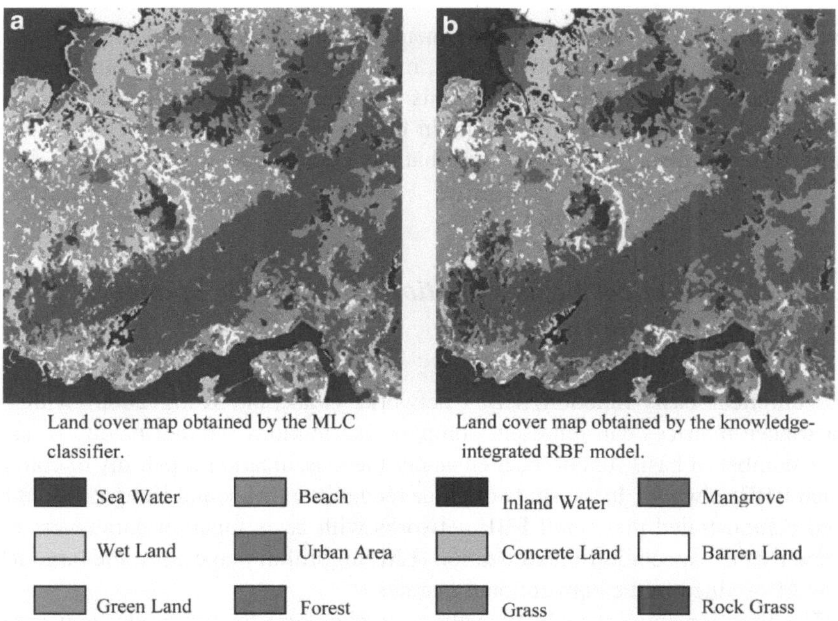

Fig. 4.9 Experimental results. (**a**) Land cover map obtained by the MLC classifier. (**b**) Land cover map obtained by the knowledge-integrated RBF model

Gomm and Yu 2000). Conventional clustering algorithms, such as K-means and K-nearest neighbors, are ordinarily employed to select the centers of the kernel functions. Such techniques, however, do not offer conventional statistical arguments. More importantly, RBF networks represent the posterior probabilities of the training data by a weighted sum of Gaussian basis functions with diagonal

Table 4.8 Error Matrix of classification by the knowledge-integrated RBF model

	C1	C2	C3	C4	C5	C6	C7	C8	C9	C10	C11	C12	Total
C1	97	0	2	0	0	0	0	0	0	0	0	0	99
C2	0	97	0	0	0	2	0	0	0	0	0	0	99
C3	3	1	94	5	0	5	0	0	0	0	0	0	108
C4	0	2	1	90	0	0	0	0	1	0	0	0	94
C5	0	0	0	0	99	0	0	0	0	0	0	0	99
C6	0	0	3	1	0	86	13	0	0	0	0	0	103
C7	0	0	0	0	0	6	84	6	5	0	0	0	101
C8	0	0	0	0	0	0	0	92	2	0	0	0	94
C9	0	0	0	4	0	1	0	2	92	0	0	0	99
C10	0	0	0	0	1	0	1	0	0	97	3	1	103
C11	0	0	0	0	0	0	0	0	0	3	95	2	100
C12	0	0	0	0	0	0	2	0	0	0	2	97	101
Total	100	100	100	100	100	100	100	100	100	100	100	100	1,200

(Accuracy = 93.17%, Kappa = 0.925)

covariance matrices. When the components of the training data vectors (and the known test data vectors) are dependent, more basis functions are required so that data in the regions covered by each basis function can still be considered to have independent components. This results in high computational cost. The following subsection describes a neural network that can circumvent such problem.

4.3.3 An Elliptical Basis Function Network for Spatial Classification

The elliptical basis function (EBF) networks (Mak and Kung 2000) with full covariance matrices can represent complex distributions without having to use a large number of basis functions. It enhances the classification capability of conventional RBF networks. In the study of voice recognition (Mak and Kung 2000), it has been demonstrated that small EBF networks with basis function parameters estimated by the expectation-maximization (EM) algorithm outperform the large RBF networks trained in the conventional manner.

The employment of the EM algorithm is to overcome the limitations in using the conventional maximum likelihood (ML) estimation to determine the parametric probability distribution characterizing, for example, remotely sensed data. Due to the complexity and randomness of real-life spatial data, mixture distributions are a common place in the feature space. Compounded by the lack of representative training samples, conventional ML estimation might introduce bias and decrease the accuracy of classification. The EM algorithm based on finite mixture density model can thus be employed for parameter estimation of the ML function (Dempster et al. 1977; McLachlan and Basford 1988; McLachlan and Krishnan 1997). The EM algorithm has been applied to multiple spatial feature extraction, multiple data fusion, and unsupervised spatial data mining (Bruzzone and Prieto 2000, 2001; Tadjudin and Landgrabe 2000). To classify more efficiently spatial data and to further extend the application of the EBF networks, Luo et al. (2004) construct an EBF network for the classification of remotely sensed images and employ the EM algorithm to estimate the parameters of the basis functions.

4.3.3.1 A Further Note on Mixture Density Distribution in Spatial Data

Spatial data in general and remote sensing data in particular often appear as a mixture of many class components which is difficult to be described by a single distribution assumed in conventional statistics. Moreover, pixels of the same class in a remotely sensed image may have complicated distributions (e.g., multiple peaks). Due to various depths, sand contents, waves of the surface and other external elements, water body in a remotely sensed image possesses, for instance, the properties of mixture density distributions in the feature space (Fig. 4.10). Furthermore, distributions of objects of different classes in remotely sensed images might overlap or look very similar to each other. Under these situations, it is very difficult to use a single probability density function (PDF) to approximate the reality. The employment of mixture density models thus become a necessity in some classification tasks.

The purpose of mixture modeling is to find a weighted mixture PDF in order to represent the distribution of the whole data set. Each component of the mixture PDF

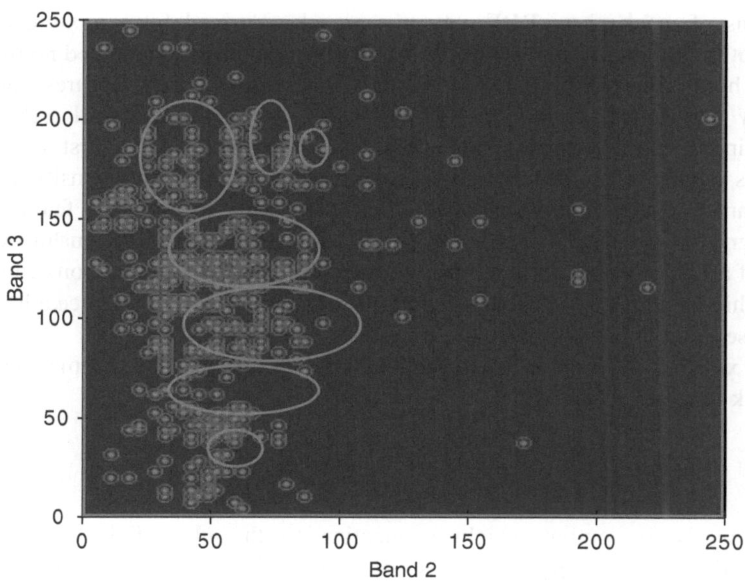

Fig. 4.10 A mixture distribution of water body sampled from a SPOT-HRV image

corresponds to a significant feature hidden in a data set. The weight of the component PDF corresponds to the proportion of points assigned to that component in the whole data set. Consequently, classification with mixture density model is transformed into the estimation of parameters and proportional weight of each component in the mixture PDF.

Gaussian mixture is often the choice in mixture modeling. Being a linear combination of Gaussian density functions, it is capable of forming smooth approximations to arbitrarily shaped densities. In classifying remote-sensing images, for instance, individual component densities can be used to model individual classes, such as urban area, barren land, grass land, and water and, constituting a genuine mixture.

Compared to neural network approaches, the statistical method holds several distinctive advantages. First, the distribution density makes the classification results more interpretable. Second, based on Bayesian theory, it can be integrated with prior probabilities obtained from other domain specific knowledge so that decision capability can be further enhanced. Third, if all parameters of the decision model can be estimated, then the computation can be simplified and easily implemented. It is sometimes advantages to embed statistical models into the neural network framework to get the best of both worlds. The EBF network by Luo et al. (2004) is constructed exactly for such purpose.

4.3.3.2 EBF Network: An Extension of the RBF Network

In terms of mechanism, RBF networks are also derived from mixture density distribution model. By the use of basis functions, patterns are mapped nonlinearly into a high-dimensional space in which the mixture density features could be linearly partitioned. Therefore, RBF networks have a feedforward architecture consisting of an input layer, a hidden layer and an output layer. At first, the feature space is separated into different hyper-regions so that the mixture density distributions can be parametrically described. Then, each hyper-region of the feature space is assigned to certain labeled category by a linear mapping relationship. Output units of a RBF network form a linear combination of the basis functions computed by the hidden units, and the basis functions in the hidden layer produce a localized response to the input vector.

Let $\mathbf{x} \in \mathbf{X} \subset \mathbf{R}^p$ be the input vector. The outputs of a non-normalized RBF network are defined as:

$$y(\mathbf{x}) = \sum\nolimits_j w_j h_j(\mathbf{x}), \tag{4.17}$$

where w_j is the output weight connecting to the jth radial basis function $h_j(\mathbf{x}) = \phi_j(||\mathbf{x} - \boldsymbol{\mu}_j||)$, $\boldsymbol{\mu}_j$ is the center of ϕ_j (or called a *prototype* of $\mathbf{x}$), and $||\cdot||$ is a norm (usually Euclidean norm) on the input space. For the Gaussian RBF networks, we have $\phi_j(t) = \exp(-t^2 / \sigma_j^2)$, in which the parameter σ_j is sometimes called the width of ϕ_j.

The basis function ϕ_j can be viewed as the activation function in the hidden layer in which each center has a localized receptive field to the input vector. It is equivalent to the mixture density model which assumes that the feature space can be separated into mixture areas according to a statistical distribution model in the feature space. The centers and parameters of the basis functions are often obtained from some clustering procedures (Lippmann 1994). Provided with a set of nonlinear basis functions, a RBF network can establish approximating relationships between the input and output spaces. Ideally, the basis functions can be chosen to meet practical needs in order to approximate the real distributions. It may however be impossible to find such subtle and complicated parametric functions based on statistical theory.

It is easily seen that the Gaussian RBF ϕ_j can be rewritten as

$$\begin{aligned}\phi_j(||\mathbf{x} - \boldsymbol{\mu}_j||) &= \exp(-||\mathbf{x} - \boldsymbol{\mu}_j||^2 / \sigma_j^2) \\ &= \exp\left\{-(\mathbf{x} - \boldsymbol{\mu}_j)^T (\sigma_j^2 \mathbf{I})^{-1} (\mathbf{x} - \boldsymbol{\mu}_j)\right\}\end{aligned} \tag{4.18}$$

where $\boldsymbol{I}$ is an identity matrix of order p. Thus the RBF network can be viewed as a weighted sum of normal basis functions with diagonal covariance matrices.

In most cases, each diagonal covariance matrix has identical elements controlling the spread of the corresponding RBF unit. As a result, the range of spread of the RBF unit is hyper-spherical. High level of accuracy in pattern recognition can be achieved when the components of the mixture density in the feature space are independent of each other. Otherwise, more basis functions are required so that data in the regions covered by each basis function can still be considered to have independent components (Mak and Kung 2000). Therefore, it would be more reasonable and beneficial if full covariance matrices can be incorporated into the basis functions of a RBF network so that complex distributions of mixture densities can be represented without the need of having to use a large number of basis functions. Under such circumstances, the range of the spread of a hidden unit is hyper-ellipsoidal, and the RBF network is then extended to the elliptical basis function (EBF) network, with the basis function taking on the following form:

$$h_j(\mathbf{x}) = \exp\left\{-\frac{1}{2\gamma_j}(\mathbf{x} - \boldsymbol{\mu}_j)^T \boldsymbol{\Sigma}_j^{-1} (\mathbf{x} - \boldsymbol{\mu}_j)\right\}, \tag{4.19}$$

where $\boldsymbol{\Sigma}_j$ is the covariance matrix corresponding to the jth basis function and γ_j is a smoothing parameter controlling the spread of the jth basis function, and it can be determined heuristically by the procedure in Mak and Kung (2000).

The connective structure of the linear perceptron is again adopted between the hidden layer and the output layer in the EBF network. If there is a connection between unit i in the hidden layer and unit j in the output layer, the weight w_{ij} is assigned to this connection. In the training phase, w_{ij} can be iteratively adjusted by the procedure outlined in (4.13) to (4.15).

4.3.3.3 EM Algorithm for the EBF Network

Selecting a clustering algorithm to determine the number of nodes in the hidden layer of a RBF or an EBF network is the most important step for their performance. The parameters of an EBF network, including centers (mean vectors) and the covariance matrices, are usually estimated by some straightforward methods such as K-means and K-nearest neighbors algorithms. The mean vector is directly estimated by averaging the sample data while the covariance matrices are approximated by the covariance of the sample data containing the center. Although it has been shown that *EBF* networks trained by such methods may perform better than *RBF* networks for the same task, it is also possible that they may lead to undesirable results if estimate of the mean vector differs significantly from the true mean. Consequently, the covariance matrices will no longer be an accurate estimate of the true covariance matrices (Mak and Kung 2000). Luo et al. (2004) adopt the mixture modeling method for estimating the parameters of the EBF network. Specifically, the EM algorithm for Gaussian mixture estimation is employed and detailed in the following discussion.

For the Gaussian density distribution:

$$f_j(\mathbf{x};\boldsymbol{\theta}_j) = (2\pi)^{-p/2}|\boldsymbol{\Sigma}_j|^{-1/2}\exp\left\{-\frac{1}{2}(\mathbf{x}-\boldsymbol{\mu}_j)^T\Sigma_j^{-1}(\mathbf{x}-\boldsymbol{\mu}_j)\right\},\ j = 1,\cdots,g, \tag{4.20}$$

where $\boldsymbol{\theta}_j$ includes mean vector $\boldsymbol{\mu}_j$ and covariance matrices $\boldsymbol{\Sigma}_j$. The estimation of the parameters $\boldsymbol{\theta}_j$ for the jth component of the mixture via the EM algorithm can be simplified as follows (Mclachlan 1988):

Step 1 (E-Step):

$$\tau_{ij}^{(t+1)} = \tau_j(\mathbf{x}_i;\boldsymbol{\theta}_j^{(t)}) = \frac{\pi_j^{(t)} f_j(\mathbf{x}_i;\boldsymbol{\theta}_j^{(t)})}{\sum\limits_{k=1}^{g}\pi_k^{(t)} f_k(\mathbf{x}_i;\boldsymbol{\theta}_k^{(t)})}$$

i.e.,

$$\tau_{ij}^{(t+1)} = \frac{\pi_j^{(t)}\left|\boldsymbol{\Sigma}_j^{(t)}\right|^{-1/2}\exp\left\{-\frac{1}{2}(\mathbf{x}_i-\boldsymbol{\mu}_j^{(t)})^T\boldsymbol{\Sigma}_j^{(t)-1}(\mathbf{x}_i-\boldsymbol{\mu}_j^{(t)})\right\}}{\sum_k \pi_k^{(t)}\left|\boldsymbol{\Sigma}_k^{(t)}\right|^{-1/2}\exp\left\{-\frac{1}{2}(\mathbf{x}_i-\boldsymbol{\mu}_k^{(t)})^T\boldsymbol{\Sigma}_k^{(t)-1}(\mathbf{x}_i-\boldsymbol{\mu}_k^{(t)})\right\}}$$

where, $\tau_{ij}^{(t)}$ is the posterior probability of the ith data point $\mathbf{x}_i$ belonging to the jth component at the tth step.

Step 2 (M-Step):

$$\pi_j^{(t+1)} = \frac{1}{n}\sum_{i=1}^{n}\tau_{ij}^{(t)},$$

$$\mu_j^{(t+1)} = \frac{1}{n\pi_j^{(t+1)}}\sum_{i=1}^{n}\tau_{ij}^{(t)}\mathbf{x}_i,$$

$$\boldsymbol{\Sigma}_j^{(t+1)} = \frac{1}{n\pi_j^{(t+1)}}\sum_{i=1}^{n}\tau_{ij}^{(t)}(\mathbf{x}_i - \boldsymbol{\mu}_j^{(t)})(\mathbf{x}_i - \boldsymbol{\mu}_j^{(t)})^T.$$

In essence, the EM algorithm firstly estimates the posterior probabilities of each sample belonging to each of the component distributions, and then computes the parameter estimates using these posterior probabilities as weights. The algorithm starts with an initial estimate $\boldsymbol{\theta}^{(0)}$ and repeats the above two steps at each iteration (time $t = 0, 1, 2, \ldots$).

Compared to conventional estimation approaches, the EM algorithm for the EBF network has several advantages:

1. Elliptical spread of centers. The degree of bias is measured by the Euclidean distance in conventional clustering algorithms. Thus, the localized response of the basis functions in the hidden layer is hyper-spherical. The EM algorithm can extend the localized response into hyper-ellipsoidal, and variances from different dimensions can be eliminated. Therefore, the estimated means and covariance matrices in Gaussian mixture densities can obtain the maximum likelihood so that all spreads of the centers can be combined to form the most appropriate relationship approximating the reality.
2. Stability and Reliability. In each iterative step, the EM algorithm guarantees that the value of the likelihood function increases continuously, resulting in stable and monotonic convergence. Moreover, it also has reliable global convergence under comparatively general initial conditions. Therefore, the requirements of initialization of the EM algorithm are low.
3. Analytical Representation and Computing Simplicity. By the EM algorithm, cluster centers are sought according to the distributed density. A center represents a certain mass of data surrounding it. Therefore, mixture centers are analytical representations of the density distributions of the feature space. In addition, the computation is comparatively simple and can easily be implemented.
4. Indication of the reasonable number of cluster centers. Conventional clustering algorithms are usually not able to give the information on the number of clusters. By the EM algorithm, the proportional and density parameters, iteratively estimated, can be regarded as important indicators to adaptively determine the

reasonable number of clusters. Accordingly, the final number of cluster centers of the EBF network can be reasonably determined.

5. Integration with knowledge. Based on the uniform framework of the Bayesian theory, other prior knowledge can be integrated into the EBF network to further improve the performance of EM algorithms.
6. Robustness. As mixture models can be viewed as contaminated densities with respect to each component in the mixture (Leung et al 2001a), the robustness of the EM algorithm should also be considered when the parameter estimates of each component are computed. It is indeed possible to construct a robust EM (REM) algorithm (Tadjudin and Landgrebe 2000).

Therefore, within the EM-based EBF network, the EM algorithm can be used to determine the status of the hidden units. By itself, the EM algorithm can be employed to discover natural clusters in spatial data.

4.3.3.4 The EBF Network with Embedded EM Algorithm for Image Classification

Based on the above analysis, information extraction and classification of remotely sensed images can be effectively carried out by an EBF network with EM algorithm. The basic mechanism of the EBF classification network can be divided into two parts. First, using the EM algorithm, the mixture density distributions of remotely sensed data in sparse feature space can be decomposed into hidden centers represented with elliptical spreads of probability distribution functions. Second, using the linear perceptron, the approximating relationship between cluster centers in the hidden layer and categorical classes in the output layer can be established. Basically, the proposed EBF network is a better model to solve classification problems especially when data are distributed as complicated mixture density.

On the implementation level, the EM-based EBF network consists of three major components (Fig. 4.11): the EM algorithm module (*A*), the EBF network training module (*B*), and the classification module (*C*).

In module *A*, the supervised sample data are first selected from remotely sensed image. Then, the status of the hidden units in the kernel layer is determined by the EM algorithm.

In module *B*, the connection weights between units in the hidden layer and the output layer are adjusted by the iterative delta rule.

Unknown vector *X* read from the image are fed into the EM-based EBF network pixel by pixel to obtain the classification results.

To evaluate the performance of the proposed EBF network, an experimental analysis on land cover classification of Hong Kong is carried out in Luo et al. (2004). The experiment used *SPOT-HRV* data with three spectral bands (*CH1*: Green Band, 0.50–0.59 μm; *CH2*: Red Band 0.61–0.68 μm; *CH3*: Near Infrared Band, 0.79–0.89 μm). The remotely sensed image was acquired on Feb. 3rd 1999, with a spatial resolution of $20 \times 20\ m^2$. The study area covers the Hong Kong

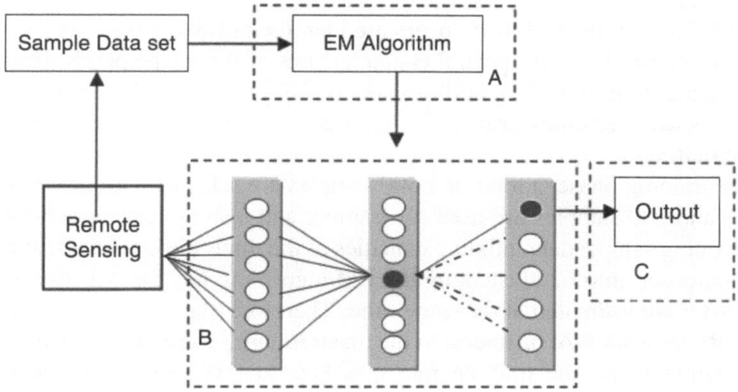

Fig. 4.11 Architecture of the EM-based EBF classification network

Fig. 4.12 Original SPOT image covering the study area

Island as shown in Fig. 4.12. The scale of the sub-image, extracted from the original image, is about 1:250,000, with a size of 600 rows by 800 columns, covering about 192 km^2 area.

Land covers in this area are very complicated. Due to its mountainous topography, the majority of the urban areas have to be built along the harbor on reclaimed land. The major part of the island is composed of rugged hills, covered with different types of vegetation.

According to the survey of the experimental area and its visual interpretation, about nine main classes of land covers are identified (Table 4.9).

Due to the similarity in spectral characteristics of the image pixels, land-covers such as water body (*C1, C2*), built-up area (*C3, C4*), vegetated area (*C7, C8, C9*), water and shadowed built-up area (*C1, C2, C3, C4*) cannot be easily separated apart in the visual map.

In the training phase, a total of 3,500 samples are selected. Out of which, 2,600 samples and 900 samples are used for training and testing, respectively. With the 2,600 training sample data, mixture densities of the three-dimensional feature space are decomposed into 62 clusters by the EM algorithm, and the ML parameters of each cluster are estimated at the same time. Therefore, the size of the hidden layer of the EBF network is of 62 nodes. In the linear training phase from the hidden layer to the output layer, the training rate η is kept at 0.02, small enough to avoid vibration commonly encountered in most of the feedforward neural networks training. The error matrix (Table 4.10) is obtained by running the trained EBF network on the testing data sets. The time for the training phase of the EBF network is about 120 s, and the overall test accuracy is 76%. The classification result is depicted in Fig. 4.13.

Table 4.9 Land covers of the study area

Type # (C)	Land covers
1	Sea water
2	Inland water
3	Urban area
4	Concrete land
5	Barren land
6	Beach
7	Grass land
8	Hilly woodland
9	Grass land after fire

Table 4.10 Error Matrix of classification by the EBF Network

Reference Result	C1	C2	C3	C4	C5	C6	C7	C8	C9	Total
C1	97	18	2	0	0	0	0	0	0	117
C2	2	62	7	0	0	1	0	0	0	72
C3	1	20	76	4	0	0	3	1	6	111
C4	0	0	10	71	6	7	0	0	12	106
C5	0	0	0	7	49	8	0	0	0	64
C6	0	0	0	7	44	71	2	0	0	124
C7	0	0	0	4	1	9	88	1	0	103
C8	0	0	1	2	0	3	5	91	1	13
C9	0	0	4	5	0	1	2	7	81	100
Total	100	100	100	100	100	100	100	100	100	900

(Time = 120 s, Accuracy = 76.11%, Kappa = 0.731)

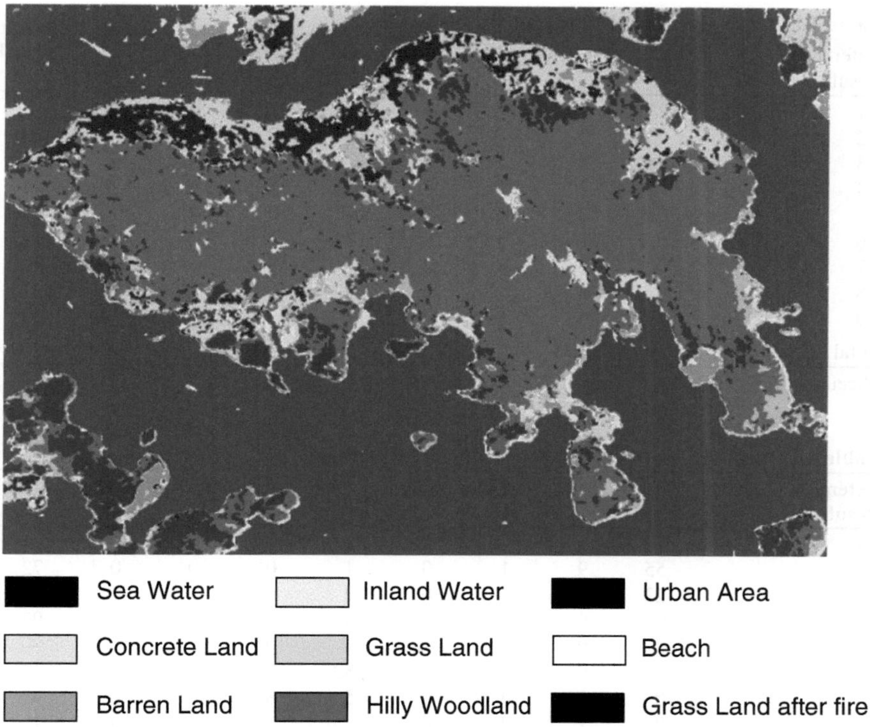

Fig. 4.13 Land covers obtained by the EBF network

As a comparison, two common classification models are also trained and tested with the same data sets. They are the maximum likelihood classifier (MLC) (elliptical, but single density model) (Table 4.11) and the conventional RBF network (mixture density model, but hyper-spherical) (Table 4.12) (Fig. 4.14).

Compared with the conventional MLC and RBF classifiers, the EBF network is simpler in structure, more accurate and easier to interpret. The relationship between accuracy and size of the hidden layer of the EBF network is given in Table 4.13. It can be observed that the accuracy levels off with the size increase of the hidden layer.

Remark 4.2. To make the proposed EBF network more effective, further research should be carried out to take advantage of the EM algorithm by integrating prior knowledge via Bayesian theory. In order to improve further the accuracy and reliability of the EBF network, robustness should be integrated into the mixture density model for the EM algorithm. To approximate reality more accurately, complexity of distributions in the feature space should be further analyzed in future research.

Table 4.11 Error matrix of classification by the MLC

Reference Result	C1	C2	C3	C4	C5	C6	C7	C8	C9	Total
C1	98	25	4	0	0	0	0	0	0	127
C2	1	23	5	0	0	0	0	0	1	30
C3	0	46	75	2	0	2	0	1	6	132
C4	0	0	7	64	4	6	0	0	13	94
C5	0	0	0	19	89	60	0	0	0	168
C6	1	5	1	5	7	24	1	0	0	44
C7	0	0	0	6	0	7	84	2	0	99
C8	0	0	2	1	0	0	10	96	11	120
C9	0	1	6	3	0	1	5	1	69	86
Total	100	100	100	100	100	100	100	100	100	900

(Accuracy = 69.11%, Kappa = 0.653)

Table 4.12 Error matrix of classification by the RBF network

Reference Result	C1	C2	C3	C4	C5	C6	C7	C8	C9	Total
C1	91	22	1	0	0	0	0	0	0	114
C2	6	55	9	1	0	1	0	0	0	72
C3	2	22	76	4	0	2	0	1	7	114
C4	0	0	6	67	4	5	0	1	14	97
C5	0	0	0	9	18	3	0	0	0	30
C6	1	0	0	7	78	82	2	0	0	170
C7	0	0	0	4	0	5	81	0	0	90
C8	0	0	2	3	0	2	12	94	9	122
C9	0	1	6	5	0	0	5	4	70	91
Total	100	100	100	100	100	100	100	100	100	900

(Training Time = 50 s, Accuracy = 70.33%, Kappa = 0.666)

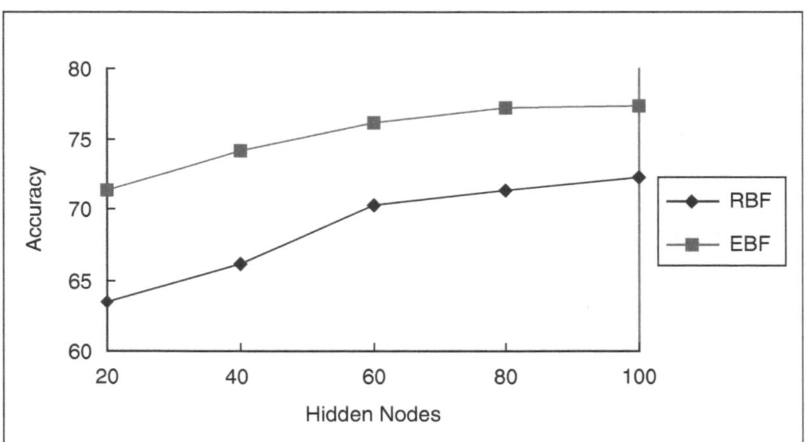

Fig. 4.14 Comparison of average accuracy between the EBF and the RBF networks. (The Curve represents the relationship between the number of hidden nodes and overall accuracy)

Table 4.13 Relationship between accuracy and size of the hidden layer

	Size of the hidden layer in the EBF network						
	20	30	40	50	60	80	100
Training time (seconds)	20	40	70	100	120	170	250
ACCURACY %							
C1	96.00	97.00	96.00	97.00	97.00	97.00	98.00
C2	39.00	40.00	52.00	64.00	62.00	62.00	63.00
C3	78.00	78.00	82.00	76.00	76.00	80.00	79.00
C4	71.00	74.00	72.00	72.00	71.00	77.00	77.00
C5	35.00	39.00	37.00	34.00	49.00	56.00	56.00
C6	71.00	71.00	70.00	78.00	71.00	69.00	69.00
C7	90.00	90.00	88.00	89.00	88.00	86.00	86.00
C8	89.00	91.00	91.00	89.00	91.00	92.00	92.00
C9	74.00	78.00	80.00	80.00	81.00	78.00	78.00
Average	71.33	73.00	74.11	75.33	76.11	77.22	77.33

4.4 Genetic Algorithms for Fuzzy Spatial Classification Systems

4.4.1 A Brief Note on Using GA to Discover Fuzzy Classification Rules

Genetic algorithms (GA) and fuzzy systems are two important areas of study in soft computing in general and classification in particular. As an extension of classical logical systems, fuzzy system methods provide an effective and flexible framework for handling problems with imprecision and have been widely applied to geographical analysis (Leung 1982, 1987, 1988a–c, 1999), control problems, pattern recognition and function approximation (Dubois and Prade 1980; Zadeh 1994). GA, as briefly discussed in Chap. 2, are adaptive and global search optimization methods that obtain solutions through the evolution of populations of encoded feasible solutions (individuals) (Holland 1975). The population is updated (evolved) by mimicking the natural evolution mechanisms such as selection, crossover and mutation.

A fuzzy classification system is composed of three parts: the fuzzy inference mechanism, the fuzzy partition of the input and output spaces, and a set of fuzzy rules whose antecedents and consequents characterizing fuzzy sets come from the fuzzy partition of the input and output spaces, respectively. In most fuzzy systems, the fuzzy partition and the fuzzy rules are determined and selected by human experts and are heuristic and subjective. Several approaches, mainly under the framework of fuzzy neural networks (Kosko 1992), have been proposed to generate fuzzy if-then rules and adjust membership functions of the fuzzy sets involved.

Recently, there is an increasing enthusiasm in training fuzzy systems via GA (Karr 1991; Feldman 1993; Park et al. 1994; Pernell et al. 1995; Ishibuchi et al. 1995). Both the fuzzy rules and parameters of the membership functions of the antecedent and consequent fuzzy sets can be selected and adjusted by GA. Parameter adjustment of the membership functions is more or less straightforward since it is a parameter optimization problem. The selection of fuzzy rules is, however, much more difficult because combinatorial optimization is involved. Two typical genetic-based approaches for selecting fuzzy rules are proposed in Ishibuchi et al. (1995) and Park et al. (1994). In the former approach, each possible set of fuzzy rules is encoded as an individual binary vector with each of its component representing a possible rule. Fuzzy rules in the latter approach appear in the form of a fuzzy relationship matrix which is learned via GA by encoding each possible fuzzy relationship matrix as an individual.

A novel genetic-based framework for the training of fuzzy classification systems developed by Leung et al. (2001b) is introduced in here. The novelty of the proposed GA, called the Genetic Algorithm with No Genetic Operators (GANGO) lies in the following facts:

1. A substantial decrease of storage requirement and computational cost in the evolution of populations. GANGO itself does not need to store the population and to use standard genetic operators required by conventional GA. So, it is almost tailor-made for large scale problems such as the training of fuzzy classification systems.
2. A new encoding scheme, based on GANGO, is quite different from the direct (but also clumsy) method used in the literature. This new encoding viewpoint also contributes to the decrease of storage requirement and results in a more natural way to eliminate irrelevant fuzzy rules.

4.4.2 A General Framework of the Fuzzy Classification System

To acquire fuzzy rules for a classification, we can consider the problem as classifying a pattern vector $\mathbf{a} = (a_1, \cdots a_d)$ from a d-dimensional pattern space $\mathbf{A} \in \mathbf{R}^d$ into M classes. The task is to design a computational device that can output a class index $i \in \{1, 2, \cdots, M\}$ for each input vector in the pattern space. The method of fuzzy logic is instrumental in pattern classification under imprecision. The construction of a fuzzy system for classification problems involves three basic aspects: (1) determination of fuzzy inference method, (2) fuzzy partition of the pattern space into fuzzy subspaces, and (3) generation of a set of fuzzy rules. We can adopt the method of simple fuzzy grid to partition the pattern space A (Ishibuchi et al. 1995). An example of a fuzzy grid partition is shown in Fig. 4.15 where the two-dimensional pattern space is divided into nine fuzzy subspaces $\mathbf{A}_{ij}, 1 \leq i, j \leq 3$. Other more complicated partitioning methods are possible. Among all fuzzy inference methods, Leung et al. (2001b) employ a variation of the so-called *new fuzzy reasoning method* (Cao et al. 1990; Park et al. 1994) as a basis of study.

The fuzzy classification system is depicted in Fig. 4.16. The component $x_i, i = 1, \cdots, N$, of the vector $\mathbf{x} = (x_1, \cdots, x_N)$ is the degree of membership of the input pattern belonging to the ith fuzzy partition space $\mathbf{A}_i$, i.e., $x_i = A_i(\mathbf{a}), N$ is the number of fuzzy subspaces of the fuzzy partition; $\mathbf{y} = (y_1, \cdots, y_M)$ is a vector

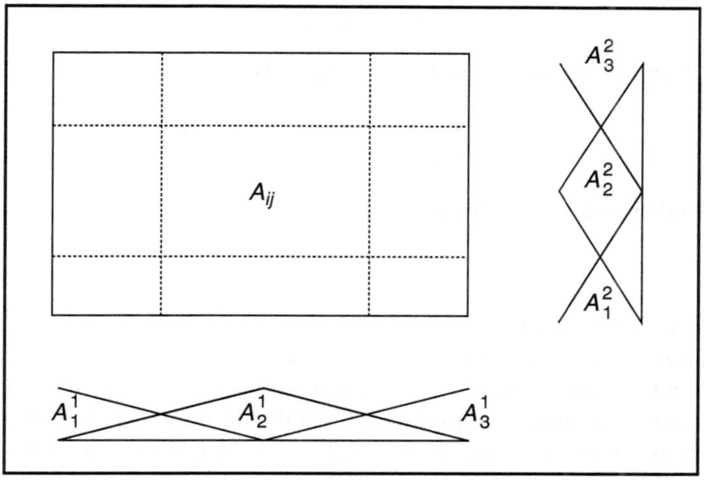

Fig. 4.15 A fuzzy grid partitioning of a pattern space

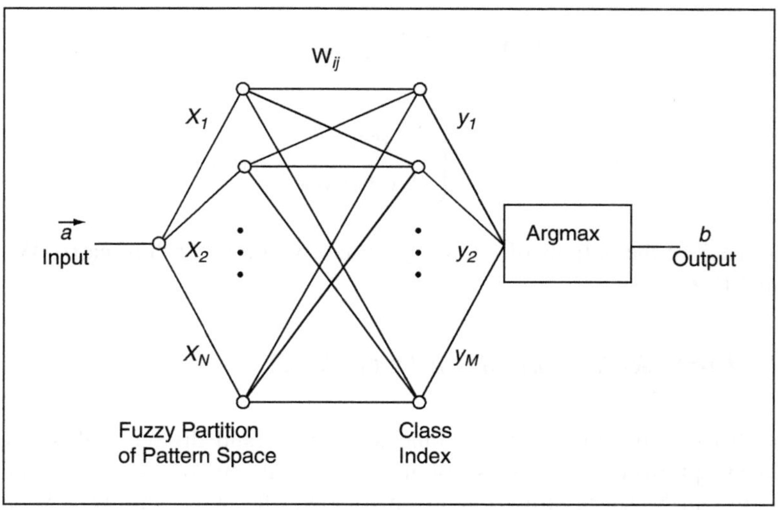

Fig. 4.16 A schema of fuzzy rule set

with y_i denoting the degree of membership of the ith class; and $\mathbf{W} = w_{ij})$ is an $N \times M$ fuzzy relationship matrix. The output b is an integer from $\{1, 2, \cdots, M\}$ indicating the class number of the input pattern. The inference algorithm is given as follows:

Step 1. For an input pattern vector $\boldsymbol{a}$, determine the membership of $\boldsymbol{a}$ for each fuzzy partition $\mathbf{A}_i$, $1 \leq i \leq N$, by

$$x_i = \mathbf{A}_i(\mathbf{a}).$$

Step 2. Calculate the vector **y** in terms of **x** and ***W***:

$$y_i = \sum_{i=1}^{N} x_i\, w_{ij}, \quad 1 \leq j \leq M.$$

Step 3. Find an index i_m such that

$$y_{i_m} = \max_{1 \leq i \leq M} y_i$$

and let the output ***b*** equal i_m

This fuzzy system actually consists of $N \times M$ fuzzy rules, each of which can be identified in the form of "IF ***a*** is $\mathbf{A}_i$, then ***a*** belongs to class j with certainty w_{ij}."

Following the method of Ishibuchi et al. (1995), the training task of the system is formulated as follows: Let $(\mathbf{a}_p, i_p)$, $p = 1, 2, \cdots, L$, be L training patterns where $\mathbf{a}_p$ is the pattern vector and i_p is the class index of $\mathbf{a}_p$. The training task of the fuzzy system described above can be formulated as the following optimization problem:

Find a fuzzy relationship matrix **W** to maximize the function

$$f(\mathbf{W}) = \sum_{p=1}^{L} I_{i_p}\left(b_W\left(\mathbf{a}_p\right)\right), \tag{4.21}$$

$$\text{where } I_i(\mathbf{b}) = \begin{cases} 1, & \text{if } \mathbf{b} = i, \\ 0, & \text{otherwise,} \end{cases} \text{ and } b_w\left(\mathbf{a}_p\right)$$

and $b_W\left(\mathbf{a}_p\right)$are the outputs of the fuzzy system with relationship matrix **W** when the input is ***a***.

4.4.3 Fuzzy Rule Acquisition by GANGO

Since $f(W)$ is a function with continuous variables and discrete values, neither numerical optimization methods nor non-numerical methods such as simulated annealing can be successfully employed to solve the above problem. So it is a natural idea to resort to the more general and powerful method of genetic algorithms. To facilitate our discussion, I first give a formalism of GA, and then a novel GA for the discovery of fuzzy rules in the classification system.

4.4.3.1 A Formalism of Canonical Genetic Algorithms

Without loss of generality, let us consider the genetic algorithm with binary string representations of length l and fixed population size N. Assume further that the algorithms use proportional selection, one-point crossover and usual bit mutation.

Each individual in the population corresponds to an element of the space $\mathbf{S} = \{0, 1\}^l$ which is called the individuals space. The population space is denoted as $\mathbf{S}^N$ and we call $\mathbf{S}^2$ the parents space. For the sake of convenience, we write the population $\overrightarrow{\mathbf{X}} \in \mathbf{S}^N$ in both vector and matrix form as follows:

$$\overrightarrow{\mathbf{X}} = (\mathbf{X}_1, \mathbf{X}_2, \cdots, \mathbf{X}_N)^T = \begin{pmatrix} x_{11} & x_{12} & \cdots & x_{11} \\ x_{21} & x_{22} & \cdots & x_{21} \\ \cdots & \cdots & \cdots & \cdots \\ x_{N1} & x_{N2} & \cdots & x_{N1} \end{pmatrix}, \tag{4.22}$$

where $\mathbf{X}_i \in \mathbf{S}$ is the ith individual of $\overrightarrow{\mathbf{X}}$, while x_{ij} is the jth component of $\mathbf{X}_i$. The fitness function f: $\mathbf{S} \rightarrow R^+$ can be derived from the objective function of the evolution, e.g., optimization, problem by a certain decoding rule.

From the mathematical point of view, genetic operators are random mappings between the space $\mathbf{S}^N$, $\mathbf{S}^2$ and $\mathbf{S}$. They are the analogous abstractions of the genetic mechanisms in the evolution of natural organisms and can be given strict probabilistic definitions as follows:

1. The proportional selection operator, T_s: $\mathbf{S}^N \rightarrow \mathbf{S}^2$, selects a couple of parents from the given population for reproduction. Given the population $\overrightarrow{\mathbf{X}}$, the probability of selecting $(\mathbf{X}_i, \mathbf{X}_j) \in \mathbf{S}^2$ as the parents is

$$P\left\{T_s\left(\overrightarrow{\mathbf{X}}\right) = (\mathbf{X}_i, \mathbf{X}_j)\right\} = \frac{f(\mathbf{X}_i)}{\sum\limits_{x \in \overrightarrow{X}} f(\mathbf{X})} \cdot \frac{f(\mathbf{X_j})}{\sum\limits_{x \in \overrightarrow{X}} f(\mathbf{X})}, \; 1 \le i \le N, \; 1 \le j \le N. \tag{4.23}$$

2. The crossover operator T_c: $\mathbf{S}^2 \rightarrow \mathbf{S}$ generates an individual from the selected parents. Given the parent $\mathbf{X}_i = (x_{il}, \cdots, x_{il}), i = 1, 2$, the probability for the one-point crossover operator to generate an individual $\boldsymbol{Y}$ is

$$P\{T_c((\mathbf{X}_1, \mathbf{X}_2)) = \mathbf{Y}\} = \begin{cases} \dfrac{k \cdot p_c}{l}, & \text{if } \mathbf{Y} \neq \mathbf{X}_1 \\ (1 - p_c) + \dfrac{k \cdot p_c}{l}, & \text{if } \mathbf{Y} = \mathbf{X}_1 \end{cases}, \tag{4.24}$$

where $0 \le p_c \le 1$ is the crossover probability, k is the number of crossover points at which the crossover of $\mathbf{X_1}$ and $\mathbf{X_2}$ can generate $\boldsymbol{Y}$.

3. The mutation operator, T_m: $\mathbf{S} \rightarrow \mathbf{S}$, operates on the individual by independently perturbing each bit string in a probabilistic manner and can be specified as follows:

$$p\{T_m(x) = \mathbf{Y}\} = p_m^{|\mathbf{X}-\mathbf{Y}|}(1 - p_m)^{l-|\mathbf{X}-\mathbf{Y}|}, \tag{4.25}$$

where $0 \leq p_m \leq 1$ is the mutation probability.

Based on the genetic operators defined above, the CGA can be represented as the following iteration of populations:

$$\overrightarrow{\mathbf{X}}(t+1) = \left\{T_m^i\left(T_c^i\left(T_S^i\left(\overrightarrow{\mathbf{X}}(t)\right)\right)\right), \quad i = 1,\cdots,N\right\}, K \geq 0, \tag{4.26}$$

where $\left(T_m^i, T_c^i, T_S^i\right), i = 1,\cdots,N$, are independent versions of (T_m, T_c, T_s). It is easy to see that the sequence of populations $\left\{\overrightarrow{\mathbf{X}}(t), t \geq 0\right\}$ is a time-homogeneous Markov chain with the state space $\mathbf{S}^N$ (henceforth it is called the population Markov chain). Similar to Rudolph (1994), it can be proved that if $P_m \geq 0$, we have for any $\overrightarrow{\mathbf{X}}$, $\overrightarrow{\mathbf{Y}}$,

$$\prod_{i=1}^{N} p\left\{T_m^i\left(T_c^i\left(T_s^i\left(\overrightarrow{\mathbf{X}}\right)\right)\right) = \mathbf{Y}_i\right\} \geq 0. \tag{4.27}$$

That is, $P\left\{\overrightarrow{\mathbf{X}}(t+1) = \overrightarrow{\mathbf{Y}}\middle|\overrightarrow{\mathbf{X}}(t) = \mathbf{x}\right\} \geq 0$. Therefore, the population Markov chain $\left\{\overrightarrow{\mathbf{X}}(t), t \geq 0\right\}$ is homogeneous, irreducible and aperiodic. Hence it can reach any state in infinite time with probability 1 regardless of the initial state. Theoretically this means that the CGA will never converge and premature convergence cannot occur provided that the mutation probability is larger than zero.

A canonical genetic algorithm (CGA) can in essence be given as follows:

Step 1. Set parameters N, l, P_c, and P_m. Set $t = 0$ and generate the initial population $\overrightarrow{\mathbf{X}}(0)$.
Step 2. Independently select N pairs of individuals from the current population for reproduction.
Step 3. Independently perform crossover to the N pairs of individuals to generate N new intermediate individuals.
Step 4. Independently mutate the N intermediate individuals to get the next generation $\overrightarrow{\mathbf{X}}(t+1) = (\mathbf{X}_1(t+1),\cdots,\mathbf{X}_N(t+1))$.
Step 5. Stop if some stopping criterion is met. Else, set $t = t+1$ and go to step 2.

So in step t, the GA generates the individuals $\mathbf{X}_i(t), 1 \leq i \leq N$, by independently performing genetic operators, which are stochastic in nature, on the population $\overrightarrow{\mathbf{X}}(t-1)$. It follows that the individual $\mathbf{X}_i(t), 1 \leq i \leq N$, are conditionally independent and identically distributed random vectors taking values in $\{0,1\}^l$. For each $1 \leq i \leq N$, the one-dimensional marginal conditional distribution of $\mathbf{X}_1(t)$ given $\overrightarrow{\mathbf{X}}(t-1)$ is

$$p_j(y_j) = (p_j(1))^{yj}(1-p_j(1))^{1-yj} + (-1)^{1-yj}\left(1 - 2\cdot\frac{F_j^1\left(\overrightarrow{\mathbf{X}}(t-1)\right)}{F\left(\overrightarrow{\mathbf{X}}(t-1)\right)}\right)\cdot p_m, \tag{4.28}$$

where p_m is a very small positive scalar called mutation probability,

$$p_j(1) = \frac{F_j^1\left(\overrightarrow{\mathbf{X}}(t-1)\right)}{F\left(\overrightarrow{\mathbf{X}}(t-1)\right)} + \left(1 - 2 \cdot \frac{F_j^1\left(\overrightarrow{\mathbf{X}}(t-1)\right)}{F\left(\overrightarrow{\mathbf{X}}(t-1)\right)}\right) \cdot p_m, \tag{4.29}$$

with

$$F\left(\overrightarrow{\mathbf{X}}(t-1)\right) = \sum_{i=1}^{N} f(\mathbf{X}_i(t-1)), \tag{4.30}$$

$$F_j^1\left(\overrightarrow{\mathbf{X}}(t-1)\right) = \sum_{i \in I_1(j)} f(\mathbf{X}_i(t-1)), \tag{4.31}$$

where $I_1(j) = \left\{1 \le i \le N; x_{ij}(t-1) = 1\right\}$.

4.4.3.2 The Genetic Algorithms with No Genetic Operators (GANGO)

It should be observed that the transition probabilities $P(\mathbf{X}, \mathbf{Y})$ of the CGA population Markov chain $\left\{\overrightarrow{\mathbf{X}}(t), t \ge 0\right\}$ cannot be given explicitly. It is because the components $\left\{x_{ij}(t+1), 1 \le i \le l\right\}$ of the individual $\mathbf{X}_i(t+1)$ generated according to (4.26) are not conditionally independent given $\mathbf{X}_i(t)$. In order to create an individual $\mathbf{Z}$ with conditionally independent components, one can proceed as follows: First generate an individual $\mathbf{Z}^1 = \left(z_1^1, \cdots, z_l^1\right)$ by (3.56) and take z_1^1 as the first component of $\mathbf{Z}$. Independent of $\mathbf{Z}^1$, create another individual $Z^2 = \left(z_1^2, z_2^2, \cdots, z_l^2\right)$ by (4.28) and take z_2^2 as the second component of $\mathbf{Z}$. Continue the process until all z_i's make it possible to construct a GA without the application of the genetic operators. Such a GA also simplifies the computation in evolving a generation.

Based on this idea, Gao et al. (1996) proposed a new algorithm, called the genetic algorithm with no genetic operations (GANGO). Instead of using genetic operators, GANGO generates new individuals by directly sampling the distribution $P(\mathbf{Y}) = \prod_{j=1}^{l} p_j(y_i)$, $\mathbf{Y} = (y_1, \cdots, y_l) \in \{0,1\}^l$ where $p_j(y_j)$ is given in (4.28). Since the distribution $P(\mathbf{Y})$ depends on the population $\overrightarrow{\mathbf{X}}(t-1)$ only through the (l+1)-tuple $\left(F\left(\overrightarrow{\mathbf{X}}(t-1)\right), F_j^1\left(\overrightarrow{\mathbf{X}}(t-1)\right), 1 \le j \le l\right)$, in GANGO we then need not to record the whole population $\overrightarrow{\mathbf{X}}(t)$. Nevertheless the ($l$+1)-tuple $\left(F\left(\overrightarrow{\mathbf{X}}(t-1)\right), F_j^1\left(\overrightarrow{\mathbf{X}}(t-1)\right), 1 \le j \le l\right)$ can be recorded in an accumulative manner so that the storage requirement of GANGO is independent of the population size N and is only $(2l+1)/(N \times l)$ that of the canonical GAs. In GANGO, the imaginary population evolves as follows: In step t, generate an individual $\boldsymbol{X}$ by sampling the conditional distribution $P^{t-1}(\mathbf{Y})$. If the fitness of $\boldsymbol{X}$ is greater than the average fitness of the population $\overrightarrow{\mathbf{X}}(t-1)$, then $\boldsymbol{X}$ is added to $\overrightarrow{\mathbf{X}}(t-1)$ to form the tth population $\vec{\mathbf{X}}(t)$. Else, set $\vec{\mathbf{X}}(t) = \vec{\mathbf{X}}(t-1)$. Of course, the imaginary population does not appear in the implementation of the GANGO because what is actually needed to be evolved is the distribution $P(\mathbf{Y}) = \prod_{j=1}^{l} p_j(y_i)$, $\mathbf{Y} = (y_1, \cdots, y_l) \in \{0,1\}^l$. The updating of $P^t(\mathbf{Y})$, which is determined by $\left(F\left(\vec{\mathbf{X}}(t-1)\right), F_j^1\left(\vec{\mathbf{X}}(t-1)\right), 1 \le j \le l\right)$, can be specified as follows:

If the fitness of $\boldsymbol{X}$ is less than the average fitness of $\vec{\mathbf{X}}(t-1)$, then

$$F\left(\vec{\mathbf{X}}(t)\right) = F\left(\vec{\mathbf{X}}(t-1)\right), \tag{4.32}$$

$$F_j^1\left(\vec{\mathbf{X}}(t)\right) = F_j^1\left(\vec{\mathbf{X}}(t-1)\right). \tag{4.33}$$

If the fitness of $\boldsymbol{X}$ is greater than the average fitness of $\vec{\mathbf{X}}(t-1)$, then

$$F\left(\vec{\mathbf{X}}(t)\right) = F\left(\vec{\mathbf{X}}(t-1)\right) + f(\mathbf{X}), \tag{4.34}$$

$$F_j^1\left(\vec{\mathbf{X}}(t)\right) = \begin{cases} F_j^1\left(\vec{\mathbf{X}}(t)\right) + f(x), & \text{if} \quad x_j = 1, \\ F_j^1\left(\vec{\mathbf{X}}(t)\right), & \text{if} \quad x_j = 0. \end{cases} \tag{4.35}$$

In summary, the algorithm of GANGO is as follows:

Step 1. Randomly generate N individuals $\vec{\mathbf{X}}(0) = (\mathbf{X}_1, \cdots, \mathbf{X}_N)^T$ and compute its characteristic $F(0), F_j^1(0), 1 \le j \le l$. Based on the characteristic of $\vec{\mathbf{X}}(0)$, compute the zero-one distributions $p_j^0(\cdot)$, $1 \le j \le l$ according to (4.29) and the threshold $a(0) = F(0)/N$. Set $t = 0$ and $kk = 1$.

Step 2. Sample the zero-one distributions $p_j^t(\cdot)$, $1 \le j \le l$, to get an individual $\mathbf{X}(N+t+1) = (x_1(N+t+1), \cdots, x_l(N+t+1))$.

Step 3. If $f(\mathbf{X}(N+t+1))a(t)$, then set $t = t+1$ and return to step 2; else update the characteristic $F(t), F_j^1(t)$, $1 \le j \le l$, according to (4.30) and (4.31) to get $F(t+1)$, $F_j^1(t+1)$, $1 \le j \le 1$. Set $kk = kk + 1$.

Step 4. Compute the zero-one distributions $p_j^{t+1}(\cdot)$, $1 \le j \le l$, and $a(t+1) = F(t+1)/N + kk$ using the updated characteristic.

Step 5. Stop or $t = t+1$ and go to step 2.

The training task of the fuzzy classification system is to find a fuzzy relationship matrix $\boldsymbol{W}$ that solves the optimization problem in (4.21). In order to use the genetic algorithm to solve the problem, we must first determine an encoding scheme that can transfer the fuzzy relationship matrix $\boldsymbol{W}$ into a binary string. A conventional (and also clumsy) encoding method is to represent each element in $\boldsymbol{W}$ in its binary form and then to combine these binary strings into a large string. Basing on the nature of GANGO and the training problem of the fuzzy system, we can however adopt the new strategy. Although the w_{ij}'s are deterministic values, it is advantageous to consider them to be the expectation of some random variables. This viewpoint has been proven to be useful in the study of complex network systems, e.g., in the study of (deterministic or stochastic) neural networks using probability models (Amari 1995).

4.4.3.3 Discovery of Classification Rules by GANGO

Two concepts are first introduced (Leung et al. 2001b):

Definition 3.1. *A fuzzy system* $\mathbf{W}$ *is called a fuzzy system with crisp fuzzy relationship matrix if all the* w_{ij}*'s take on the value 0 or 1.*

Definition 3.2. *A fuzzy system* $\mathbf{V}$ *is called a fuzzy system with random and crisp fuzzy relationship matrix if each element* v_{ij} *of* $\mathbf{V}$ *is a 0–1 random variable.*

Given a fuzzy system $\boldsymbol{V}$ with random and crisp fuzzy relationship matrix, let $w_{ij} = E(v_{ij}) = P\{v_{ij} = 1\}$ (here E denotes expected value and P denotes probability). The fuzzy system $\mathbf{W} = (w_{ij})$ is called the mean fuzzy system of $\boldsymbol{V}$. Conversely, any fuzzy system $\boldsymbol{W}$ can be treated as the mean fuzzy system of some fuzzy system with random and crisp fuzzy relationship matrix. For convenience, we treat the $N \times M$ matrices $\boldsymbol{W}$ and $\boldsymbol{V}$ as vectors whose components are still indexed by subscript i, j. For example, we treat $\mathbf{W} = (w_{ij}, 1 \le i \le N, 1 \le j \le M)$ as a vector of $N \times M$ dimension with the $((i-1)M + j)$th components being w_{ij}. In this way, any (random) and crisp relationship matrix may be regarded as a (random) binary string. Returning to the problem of encoding a fuzzy system $\boldsymbol{W}$ for GANGO, We treat the fuzzy system $\boldsymbol{W}$ involved in the training task in (4.21) as the mean fuzzy system of a fuzzy system $\boldsymbol{V}$ with random and crisp fuzzy relationship matrix. To find w_{ij} is equivalent to finding the parameter of the 0–1 distribution of $w_{ij}, P\{v_{ij} = 1\}$. In their algorithm for the training of the fuzzy system, Leung et al. (2001) use the crisp relationship matrices $\boldsymbol{V}$'s as the individuals while the corresponding fuzzy relationship matrices $\boldsymbol{W}$'s are given by the expectations of the random and crisp relationship matrices corresponding to the individuals in the algorithm, that is indeed the parameters of the 0–1 distribution of v_{ij}'s, $P\{v_{ij} = 1\}$.

Having specified the encoding scheme, the algorithm for training the fuzzy classification systems can be summarized as follows:

Step 1. Randomly generate T fuzzy systems with crisp relationship matrix $\{\mathbf{V}(t)\}_1^T$. Compute the fitness of $f(\mathbf{V}(t))$ according to (4.21). Compute the characteristic $\left(F(0),\ F_{ij}^1(0)\right)$ of the population $\{\mathbf{V}(t)\}_1^T$ according to:

$$F\left(\overrightarrow{\mathbf{X}}(t-1)\right) = \sum_{i=1}^{N} f(\mathbf{X}_i(t-1)),$$

$$F_j^1\left(\overrightarrow{\mathbf{X}}(t-1)\right) = \sum_{i \in I_1(j)} f(\mathbf{X}_i(t-1)),$$

where $I_1(j) = \left\{1 \le i \le N;\ x_{ij}(t-1) = 1\right\}$.

For each pair (i, j), compute

$$p_{ij}^{(0)}(1) = \frac{F_{ij}^{1}(0)}{F(0)} + \left(1 - 2\frac{F_{ij}^{1}(0)}{F(0)}\right) p_m$$

and let $w_{ij}(0) = P_{ij}^{(0)}(1)$. Set $t = T$ and $k = T$.
(here p denotes probability and m denotes mutation)

Step 2. Sample the zero-one distribution $p_{ij}^{(t)}(\cdot)$, $1 \le i \le M$, $1 \le j \le N$, with parameters $p_{ij}^{(t)}(1)$ to get an individual $\mathbf{V}(t+1) = (v_{ij}(t+1))$.

Step 3. If $f(\mathbf{V}(t+1)) < F(0)/k$, set $t = t+1$ and return to *step 2*; else update the characteristic $\left(F(t),\ F_{ij}^{1}(t)\right)$, according to:

$$F\left(\vec{\mathbf{X}}(t)\right) = F\left(\vec{\mathbf{X}}(t-1)\right) + f(\mathbf{X})$$

$$F_j^1\left(\vec{\mathbf{X}}(t)\right) = \begin{cases} F_j^1\left(\vec{\mathbf{X}}(t-1)\right) + f(\mathbf{X}), & \text{if} x_j = 1, \\ F_j^1\left(\vec{\mathbf{X}}(t-1)\right), & \text{if}\, x_j = 0, \end{cases}$$

to get the new characteristic $\left(F(t+1),\ F_{ij}^{1}(t+1)\right)$. Set $k = k+1$.

Step 4. For each pair (i,j) compute

$$p_{ij}^{(t+1)}(1) = \frac{F_{ij}^{1}(t+1)}{F(t+1)} + \left(1 - 2\frac{F_{ij}^{1}(t+1)}{F(t+1)}\right) p_m$$

and let $w_{ij}^{(t+1)} = P_{ij}^{(t+1)}(1)$. Set $t = t+1$.

Step 5. Repeat *step 2-step 4* until the stopping criterion is met.

4.4.3.4 The Reduction of the Number of Fuzzy Rules in the GANGO Trained Fuzzy System

In real-world problems, the number of possible rules may be huge due to the high dimension of the pattern space. To improve computational efficiency and to obtain a practical system, we need methods to eliminate some irrelevant rules to derive a compact fuzzy system. The irrelevant rules essentially fall into two categories: the dummy rules and the inactive rules. Recall that the fuzzy rules in the fuzzy system take the form:

If $\vec{\mathbf{a}} \in \mathbf{A}_i$, then $\vec{\mathbf{a}}$ belong to class j with certainty w_{ij}.

Let $(\vec{\mathbf{a}}_p,\ i_p)$, $p = 1, \cdots, L$, be the training patterns. A fuzzy rule is called a α-level dummy rule if $\sum_{i_p=j} \mathbf{A}_i\left(\vec{\mathbf{a}}_p\right) < \alpha$. A fuzzy rule is called a β-level inactive rule if $w_{ij} < \beta$. Both the dummy rules and the inactive rules have little or no effect on the performance of the fuzzy systems, and should be eliminated.

1. Fitness Reassignment Strategy for the Elimination of Dummy Rules

   Though a fuzzy system has an overall fitness, different fuzzy rules in the fuzzy system have different contributions to that overall fitness. For example, the dummy rules have no or little contributions to the performance (fitness) of a fuzzy system. The strategy for the elimination of the dummy rules is to discourage dummy rules in the course of evolution by reassigning the fitness to the dummy rules. This is possible only in the GANGO framework since it operates on the components (i.e., genes or fuzzy rules) level, while selection in conventional genetic algorithms is done on the individuals (i.e., fuzzy systems) level. To implement the reassignment of fitness to the dummy rules in the training algorithm, all that need to be changed is the updating scheme of $F^1_{ij}(t+1)$ in *step 3*. For each $1 \le i \le M,\ 1 \le j \le N$, define the weight of reassignment as

$$r_{ij} = \begin{cases} 0, & \text{if} \sum\limits_{i_p = j} A_i(\vec{a}_p) < \alpha, \\ 1, & \text{otherwise}, \end{cases} \tag{4.36}$$

   where α is a small scalar. The updating scheme of $F^1_{ij}(t+1)$ becomes

$$F^1_{ij}(t+1) = \begin{cases} F^1_{ij}(t) + r_{ij} f(\mathbf{V}(t+1)), & \textit{if } v_{ij} = 1, \\ F^1_{ij}(t), & \textit{if } v_{ij} = 0. \end{cases} \tag{4.37}$$

2. Weight Truncation Strategy for the Reduction of Inactive Rules

   Let $\mathbf{W} = (w_{ij})$ be the fuzzy relationship matrix of a trained fuzzy system. As has been explained previously, this fuzzy system consists of $N \times M$ fuzzy rules of the form

$$\textit{If } \vec{A} \text{ is } A_i, \vec{a} \text{ belongs to class } j \text{ with certainty } w_{ij}.$$

   Moreover, based on the probabilistic interpretation of w_{ij} in the encoding scheme, w_{ij} can be view as the conditional probability that a pattern belongs to class j given that the pattern is in A_i, or w_{ij} can also be viewed as the probability that the rule "If A_i then j" is active in the fuzzy system. We can reduce the number of fuzzy rules by eliminating those rules whose active probability w_{ij} is small. This is done by setting the w_{ij}'s that are smaller than a small scalar (threshold) to be zero. Formally, let $\mathbf{W} = (w_{ij})$ be the fuzzy relationship matrix of a trained fuzzy system and let $0 < \alpha < 1$ be the threshold. Define a new fuzzy relationship matrix $\mathbf{W}^{\alpha} = \left(w^{\alpha}_{ij}\right)$ by

$$w^{\alpha}_{ij} = \begin{cases} w_{ij}, & \text{if } w_{ij} > \alpha, \\ 0, & \text{if } w_{ij} \le \alpha. \end{cases} \tag{4.38}$$

The number of active fuzzy rules in the fuzzy system with fuzzy relationship matrix $\mathbf{W}^{\alpha}$ is thus less than that in the original fuzzy system. This gives a tight set of rules with sensible interpretation.

4.4.4 An Application in the Classification of Remote Sensing Data

As an application, the GANGO system has been employed to discover rules and form a classification system for remotely sensed data. The automatic knowledge discovery process was completed in relatively short training time. The data is the TM data of Hong Kong, 1993. The data are vectors of three dimensions with each component representing one band of spectrum within a pixel ranging from 0 to 255. Each datum comes from one of the three classes: water (0), plant (1), and building (2). The 150 training data consist of 50 spectral data from each class. There are also 600 spectral data (200 from each class) to be used as the testing data to examine the ability of generalization of the trained fuzzy classification system.

To partition the pattern space $[0, 255] \times [0, 255] \times [0, 255]$ into fuzzy subspaces, the kth axis ($k = 1, 2, 3$) $[0, 255]$ is partitioned using 6 triangular fuzzy sets U_i^k, $0 \leq i \leq 5$. We may name the 6 fuzzy sets as *very small* $\left(U_0^k\right)$, *small* $\left(U_1^k\right)$, *moderately small* $\left(U_2^k\right)$, *moderately large* $\left(U_3^k\right)$, *large* $\left(U_4^k\right)$, and *very large* $\left(U_5^k\right)$(Fig. 4.17). After each axis has been partitioned, the fuzzy partitions A_i, $0 \leq i \leq 215$, of the entire pattern space are formed by re-indexing the set of 216 fuzzy subsets $U_i^1 \cup U_j^2 \cap U_p^3$, $0 \leq i, j, p \leq 5$.

With this fuzzy partition, the parameters $M \times N$ in the training algorithm for the fuzzy classification system are given as $M = 216$, $N = 3$. Since GANGO does not have the concept of generation, then the number of function evaluations is employed as the stopping criterion. The maximum number of function evaluations is specified as 5000, which is equivalent to 1,000 generations with a population of

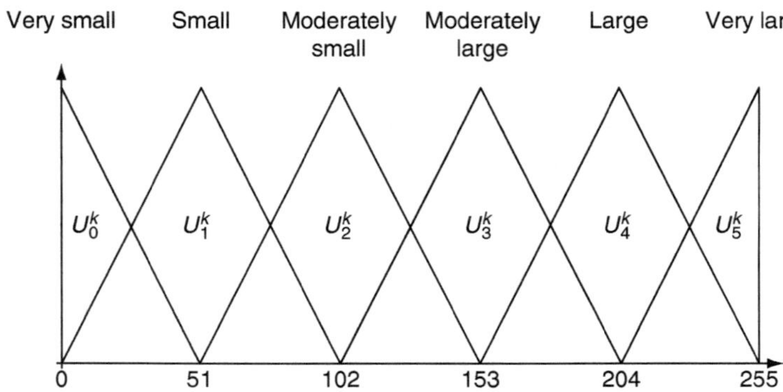

Fig. 4.17 A fuzzy partition of an axis of spectrum

Table 4.14 The performance of the proposed training algorithms in five independent runs with $p_m = 0.00$

Run No.	1	2	3	4	5	Aver.
No. Rules	19	18	15	18	17	17.4
TRCR(%)	100.0	100.0	98.0	100.0	98.0	99.2
TECR(%)	99.8	99.7	97.2	99.3	96.7	98.5

"Run No." indicates the five different runs, "No. Rules" is the number of fuzzy rules in the trained fuzzy system. TRCR and TECR denote the classification rates of the trained fuzzy system on the training data and on the test data, respectively

Table 4.15 The performance of the proposed training algorithms in five independent runs with $p_m = 0.01$

Run No.	1	2	3	4	5	Aver.
No. Rules	21	17	18	20	19	19
TRCR(%)	100.0	99.3	100.0	100.0	100.0	99.8
TECR(%)	99.7	99.5	99.8	99.7	99.7	99.6

size 5, 500 generations with a population of size 10, and 100 generations with a population of size 50 in conventional GAs,

The training algorithm is run five times for each of the two typical mutation probabilities $p_m = 0.00$ and $p_m = 0.01$. The results (Tables 4.14 and 4.15) are very encouraging.

For the case of $p_m = 0.00$, the average classification rate on the training data is 99.2% and the average classification rate on the test data (the generalization ability) is 98.5%. For the case of $p_m = 0.01$, the average classification rates on the training data and the test data are, respectively, 99.8% and 99.6%.

In both cases, our algorithm outperforms previous research reviewed in this study. Though direct comparisons cannot be drawn because the application problems are different, we can observe the advantages of the GANGO system by comparing some statistics. For example, the classification rate of 99.47% is obtained on the training data and 96.67% is obtained on the test data in using the conventional GAs in Ishibuchi et al. (1995). Moreover, the GANGO results are obtained within the framework of a simple fuzzy grid partition while in several fuzzy grid partitions of different levels of granularity are simultaneously used in Ishibuchi et al. (1995). With regard to the computational cost, the maximum number of function evaluations used in our algorithm is 5,000 while the maximum number of function evaluations of 10,000 is adopted in Ishibuchi et al. (1995).

Figure 4.18 shows the dynamics of the classification rate as function of the number of function evaluations in the GANGO system. It can be observed that the convergence rate is very high. It has been demonstrated that GA can be employed to discover classification rules for spatial data. In particular, the novel encoding scheme, together with the no-population-storage and no-genetic-operators nature of the GANGO, contributes to a dramatic decrease in storage requirement and computational coast. The results of training a fuzzy classification system for remote

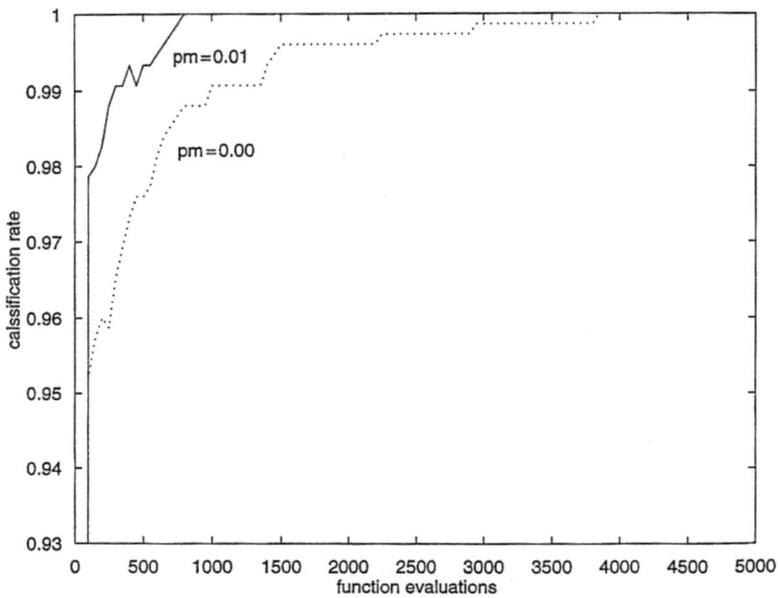

Fig. 4.18 Classification rate of GANGO

sensing data are encouraging. It is found from the experiments that the GANGO method outperforms the conventional GA-based approaches in convergence speed, classification rate, and generalization ability. The novelty of the GANGO also lies in the way that the irrelevant fuzzy rules are eliminated automatically throughout the evolution.

4.5 The Rough Set Approach to the Discovery of Classification Rules in Spatial Data

4.5.1 Basic Ideas of the Rough Set Methodology for Knowledge Discovery

The basic issue of rule-based system is the determination of a minimal set of features (and feature values) and the optimal (usually the minimal) set of consistent rules for classification. All of this has to be achieved with data available. Rough set theory, proposed by Pawlak (1982, 1991), is an extension of set theory for the study of information systems characterized by insufficient and incomplete information and has been demonstrated to be useful in fields such as pattern recognition, machine learning, and automated knowledge acquisition (see, e.g., Yasdi 1996, Polkowski and Skowron 1998, Polkowski et al. 2000, Leung and Li 2003, Leung et al. 2006a). Its basic idea is to unravel an optimal set of decision rules from an information system (basically a feature-value table) via an objective knowledge induction process which determines the necessary and sufficient features constituting the minimal rule set for classification.

Unlike the statistical approach, such as the maximum likelihood classifier, which is restricted by parametric assumptions, the fuzzy sets approach which relies on the definition of a membership function, and the neural networks approach, such as the multilayered feedforward network, which depends on the specification of network architecture and learning parameters, rough set approach works directly on data without making any assumptions about them. It is a non-presumptions bottom-up method for the discovery of classification rules in data.

Though rough set theory has not been commonly applied to the analysis of spatial databases such as vector-based GIS and remotely sensed data, recent works by some researchers have argued for the advantages of using a rough set approach to geo-referenced data, specifically qualitative data (see Stell and Worboys 1998; Worboys 1998a, b; Bittner and Stell 2002; Wang et al. 2002). For data mining in spatial databases, Aldridge (1998) has developed a rough-set methodology for obtaining knowledge from multi-theme geographic data, and applied the classical rough set method to estimate landslide hazards in New Zealand. Wang et al. (2001) have employed the rough set method to discover land control knowledge, with a case study indicating its feasibility. Ahlqvist et al. (2000, 2003) and Ahlqvist (2005) have also applied the rough set method for spatial classification and uncertainty analysis. These studies, however, have not explicitly studied the mining of rules, an important undertaking in rough set research, for the classification of spatial data, particularly remotely sensed data.

Though using the rough set approach for knowledge discovery in spatial databases is still in its early stage, we can see its potential in spatial data mining, particularly when data are discrete and qualitative. It is an objective way to unravel decision rules from information systems with incomplete and qualitative data. It renders an effective methodology to optimally select features, e.g., selection of the most relevant spectral bands, constituting an optimal rule set necessary and sufficient for a classification task. However, the standard Pawlak's rough set model that has been applied to discover knowledge in databases so far is generally not appropriate in handling spatial information, particularly remotely sensed data, which is real-valued or integer-valued in nature. It should be noted that equivalence classes is a key notion in Pawlak's rough set model. It is the basic building block for the knowledge induction procedure. With real-valued or integer-valued (in large range) information, we most likely will have way too many equivalence classes which will eventually lead to too large a number of classification rules. Though such classification rules may fit the training data, their generalization capability will be rather low since perfect match of the real-valued or integer-valued condition parts of the rules will be difficult if not impossible to realize.

To make the rough set approach effective and efficient for knowledge discovery in spatial databases, it is thus essential to develop novel rough set models for real-valued or integer-valued information. Since integer-valued information is a particular class of real-valued information and the method to be discussed applied to both, the term "real-valued" is henceforth used for simplicity of presentation. Our discussion is based on the study by Leung et al. (2007) which proposes to first transform a real-valued information system into an interval-valued information system, and then construct a new rough-set knowledge induction method to select optimal decision rules with a minimal set of features necessary and sufficient for the classification of real-valued spatial information in general, and remotely sensed data in particular.

4.5.2 Basic Notions Related to Spatial Information Systems and Rough Sets

The notion of an information system provides a convenient representation of objects in terms of their attributes. An (complete) information system can be defined by a pair $S = (U, A)$, where U is a nonempty finite set of objects called the universe of discourse, and A is a nonempty finite set of attributes, i.e., $a : U \to V_a$ is an information function for $a \in A$, where V_a is called domain of a.

Elements of U are called objects which, in spatial context, may be cites, states, processes, pixels, points, lines, and polygons. Attributes can be features, variables, spectral bands, and socio-economic characteristics.

For an information system $S = (U, A)$, one can describe relationships between objects through their attribute values. With respect to an attribute subset $B \subseteq A$, a binary equivalence relation R_B can be defined as

$$x, y \in U, (x, y) \in R_B \Leftrightarrow a(x) = a(y), \forall a \in B. \tag{4.39}$$

The term R_B is the relation with respect to B derived from the information system S, and we call (U, R_B) the Pawlak approximation space with respect to B induced from S. With the relation B, two objects are considered to be indiscernible if and only if they have the same value on each $a \in B$. Based on the approximation space (U, R_B), one can derive the lower and upper approximations of an arbitrary subset X of U defined, respectively, as

$$\underline{B}(X) = \{x \in U : [x]_B \subseteq X\}, \overline{B}(X) = \{x \in U : [x]_B \cap X \neq \emptyset\}, \tag{4.40}$$

where $[x]_B = \{y \in U : (x, y) \in R_B\}$ is the B-equivalence class containing x (Fig. 4.19). The pair B is the representation of B in the Pawlak approximation space (U, R_B), or is referred to as the Pawlak rough set of X with respect to (U, R_B). The boundary of X, $Bd(X)$, is thus

$$Bd(X) = \bar{B}(X) - \underline{B}(X). \tag{4.41}$$

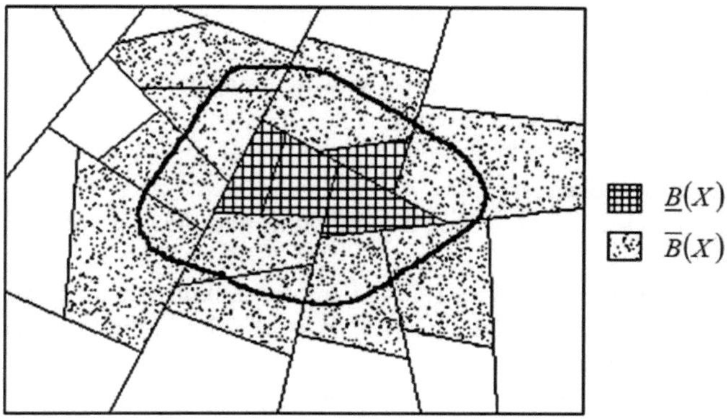

Fig. 4.19 Lower and upper approximations of a rough concept

Table 4.16 A simple decision table

U	Slope (a_1)	Altitude (a_2)	d(hill fire)
x_1	0	L	0
x_2	0	H	0
x_3	0	H	0
x_4	1	L	0
x_5	2	L	1
x_6	1	H	1
x_7	1	H	1
x_8	1	L	0
x_9	2	L	1
x_{10}	2	H	1

If $\bar{B}(X) = \underline{B}(X)$, then $Bd(X) = \phi$. This implies that X is precise.
A special case of an information system is a decision table. A decision table is an information system of the form $S = (U, A \cup \{d\})$, where $d \notin A$ is a distinguishing attribute called decision. The elements of A are called conditional attributes. We can interpret the decision attribute as a kind of classifier on the universe of objects given, for example, by an expert or a decision-maker. In machine learning, decision tables are called sets of training examples.

Without loss of generality, we assume that $V_d = \{1, 2, \ldots, I\}$. It can be observed that the decision d determines a partition of the universe of discourse, $U/d = \{ [x]_d : x \in U\} = \{X_1^d, X_2^d, \ldots, X_I^d\}$, where $X_i^d = \{x \in U : d(x) = i\}, i = 1, 2, \ldots, I$. The set X_i^d is called the ith decision class of the decision table $S = (U, A \cup \{d\})$. Thus i may be treated as the label of the class X_i^d.

Table 4.16 is an information system (without the decision column d), where $U = \{x_1, \cdots, x_{10}\}$, $A = \{a_1, a_2\}$, $V_a\{0,\ 1,\ 2\}$ and $Va_2 = \{L, H\}$. With respect to A, the equivalence classes are: $\{\{x_1\}, \{x_2, x_3\}, \{x_4, x_8\}, \{x_5, x_9\}, \{x_6, x_7\}, \{x_{10}\}\}$. Augmenting d as a decision, where $V_d = \{0,\ 1\}$, Table 4.16 then becomes a decision table. The equivalence classes with respect to d are $\{\{x_1, x_2, x_3, x_4, x_8\}, \{x_5, x_6, x_7, x_9, x_{10}\}\}$.

Taking the classification of remote sensing imagery as an example, we can formalize the decision table as follows:

Let

$d_1, d_2, \ldots, d_I$ be I classes;

$O_i = \{o_{ij} : j = 1, 2, \ldots, J_i\}$ be the random sample set of the ith class, $i = 1, 2, \ldots, I$;

$A = \{a_1, a_2, \ldots, a_m\} = \{a_k : k = 1, 2, \ldots, m\}$ be a finite set of attributes which represent m spectral bands, and $a_k(o_{ij}) = v_{ij}^k \in [0, 255]$ is the gray scale value of o_{ij} measured by spectral band a_k.

Such a (training) data set can be represented by a decision table $(O, A \cup \{d\})$, where $O = \{o_{ij} : i = 1, 2, \ldots, I, j = 1, 2, \ldots, J_i\}$ is a finite set of objects $A = \{a_1, a_2, \ldots, a_m\}$ is an attribute (spectral band) set, such that $a_k(o_{ij}) = v_{ij}^k \in R_+$ for all $j = 1, 2, \ldots, J_i$; $i = 1, 2, \ldots, I$; $k = 1, 2, \ldots, m$; d is the decision attribute; $V_d = \{1, 2, \ldots, I\}$ is the value set of decision such that $d(o_{ij}) = i$, $\forall j = 1, 2, \ldots, J_i$, $i = 1, 2, \ldots, I$;, and $O_i = \{o_{ij} : j = 1, 2, \ldots, J_i\}$is the random sample set of the ith class of objects.

Based on the lower and upper approximations of the decision classes X_i^d, $i = 1, 2, \ldots, I$, with respect to (U, R_A) in the decision table $(U, A \cup \{d\})$, all the certain and possible decision rules can be unraveled (Pawlak 1991).

4.5.3 *Interval-Valued Information Systems and Data Transformation*

Given a number of facts, generalization can be performed in many different directions. In order to extract interesting rules from databases, learning should be directed by some background knowledge. To discover patterns in remotely sensed data, we, for example, need to know initially the classes of interest and the plausible spectral bands which might be relevant to the classification task.

Differing from most rough-set applications, integer-valued attributes (spectral bands) need to be employed to discover knowledge in remotely sensed data. That is, we are to classify objects by integer-valued spectral reflectance. A direct application of conventional rough-set models to such a data base will most likely lead to a huge number of equivalence classes on which knowledge induction is based. Consequently, a large number of decision rules will be discovered with low generalization capability.

To make the rough-set approach effective, efficient and practical, and to achieve higher level of generalization, a novel rough-set framework is formulated by Leung et al. (2007). Their proposed approach is to first convert the real-valued information system into an interval-valued information system through a simple manipulation of the data. Then, a new rough-set model is constructed for knowledge induction in interval-valued databases.

An interval-valued information system can be defined by a pair $K = (U, A)$, where the universe of discourse $U = \{u_i : i = 1, 2, \ldots, I\}$ is represented as a set of distinct I classes, $A = \{a_1, a_2, \ldots, a_m\}$ is the attribute (spectral band) set such that $a_k(u_i) = [l_i^k, u_i^k]$, for all $i = 1, 2, \ldots, I$ and $k = 1, 2, \ldots, m$, where l_i^k and u_i^k are the lower and upper limits of the interval for class i under attribute a_k. Specially, each class i is signified by a value range, an interval, under spectral band k. Signifying a class by an interval of spectral values under a spectral band is evident in theory and applications (Jenson 1996; Jenson and Langari 1999; Ji 2003). Due to the variation of values of the sample points belonging to a class under a specific spectral band, taking interval-values will not result in information loss but will actually make it more true-to-life and representative. This is particularly relevant in region-based classification. This justifies the conversion of integer-valued remote sensing database into the one with interval values.

We can transform a real-valued information system $S = (O, A \cup \{d\})$ into an interval-valued information system $K = (U, A)$ by methods such as statistics, discretization, or expert opinions. Discretization may be based on experience or specification of arbitrary cut-off points. Also, whether a determination could be made to select a given discretization method based solely on the data characteristics of an attribute or a data set (Chmielewski and Grzymala-Busse 1996). Sometimes expert opinions are very useful and reliable in the identification of cut-off points demarcating the intervals (Leung and Leung 1993a, b). Statistical methods, on the other hand, may be used to capture most of the data variation under some probability density function fitting the data. A simple statistical method is that for each attribute, we only include values that fall within an interval, say $\mu \pm 2\sigma$, under a particular probability density function, say normal distribution with parameters μ and σ^2. Taking randomness into account, such an interval would be a good representation of the data since it accounts for 95.6% of the variation in the normal distribution case.

Formally, we let $(O, A \cup \{d\})$ be the information system obtained from the randomly selected training samples. We assumed that for each sample set $O_i = \{o_{ij} : j = 1, 2, \ldots, J_i\}$, and each attribute (spectral band) a_k, the gray values $\left\{v_{ij}^k : j = 1, 2, \ldots, J_i\right\}$ satisfy a normal distribution $N\left(\mu_i^k, (\sigma_i^k)^2\right)$. Such an observation is also made in studies such as Jenson (1996), Jenson and Langari (1999), and Ji (2003). Again, the principle of the transformation method can likewise be applied to other probability distributions.

We then transformed $S = (O, A \cup \{d\})$ into an interval-valued information system $K = (U, A)$, where $U = \{u_i : i = 1, 2, \ldots, 5\}$ is represented as classes (the distinct five classes) called the universe of discourse, $A = \{a_1, a_2, a_3, a_4\}$ is the attribute (spectral band) set such that $a_k(u_i) = [l_i^k, u_i^k]$, for all $i = 1, 2, \ldots, 5$ and $k = 1, 2, 3, 4$, where

$$l_i^k = \mathrm{int}\left(\max\{\mu_i^k - 2\sigma_i^k,\ 0\}\right) + 1,\ u_i^k = \mathrm{int}\left(\min\{\mu_i^k + 2\sigma_i^k,\ 255\}\right). \tag{4.42}$$

Similar method can be used in other probability distributions fitting the data. It should be noted that in the discretization method, interval-valued set $\{a_k(u_i) : u_i \in U\}$ forms a partition of a set for the same attribute a_k, but in the distribution-based statistical method, the value intervals may have non-empty intersections for distinct objects in the universe of discourse. This is rather natural, since the gray values of different objects might have rather close spectral signature in the same spectral band.

Remark 4.3. It should be pointed out that we only use the statistical method to preprocess the data, i.e., by transforming real-valued attributes into interval-valued attributes. Other than that, the rough set knowledge induction method to be discussed has nothing to do with any statistical arguments. That is, the knowledge induction process is independent of the way the intervals are formed by either the statistical method, the discretization method, or expert opinion.

4.5.4 Knowledge Discovery in Interval-Valued Information Systems

Let $K = (U, A)$ be an interval-valued information system and $B \subseteq A$. We can define a binary relation, denoted by R_B, on U as:

$$R_B = \{(u_i, u_j) \in U \times U : a_k(u_i) \cap a_k(u_j) \neq \emptyset, \forall a_k \in B\}. \tag{4.43}$$

Two classes u_i and u_j have relation R_B if and only if they cannot be separated by the attribute set B. Obviously, R_B is reflexive and symmetric, but may not be transitive. So R_B is a tolerance relation which satisfies

$$R_B = \bigcap_{b \in B} R_{\{b\}}. \tag{4.44}$$

Denote $S_B(u_i) = \{u_j \in U : (u_i, u_j) \in R_B\}$, $u_i \in U$. Then $S_B(u_i)$ is the tolerance classes of u_i with respect to R_B, $u_j \in S_B(u_i)$ if and only if u_i and u_j cannot be separated by the attribute set B.

One fundamental aspect of rough set theory involves the search for particular subsets of attributes which provide the same information for classification purposes as the full set of attributes. Such subsets are called attribute reducts. To acquire concise decision rules from the information systems, knowledge reduction is needed. Many types of attribute reducts and decision rules have been proposed in rough set research. For example, Kryszkiewicz (2001) has established static relationships among conventional types of knowledge reduction in inconsistent complete decision tables. Zhang et al. (2003a,b; 2004) have introduced a new kind of knowledge reduction called a maximum distribution reduct which preserves all maximum decision rules. Mi et al. (2004) have proposed approaches to knowledge reduction based on variable precision rough set model. Wu et al. (2005) have investigated knowledge reduction via Dempster-Shafer theory of evidence in information systems. Leung et al. (2008a) study knowledge reduction in interval-valued information systems so that optimal classification rules can be discovered.

Let $K = (U, A)$ be an interval valued-information system and $B \subseteq A$. If $R_B = R_A$, then B is referred to as a classification consistent set in K. If B is a classification consistent set in K, $B - \{b\}$ is not a classification consistent set in K for all $b \in B$, i.e., $R_{B-\{b\}} \neq R_A$, then B is called a (global) classification reduct in K. The set of all classification reducts in K is denoted by $re(K)$. The intersection of all classification reducts is called the classification core in K, the elements of which are those attributes that cannot be eliminated without introducing contradictions to the data set.

On the other hand, for $u \in U$, $B \subseteq A$, if $S_B(u) = S_A(u)$, then B is referred to as a classification consistent set of u in K. If B is a classification consistent set of u in K, $B - \{b\}$ is not a classification consistent set of u in K for all $b \in B$, i.e., $S_{B-\{b\}}(u) \neq S_A(u)$, then B is called a (local) classification reduct of u in K. The set of all classification reducts of u in K is denoted by $re(u)$. The intersection of all classification reducts of u is called the classification core of u in K.

It should be noted that, in general, a local reduct may not necessarily be included in any global reducts. However, if B is a global reduct, for any class u, there must exist a local reduct of u such that the local reduct is a subset of B, and such a local reduct can simplify a classification rule generated from the global reduct and may be of higher generalization capability.

A classification consistent set in K is a subset of the attribute set that preserves the tolerance classes of objects. A classification reduct is a minimal consistent set that preserves the tolerance relation and, consequently, leads to the same classification. The remaining attributes are then redundant, and their removal does not affect (e.g., worsen) the classification.

In what follows, we propose a Boolean reasoning method to calculate the attribute reducts by introducing a discernibility matrix.

Let $K = (U, A)$ be an interval-valued information system. Denote

$D_{ij} = \{a_k \in A : a_k(u_i) \cap a_k(u_j) = \emptyset\}$, $i \neq j$, and $D_{ii} = \emptyset$ for all $i = 1, 2, \ldots, I$.

The term D_{ij} is called the discernibility set of classes u_i and u_j in K, containing attributes separating classes u_i and u_j. Denote $\mathbf{M} = (D_{ij} : i, j = 1, 2, \ldots, I)$. $\mathbf{M}$ is referred to as the discernibility matrix of K. Let $\mathrm{M}_0 = \{D_{ij} : D_{ij} \neq \emptyset\}$.

Theorem 4.1. *(Judgment Theorem). Let $K = (U, A)$ be an interval-valued information system. Then $B \subseteq A$ is a classification consistent set in K, i.e., $R_B = R_A$, iff $B \cap D \neq \emptyset$, $\forall D \in \mathrm{M}_0$ (See Leung et al. (2007) for the proof).*

According to Theorem 3.1, $B \subseteq A$ is a classification reduct in K iff B is the minimal set satisfying $B \cap D \neq \emptyset$, $\forall D \in \mathrm{M}_0$. Since reducts are not unique, it is useful to identify the core (attribute(s)) common to all reducts. It contains attribute (s) that is essential to the classification rules, and the classification result will be significantly affected without it. The following theorem stipulates the criteria for the identification of the classification core in interval-valued information systems.

Theorem 4.2. *Let $K = (U,A)$ be an interval-valued information system. Then $a_k \in A$ is an element of classification core in K iff there exists $D \in \mathrm{M}_0$ such that $D = \{a_k\}$. (See Leung et al. (2007) for the proof).*

Reduct computation can be translated into the computation of prime implicants of a Boolean function. It was shown in (Skowron and Rauszer 1992) that the problem of finding reducts of a given Pawlak (complete) information system may be solved as a case in Boolean reasoning. The idea of Boolean reasoning is to represent a problem with a Boolean function and to interpret its prime implicants (an implicant of a Boolean function f is any conjunction of literals (variables or their negations) such that for each valuation v of variables, if the values of these literals are true under v then the value of the function f under v is also true; a prime implicant is a minimal implicant) as solutions to the problem. This is a useful approach to the calculation of the reducts of classical information systems. It can be generalized to the interval-valued information systems. It should be pointed out that we are interested in implicants of monotone Boolean functions only, i.e., functions constructed without negation.

Let $K = (U,A)$ be an interval-valued information system. A discernibility function f_K for the system K is a Boolean function of m Boolean variables $\overline{a}_1, \overline{a}_2, \ldots, \overline{a}_m$ corresponding to the attributes $a_1, a_2, \ldots, a_m$, respectively, and is defined as follows:

$$f_K(\bar{a}_1, \bar{a}_2 \ldots, \bar{a}_m) = \wedge\{\vee D_{ij} : D_{ij} \in M_0\}, \tag{4.44}$$

where $\vee D_{ij}$ is the disjunction of all variables $\overline{a}$ such that $a \in D_{ij}$.

Theorem 4.3. Let *$K = (U,A)$ be an interval-valued information system. Then an attribute subset $B \subseteq A$ is a classification reduct in K iff $\bigwedge_{a_k \in B} \overline{a_k}$ is a prime implicant of the discernibility function f_K (See Leung et al. (2007) for the proof).*

Without causing any confusion, we shall write a_k instead of $\overline{a_k}$ in the discussion to follow.

If we instead construct a Boolean function by restricting the conjunction to run over only column i (instead of over all columns) in the discernibility matrix, we obtain the so-called i discernibility function, denoted by f_i, that is,

$$f_i(\overline{a}_1, \overline{a}_2, \ldots, \overline{a}_m) = \bigwedge_{\{j: D_{ij} \in \mathrm{M}_0\}} (\vee D_{ij}). \tag{4.45}$$

The set of all prime implicants of function f_i determines the set of all (local) classification reducts of u_i in K. These classification reducts reveal the minimum amount of information needed to discern the class u_i from all other classes which are not included in the tolerance classes of u_i. This can be summarized in the following theorem:

Theorem 4.4. *Let $K = (U, A)$ be an interval-valued information system, $u_i \in U$. Then an attribute subset $B \subseteq A$ is a classification reduct of u_i in K iff $\bigwedge_{a_k \in B} \overline{a_k}$ is a prime implicant of the discernibility function f_i.*

After a classification reduct B has been calculated, classification knowledge hidden in an interval-valued information system may be discovered and expressed in the form of classification rules as follows:

If $a_k(x) \in [l_i^k, u_i^k]$, for all $a_k \in B$, then the sample x should be classified into the class u_i.

4.5.5 Discovery of Classification Rules for Remotely Sensed Data

A SPOT-4 multispectral data acquired on November 22, 2000 in the northwestern part of Hong Kong are used in the experiment. The data were acquired in four multispectral bands (green, red, near-infrared and short wave infrared) at 20 m spatial resolution. A 256×256 image was extracted covering the Maipo Ramsar Wetland site. The Ramsar Site covers an area of 1,500 ha and is rich in both flora and fauna (Fung 2003). It hosts over 50,000 migrant birds annually with 340 different bird species among which 23 are rare species. Fringing the Deep Bay coastline, mangrove and mudflats form the major conservation foci in the Ramsar site. Further inland are fish ponds and shrimp ponds (named gei wai locally) which are noted as artificial wetland. Other than the natural landscape, low density residential estates and the Yuen Long industrial estate form the major urban land covers. In their study, Leung et al. (2007) use five general land covers to test and illustrate the effectiveness of rough set concepts in classification. The land covers are water, vegetation, mudflat, residential land and industrial land.

The pedagogical experiment demonstrates the capability of the interval-valued rough set method in the discovery from remotely sensed data optimal spectral bands and optimal rule set necessary and sufficient for the classification of land covers. The method is also capable of discovering "the" spectral band(s) discerning certain classes.

Basing on field experience aided by high resolution aerial photographs, two sets of independent samples are extracted in the experiment. The first set is used for training purpose with each class comprises of 30–60 independent pixels randomly selected (Table 4.17). The second set is used for testing and the number of independent samples ranges from 30 to 36 (Table 4.18).

Table 4.19 depicts the interval-valued information system transformed from the original data matrix summarized in Table 4.17. The term $U = \{u_1, u_2, \ldots, u_5\}$ is the universe of discourse containing the five land covers, and $A = \{a_1, a_2, a_3, a_4\}$ is the set of four attributes (e.g., spectral bands), with each of its attribute value $a_k(u_j)$ being an interval obtained by (4.42). In this case, only those values that fall within $\mu \pm 2\sigma$ under the density function are included.

Table 4.17 A description of the training samples

Land cover	No. of samples	Spectral band															
		Green (a_1)				Red (a_2)				NIR (a_3)				SWIR (a_4)			
		mean	variance	min	max	mean	variance	min	max	mean	variance	min	max	mean	variance	min	max
Water (u_1)	60	68.45	20.05	60	77	56.13	37.13	44	68	24.25	8.70	19	30	10.73	8.37	5	16
Mudflat (u_2)	60	79.22	1.43	77	81	77.13	3.64	74	80	56.47	4.80	53	60	12.17	13.60	5	19
Residential land (u_3)	60	85.02	36.08	74	97	90.40	171.36	65	116	84.92	129.54	63	107	81.57	172.32	56	107
Industrial land (u_4)	30	146.20	2300.79	51	242	174.73	2590.20	73	276	156.30	597.94	108	205	154.23	631.22	104	204
Vegetation (u_5)	60	55.60	1.26	54	57	40.05	1.71	38	42	116.38	133.66	94	139	34.95	10.90	29	41

Table 4.18 A description of the test samples

Land cover	No. of samples	Spectral band															
		Green (a_1)				Red (a_2)				NIR (a_3)				SWIR (a_4)			
		mean	variance	min	max	mean	variance	min	max	mean	variance	min	max	mean	variance	min	max
water	35	65.94	12.35	59	72	52.20	13.11	45	59	24.37	24.83	15	34	11.20	7.05	6	16
mudflat	34	78.74	4.38	75	82	76.35	9.69	71	82	56.82	12.82	50	63	12.79	18.23	5	21
residential land	33	82.88	46.61	70	96	82.91	72.27	66	99	78.73	42.83	66	91	75.36	29.49	65	86
industrial land	30	135.47	2034.60	46	225	156.70	2648.01	54	259	139.07	1441.38	64	215	135.77	984.19	74	198
vegetation	36	56.08	1.91	54	58	40.42	3.34	37	44	115.97	159.28	91	141	35.28	10.66	29	41

Here, the gray value $a_k(u_i)$ is a positive integer between 0 and 255 for each $i = 1, 2, \ldots, 5$ and $k = 1, 2, 3, 4$, and none of the $\max\{\mu_i^k - 2\sigma_i^k\}$, $i = 1, 2, \ldots, 5$, $k = 1, 2, 3, 4$, is an integer. The integer function int(x)in (4.42), for example int $(6.56) = 6$, needs to be employed. We can observe that $a_k(u_i) \in [\max\{\mu_i^k - 2\sigma_i^k, 0\}, \min\{\mu_i^k + 2\sigma_i^k, 255\}]$ if and only if $a_k(u_i) \in [l_i^k, u_i^k]$. Hence, the raw data set is transformed into an interval-valued information system as shown in Table 4.19.

Accordingly, the discernibility sets can be obtained (Table 4.20). Since $D_{ij} = D_{ji}$, for simplicity, we only list D_{ij}'s with $1 \le j < i < I$. By Theorems 3.1 and 3.3, we obtain the Boolean function:

$$f_K(a_1, a_2, a_3, a_4) = (a_2 \vee a_3) \wedge (a_3 \vee a_4) \wedge (a_2 \vee a_3 \vee a_4) \wedge (a_1 \vee a_2 \vee a_3 \vee a_4) \\ \wedge a_3 \wedge (a_2 \vee a_4) \wedge (a_1 \vee a_2 \vee a_4).$$

After simplification (using the absorption laws), we obtain the prime implicants representation of the Boolean function:

$$f_K(a_1, a_2, a_3, a_4) = (a_2 \vee a_4) \wedge a_3 = (a_2 \wedge a_3) \vee (a_3 \wedge a_4).$$

Hence there are two classification reducts in the system: $B_1 = \{a_2\,(\text{red}),'' a_3\,(\text{NIR})\}$ and $B_2 = \{a_3\,(\text{NIR}), a_4\,(\text{SWIR})\}$, and the classification core is $\{a_3\}$. The remaining attribute a_1 is then redundant, and its removal does not worsen the classification.

Therefore, to obtain the classification rules that discriminate one class from the others, at most two bands, $\{a_2, a_3\}$ or $\{a_3, a_4\}$, are necessary. That means, the proposed method reduces the number of spectral bands (attributes) by 50%. The

Table 4.19 An interval-valued information system

U	a_1	a_2	a_3	a_4
u_1	[60, 77]	[44, 68]	[19, 30]	[5, 16]
u_2	[77, 81]	[74, 80]	[53, 60]	[5, 19]
u_3	[74, 97]	[65, 116]	[63, 107]	[56, 107]
u_4	[51, 242]	[73, 276]	[108, 205]	[104, 204]
u_5	[54, 57]	[38, 42]	[94, 139]	[29, 41]

Note For simplicity, attributes are coded as a_k, $k = 1, 2, \ldots, 4$, and classes are coded as u_j, $j = 1, 2, \ldots, 5$ in here

Table 4.20 Discernibility set

	u_1	u_2	u_3	u_4	u_5
u_1					
u_2	a_2a_3				
u_3	a_3a_4	a_3a_4			
u_4	$a_2a_3a_4$	a_3a_4	a_3		
u_5	A	A	$a_1a_2a_4$	a_2a_4	

green band (a_1), does not appear in any reduct. Since the green band (a_1) and the red band (a_2) have a very high correlation coefficient of 0.96, they are more or less identical in information content. Thus, only one of them is needed in classification and removal of the green band will not worsen the classification. The two reducts share a common spectral band a_3 (the near-infrared band), the classification core by Theorem 3.2. It demonstrates the importance of the near-infrared band for delineating land from water, and vegetation from non-vegetation land covers. Its elimination will affect the classification results significantly. Therefore, the proposed method manages to identify which spectral band is necessary and which spectral band is redundant for a classification task. It produces a sound result for feature selection highlighting the discriminatory power in different combinations of spectral bands (or attributes). It also sheds light on the use of appropriate spectral band (s) in each level of a hierarchical classification should such a procedure be preferable. That is, we may, for example, want to use a particular band to separate major land covers first, and then use relevant band(s) to separate sub-covers.

To obtain the classification reducts for each individual land cover, we can obtain the Boolean function with respect to u_i for $i = 1, 2, \ldots, 5$, and then obtain the classification reduct of each class (local reduct) as follows:

Since

$$\begin{aligned} f_1(a_1, a_2, a_3, a_4) &= (a_2 \vee a_3) \wedge (a_3 \vee a_4) \wedge (a_2 \vee a_3 \vee a_4) \wedge (a_1 \vee a_2 \vee a_3 \vee a_4) \\ &= (a_2 \vee a_3) \wedge (a_3 \vee a_4) = (a_2 \wedge a_4) \vee a_3, \end{aligned}$$

$\{a_3\}$ and $\{a_2, a_4\}$ are the classification reducts of water(u_1), i.e., $re(u_1) = \{\{a_3\}, \{a_2, a_4\}\}$.

Similarly, the classification reducts for mudflat (u_2), residential land (u_3), industrial land (u_4), and vegetation (u_5) are, respectively:

$$f_2(a_1, a_2, a_3, a_4) = (a_2 \wedge a_4) \vee a_3, \ re(u_2) = \{\{a_3\}, \{a_2, a_4\}\}.$$

$$\begin{aligned} f_3(a_1, a_2, a_3, a_4) &= (a_1 \wedge a_3) \vee (a_2 \wedge a_3) \vee (a_3 \wedge a_4), \\ re(u_3) &= \{\{a_1, a_3\}, \{a_2, a_3\}, \{a_3, a_4\}\}. \end{aligned}$$

$$f_4(a_1, a_2, a_3, a_4) = (a_2 \wedge a_3) \vee (a_3 \wedge a_4), \ re(u_4) = \{\{a_2, a_3\}, \{a_3, a_4\}\}.$$

$$f_5(a_1, a_2, a_3, a_4) = a_2 \vee a_4, \ re(u_5) = \{\{a_2\}, \{a_4\}\}.$$

We can see that the local attribute reducts $\{a_1, a_3\}$ and $\{a_2, a_4\}$ are not included in any global reduct. Based on the classification reduct of each class, all classification rules hidden in the interval-valued information system can be discovered and expressed as follows:

$$\mathrm{r}_1 : \text{If } a_3(x) \in [19, 30], \text{ then } x \in u_1.$$

$$r_1' : \text{If } a_2(x) \in [44, 68] \text{ and } a_4(x) \in [5, 16], \text{ then} \quad x \in u_1.$$

$$r_2 : \text{If } a_3(x) \in [53, 60], \text{ then} \quad x \in u_2.$$

$$r_2' : \text{If } a_2(x) \in [74, 80] \text{ and } a_4(x) \in [5, 19], \text{ then} \quad x \in u_2.$$

$$r_3 : \textit{If } a_1(x) \in [74, 97] \text{ and } a_3(x) \in [63, 107], \text{ then} \quad x \in u_3.$$

$$r_3' : \text{If } a_2(x) \in [65, 116] \text{ and } a_3(x) \in [63, 107], \text{ then} \quad x \in u_3.$$

$$r_3'' : \text{If } a_3(x) \in [63, 107] \text{ and } a_4(x) \in [56, 107], \text{ then} \quad x \in u_3.$$

$$r_4 : \text{If } a_2(x) \in [73, 276] \text{ and } a_3(x) \in [108, 205], \text{ then} \quad x \in u_4.$$

$$r_4' : \text{If } a_3(x) \in [108,\ 205] \text{ and } a_4(x) \in [104, 204], \text{ then} \quad x \in u_4.$$

$$r_5 : \text{If } a_2(x) \in [38, 42], \text{ then} \quad x \in u_5.$$

$$r_5 : \text{If } a_4(x) \in [29, 41], \text{ then} \quad x \in u_5.$$

This is actually the answer to the problem involving Table 1.1.

While classification rules are derived from a sample data set, an independent set of samples is used as reference data for accuracy verification to test the effectiveness of the proposed rough set method. The composition of error matrix helps generate standard accuracy indices, including producer's accuracy and user's accuracy for individual classes, as well as overall accuracy and Kappa coefficient of agreement for the entire data set (Congalton and Green 1999).

Since the two classification reducts $\{a_2, a_3\}$ and $\{a_3, a_4\}$ all have two spectral bands, to search for the optimal reduct, we have to compare the overall accuracies of classification. From the classification reduct $\{a_2, a_3\}$, and from using the local reducts, we obtain five classification rules: r_1, r_2, r_3', r_4, r_5. Similarly, another five classification rules: r_1, r_2, r_3'', r_4', r_5' are obtained from using the classification reduct $\{a_3, a_4\}$ and the corresponding local reducts. The corresponding error matrices, user's accuracies, producer's accuracies, overall accuracies, and $\hat{K}$ value of the two classifications for the training samples are depicted in Tables 4.21 and 4.22, respectively. Since the overall accuracy of the first classification (0.944) is greater than that of the second one (0.896), so is the $\hat{K}$ value, we can assert that the spectral band set $\{a_2, a_3\}$ is the optimal reduct, and the optimal classification rules are: r_1, r_2, r_3', r_4, r_5. The combination of red and NIR bands tend to provide a good result with all classes having both the producer's and user's accuracies greater than 0.90. Only mudflat and vegetation have user's accuracy less than 0.95 with four samples being unrecognizable with the classification rules. The overall accuracy of classification corresponding to the optimal reduct for the training samples is 0.944.

The corresponding results for the test samples with respect to the two classifications are summarized in Tables 4.23 and 4.24, respectively. Again, most classes

Table 4.21 Classification accuracy from applying classification reduct $B_1 = \{a_2, a_3\}$ and five rules: r_1, r_2, r_3', r_4, r_5 to the training samples

Training samples	Water	Mudflat	Resident	Industry	Vegetation	Unrecognizable	User accuracy
Water (60samples)	57	0	0	0	0	3	57/60 = 0.95
Mudflat (60samples)	0	55	1	0	0	4	55/60 = 0.917
Resident (60samples)	0	0	58	2	0	0	58/60 = 0.967
Industry (30samples)	0	0	0	29	0	1	29/30 = 0.967
Vegetation (60samples)	0	0	0	0	56	4	56/60 = 0.933
Producer accuracy	57/57 = 1.0	55/55 = 1.0	58/59 = 0.983	29/31 = 0.935	56/56 = 1.0	Overall accuracy = 255/270 = 0.944	$\hat{K}$ = 0.931351

Table 4.22 Classification accuracy from applying classification reduct $B_2 = \{a_3, a_4\}$ and five rules: r_1, r_2, r_3'', r_4', r_5' to training samples

Training samples	Water	Mudflat	Resident	Industry	Vegetation	Unrecognizable	User accuracy
Water (60samples)	57	0	0	0	0	3	57/60 = 0.95
Mudflat (60samples)	0	55	0	0	0	5	55/60 = 0.917
Resident (60samples)	0	0	56	4	0	0	56/60 = 0.933
Industry (30samples)	0	0	0	16	0	14	16/30 = 0.533
Vegetation (60samples)	0	0	0	0	58	2	58/60 = 0.967
Producer accuracy	57/57 = 1.0	55/55 = 1.0	56/56 = 1.0	16/20 = 0.80	58/58 = 1.0	Overall accuracy = 242/270 = 0.896	$\hat{K}$ = 0.873116

Table 4.23 Classification accuracy from applying classification reduct $B_1 = \{a_2, a_3\}$ and five rules: r_1, r_2, r_3', r_4, r_5 to the test samples

Test samples	Water	Mudflat	Resident	Industry	Vegetation	Unrecognizable	User accuracy
Water (35samples)	30	0	0	0	0	5	30/35 = 0.857
Mudflat (34samples)	0	28	1	0	0	5	28/34 = 0.824
Rresident (33samples)	0	0	33	0	0	0	33/33 = 1.0
Industry (30samples)	0	0	7	20	0	3	20/30 = 0.667
Vegetation (36samples)	0	0	0	0	33	3	33/36 = 0.917
Producer accuracy	30/30 = 1.0	28/28 = 1.0	33/41 = 0.805	20/20 = 1.0	33/33 = 1.0	Overall accuracy = 144/168 = 0.857 $\hat{K}$ = 0.828644	

Table 4.24 Classification accuracy from applying classification reduct $B_2 = \{a_3, a_4\}$ and five rules: r_1, r_2, r_3'', r_4', r_5' to the test samples

Test samples	Water	Mudflat	Resident	Industry	Vegetation	Unrecognizable	User accuracy
Water (35samples)	30	0	0	0	0	5	30/35 = 0.857
Mudflat (34samples)	0	28	0	0	0	6	28/34 = 0.824
Resident (33samples)	0	0	33	0	0	0	33/33 = 1.0
Industry (30samples)	0	0	6	13	0	11	13/30 = 0.433
Vegetation (36samples)	0	0	0	0	33	3	33/36 = 0.917
Producer accuracy	30/30 = 1.0	28/28 = 1.0	33/39 = 0.846	13/13 = 1.0	33/33 = 1.0	overall accuracy = 137/168 = 0.815 $\hat{K}$ = 0. 782247	

have their user's and producer's accuracies greater than 0.80. Only industrial land has a relatively poor user's accuracy (0.667 for $B_2 = \{a_2, a_3\}$ and 0.433 for $B_2 = \{a_3, a_4\}$) showing confusion with residential land. Clearly, these results show that the classification derived from the reduct $\{a_2, a_3\}$ (overall accuracy of 0.857) is more efficient and effective than the one derived from reduct $\{a_3, a_4\}$ (overall accuracy being 0.811). Also, the result shows that the generalization of our method is reasonably good, or alternatively, the situation of over-generalization will not occur.

One might suspect that a higher level of accuracy could be achieved if more spectral bands are employed to classify images. For the present application, if we use three spectral bands, namely red, NIR and SWIR, i.e., $\{a_2, a_3, a_4\}$, for classification, we can generate ten classification rules (i.e., r_1, r_1', r_2, r_2', r_3', r_3'', r_4, r_4', r_5, r_5') on the basis of local reducts. The results obtained from applying them to the training and test samples are summarized in Tables 4.25 and 4.26, respectively. It should be pointed out that the green band (band a_1) is again redundant, so rule r_3 is omitted (we have checked that the overall accuracies of the classifications corresponding, respectively, to the ten rules (i.e., r_1, r_1', r_2, r_2', r_3', r_3'', r_4, r_4', r_5, r_5') and the eleven rules (i.e., r_1, r_1', r_2, r_2', r_3, r_3', r_3'', r_4, r_4', r_5, r_5') are the same). Compared to the results obtained from applying the classification rules derived from the optimal classification reduct (Tables 4.23 and 4.24), the improvement in overall accuracy are only 1.9 and 5.4%, with reference to training and test samples, respectively. Confusion between industrial and residential land still remains. In this regard, less confusion is found in the two band $\{a_2, a_3\}$ classification. It means that, comparing to the use of the whole set of attributes (original spectral bands), if we use the optimal reduct for classification, the decrease in classification accuracy is rather small. In other words, the loss of information is almost negligible by using only spectral bands that really matter. It should further be pointed out that, if we only want to discern the special class of "water" (respectively, mudflat, vegetation) from other classes, one and only one band is sufficient. That is, a further parsimony in the use of spectral bands is achieved. Such discriminatory power of the proposed approach will prove to be important in knowledge discovery in hyperspectral data. Under that situation, our ability to minimize the number of spectral bands used becomes pertinent.

Remark 4.4. A general framework for the discovery of classification rules in real-valued or integer-valued information system has been introduced in this section. Particular emphasis has been placed on the analysis of remotely sensed data which are integer-valued in nature. The approach involves the transformation of real-valued or integer-valued decision table into interval-valued information system in the data preprocessing step and the construction of a rough-set based knowledge induction procedure to discover rules necessary and sufficient for a classification task. I have also introduced several useful concepts such as local and global classification reducts as well as classification core pertinent to data analysis in interval-valued information systems. A method by Boolean functions to compute the classification reducts in the interval-valued information system has also been

Table 4.25 Classification accuracy from applying ten rules and three bands (a_2, a_3, a_4) to the training samples

Training samples	Water	Mudflat	Resident	Industry	Vegetation	Unrecognizable	User accuracy
Water (60samples)	59	0	0	0	0	1	59/60 = 0.983
Mudflat (60samples)	0	56	1	0	0	3	56/60 = 0.933
Resident (60samples)	0	0	56	4	0	0	56/60 = 0.933
Industry (30samples)	0	0	0	29	0	1	29/30 = 0.967
Vegetation (60samples)	0	0	0	0	60	0	60/60 = 1.0
Producer accuracy	59/59 = 1.0	56/56 = 1.0	56/57 = 0.982	29/33 = 0.879	60/60 = 1	Overall accuracy = 260/270 = 0.963	
						$\hat{K} = 0.959081$	

Table 4.26 Classification accuracy from applying ten rules and three bands (a_2, a_3, a_4) to the test samples

Test samples	Water	Mudflat	Resident	Industry	Vegetation	Unrecognizable	User accuracy
Water (35 samples)	35	0	0	0	0	0	35/35 = 1.0
Mudflat (34 samples)	0	29	1	0	0	4	29/34 = 0.853
Resident (33 samples)	0	0	33	0	0	0	33/33 = 1.0
Industry (30 samples)	0	0	8	20	0	2	20/30 = 0.667
Vegetation (36 samples)	0	0	0	0	36	0	36/36 = 1.0
Producer accuracy	35/35 = 1.0	29/29 = 1.0	33/42 = 0.786	20/20 = 1.0	36/36 = 1.0	overall accuracy = 153/168 = 0.911	
						$\hat{K} = 0.889894$	

proposed. Theoretical analysis and the real-life experiment all show that the proposed approach is effective in discovering classification rules hidden in remotely sensed data. It is also instrumental in dimension reduction by unraveling the minimal number of features (spectral bands) and the optimal number of rules for a classification task. Furthermore, critical features for differentiating specific classes can also be discovered. Such ability can facilitate the orderly use of relevant features to classify remotely sensed data in a hierarchical manner. We, for example, can use only a few key spectral bands to classify broad types and then use other relevant spectral bands to classify subtypes. All these spectral bands can automatically be discovered from the data by the proposed method.

Though the emphasis has been placed on knowledge discovery in remotely sensed data, the proposed approach is general enough to mine knowledge from any real-valued or integer-valued spatial information system. As aforementioned, Pawlak's rough set model is essentially catered for qualitative data. It is ineffective and inefficient in analyzing quantitative, e.g., real-valued and or integer-valued, data commonly encountered in real-life problems. The extension of the conventional rough set model by Leung et al. (2007, 2008a) has greatly extended its applicability. Furthermore, it has built a basis for knowledge discovery in mixed, e.g., qualitative and quantitative, databases. As is well-known, the use of spectral signatures alone is not sufficient to classify complex remotely sensed images, rich high order image characteristics such as shape, shadow, size, texture, pattern, site and association should be used to perform classification with higher level of accuracy. The rough set approach to knowledge discovery in such a mixture of qualitative and quantitative information deserves further investigation. Moreover, the dimension reduction capability of the proposed method will be very useful in the analysis of hyperspectral data. All of these problems can be solved by further extending the current framework in future studies.

4.5.6 Classification of Tree Species with Hyperspectral Data

The rough set approach has been employed to classify 15 tree species with hyperspectral data (Leung et al. 2007). There were totally 689 bands within the 400–900 nm region in the experiment. Fifteen tree species commonly found in Hong Kong were selected for the study. They are listed as follows: *Acacia confusa* (u_1), *Araucaria heterophylla* (u_2), *Acacia mangium* (u_3), *Bauhinia variegata* (u_4), *Cinnamomum camphora* (u_5), *Casuarina equisetifolia* (u_6), *Aleurites moluccana* (u_7), *Ficus microcarpa* (u_8), *Firmiana simplex* (u_9), *Ficus variegata* (u_{10}), *Hibiscus tiliaceus* (u_{11}), *Melaleuca quanqueenervia* (u_{12}), *Pinus elliottii* (u_{13}), *Schima superba* (u_{14}), *Sapium sebiferum* (u_{15}).

For each type of tree, 36 sample spectra were taken in the laboratory. They are separated randomly into two independent sets. The first set was used for training purpose with each class comprises of 18 independent samples. The second set was used for testing and the number of samples is 18 also.

By transforming the original information system into the interval-valued information system and by applying the methods discussed in the previous subsections, a reduct $B = \{a_{39}, a_{56}, a_{89}, a_{107}, a_{164}, a_{203}, a_{295}, a_{336}, a_{368}, a_{377}, a_{412}, a_{420}, a_{434}, a_{452}, a_{540}\}$ containing 15 spectral bands was obtained from the interval-valued information system (Table 4.27).

Result of this feature selection process selects four blue bands, two green bands (including the green peak at 550 nm), four red band, four bands along the red edge and 1 NIR band. The four bands selected along the red edge echo earlier work that these bands possess strong discriminatory power for tree species identification (Fung et al. 2003).

The reduct can be used to obtain the classification rules which discriminate the tree species from each other. That means the proposed method significantly reduces the number of spectral bands (attributes) by 97.8%. It gives a sound result for feature selection by highlighting the discriminatory power in different combinations of the spectral bands (or attributes).

While classification rules are derived from a sample data set, an independent set of samples is used as reference data for accuracy verification to test the effectiveness of the proposed rough set method. The corresponding error matrices, user's accuracies, overall accuracies of the classifications for the training samples and the test samples are depicted in Table 4.28.

Comparing to the use of the whole set of attributes (original 689 spectral bands), it is noticed that the decrease in classification accuracy is rather small when the reduct (15 spectral bands) is employed. In other words, the loss of information is almost negligible if we only use the spectral bands that really matter. This experiment demonstrates that the proposed approach significantly minimizes the number of spectral bands necessary for a classification task.

Table 4.27 Spectral bands selected for classification

Spectral band	Description
a_{39} = 428.71 nm	Blue band
a_{56} = 441.36 nm	
a_{89} = 465.87 nm	
a_{107}= 479.2 nm	
a_{164}= 521.31nm	Green band
a_{203}= 550 nm	
a_{295}= 617.32nm	Red band
a_{336}= 647.16 nm	
a_{368}= 669.65 nm	
a_{377}= 676.89 nm	
a_{412}= 701.47nm	Red edge
a_{420}= 707.96nm	
a_{434}= 717.34 nm	
a_{452}= 731.01 nm	
a_{540}= 793.37nm	Near infrared band

Table 4.28 Comparison of classification accuracies from applying classification reduct B to the training and test tree samples

U	No. of training samples	No. of training samples correctly identified	Training accuracy	No. of test samples	No. of test samples correctly identified	Test accuracy
u_1	18	12	0.666667	18	13	0.722222
u_2	18	16	0.888889	18	14	0.777778
u_3	18	15	0.833333	18	15	0.833333
u_4	18	16	0.888889	18	16	0.888889
u_5	18	16	0.888889	18	16	0.888889
u_6	18	17	0.944444	18	15	0.833333
u_7	18	15	0.833333	18	15	0.833333
u_8	18	17	0.944444	18	14	0.777778
u_9	18	16	0.888889	18	14	0.777778
u_{10}	18	14	0.777778	18	14	0.777778
u_{11}	18	16	0.888889	18	13	0.722222
u_{12}	18	17	0.944444	18	16	0.888889
u_{13}	18	14	0.777778	18	14	0.777778
u_{14}	18	16	0.888889	18	15	0.833333
u_{15}	18	16	0.888889	18	15	0.833333
Overall	270	233	0.862963	270	219	0.811111

4.6 A Vision-Based Approach to Spatial Classification

4.6.1 On Scale and Noise in Spatial Data Classification

Robustness and model selection are problems surrounding many of the classification methods discussed so far. Most algorithms are very sensitive to sample neatness, i.e., low tolerance of noise and/or outliers, and heavy dependence on the tuning of model parameters. The neural network approach generally needs a parametric network topology. The evolutionary approach usually depends on genetic operators with subjectively selected probabilities. The statistical approach often relies on some kinds of assumptions about the probability distribution of data, and the fuzzy sets approach generally needs the notion of a membership function. However, the selection and specification of all these models and parameters often lack general rules. Additionally, none of the algorithms have explicit considerations of scale which is important in spatial analysis in general and image classification in particular. Almost all of the classification methods operate on a fixed scale defined by the spatial resolution of the data. Although there are studies (Atkinson and Curran 1997; Ferro and Warner 2002) on how information and accuracy vary in scale by making use of a collection of images acquired from satellite sensors of different resolutions, the classification algorithms per se do not address this problem.

Similar to the discovery of natural clusters, human beings, with natural coordination of eyes and brains, are excellent classifiers of objects/data. Classification, from the physiological point of view, may then be modeled after our senses and perception. Thus, it might be beneficial to mimic how our eyes and brains sense and perceive objects/data in order to come up with an efficient and effective classification method. In this classification, scale becomes a natural parameter and the algorithm can automatically select the right scale for specific information. Unlike existing methods, algorithms thus derived will be less mechanical.

According to physiological experiments, human can sense and perceive the change of light. In the retina, there are only three types of cell responses (Coren et al. 1994). The "ON" response is the response to the arrival of a light stimulus. The "OFF" response is the response to the removal of a light stimulus, and the "ON-OFF" response is the response to the hybrids of "ON" and "OFF" (because both presentation and removal of the stimulus may simultaneously occur). For a single small spot of light (a fixed light spot), if it causes "ON"/"OFF" response on the retina, the cells in the retina with "ON"/"OFF" response forms a Gaussian-like region (called "ON"/"OFF" region) and all cells outside of the region then forms an "OFF"/"ON" region. Consequently, for multiple spots of lights, different "ON," "OFF," and "ON-OFF" regions may coexist in the retina. Particularly, the "ON-OFF" region intuitively forms narrow boundary between "ON" region and "OFF" region. By treating a multidimensional data point as a light source, we can thus develop a vision-based classification method which identifies classes through the analysis of the blurring process of the light sources along a scale (Fig. 4.20).

The advantages of the proposed method are: (1) the explicit consideration of scale in image classification; (2) the physiological basis for classification and its interpretation; (3) free of assumption about the distribution of the underlying data; (4) computationally stable and robust algorithm for noisy data; and (5) efficient for high dimensional image classification with very large training data set.

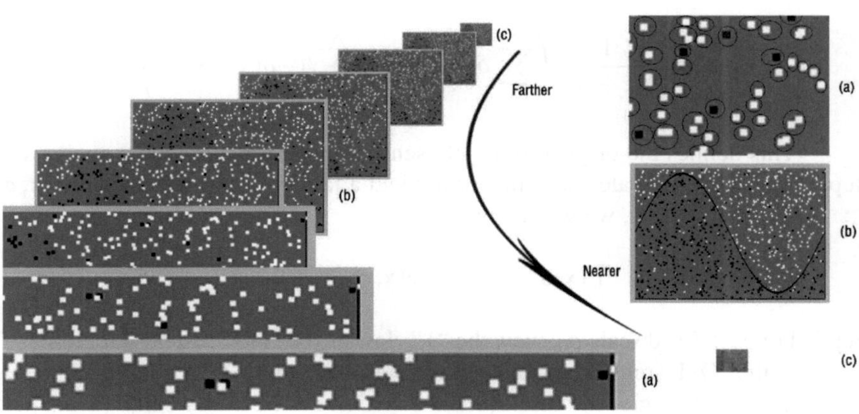

Fig. 4.20 Discovery of the optimal discriminant function through a blurring process. (**a**) Observing the data set from a very close distance, a discriminant function consisting of the disconnected circles surrounding each datum is perceived. (**b**) Observing the data set from a proper distance, a discriminant function that optimally compromises approximation and generalization performance is perceived. (**c**) Observing the data set from far away, no discriminant function is perceived

4.6.2 The Vision-Based Classification Method

Based on the study of Meng and Xu (2006), the vision-based classification method, without loss of generality, is proposed by Fung et al. (2007) for the classification of remotely sensed images. The fundamental mechanism on which the vision-based classification method models after is the blurring of images on the retina of human eyes at different scales. An existing model that captures such a process is the scale space theory (Witkin 1983, 1984; Koenderink 1984; Hummel and Moniot 1989; Leung et al. 2000a). As discussed in Chap. 2, in scale space theory, an n-dimensional scale-space image, given by a mapping $p(\mathbf{x}) = \mathbf{R}^n \to \mathbf{R}$, can be embedded into a continuous family $P(\mathbf{x}, \sigma)$ of gradually smoother versions of it. The original image corresponds to the scale $\sigma = 0$ and increasing the scale (can be interpreted as the distance between our eyes and an object) should simplify the image without creating spurious structures. If there are no prior assumptions that are specific to the image, then the image can be blurred in a unique and sensible way in which the scale space, $P(\mathbf{x}, \sigma)$, is the convolution of $p(\mathbf{x})$ with the Gaussian kernel, (2.1), obeying to the heat diffusion equation, (2.3). Since any multiple-label classification problem can be directly deduced to a series of two-label problems, it is sufficient to explain the vision-based method for the two-label classification case.

Given a two-label training data set $D = \{\mathbf{x}_i{}^+\}_{i=1}^{N_+} \cup \{\mathbf{x}_i^-\}_{i=1}^{N_-}$ that is generated from an unknown but fixed distribution. The proposed vision-based classification method determines a discriminant rule through the following steps:

Step 1. To view every positive (negative) training samples $\mathbf{x}_i^+ (\mathbf{x}_i^-)$as a spot light with unit strength $\delta(\mathbf{x} - \mathbf{x}_i)(-\delta(\mathbf{x} - \mathbf{x}_i))$, causing an "ON"("OFF") response in the retina (where $\delta(\mathbf{x})$ is a Dirac function). Consequently, all the data form an image

$$p(\mathbf{x}) = \frac{1}{N_+ + N_-}\left(\sum_{i=1}^{N_+} \delta(\mathbf{x} - \mathbf{x}_i^+) + \sum_{i=1}^{N_-} (-\delta(\mathbf{x} - \mathbf{x}_i^-))\right). \tag{4.46}$$

This defines the original remotely sensed image for subsequent analysis.

Step 2. To apply the scale space theory to yield a family of blurred images $P(\mathbf{x}, \sigma)$ of $p(\mathbf{x})$. That is, we will define

$$P(\mathbf{x}, \sigma) = p(\mathbf{x}) * g(\mathbf{x}, \sigma), \sigma \geq 0. \tag{4.47}$$

Step 3. For each fixed scale σ_0, view the "+" class as the "ON" response region, "$-$" as the "OFF" region, and the boundary as the "ON-OFF" region in the retina. Correspondingly, the discriminant function is defined by $\text{sgn}(P(\mathbf{x}, \sigma_0))$and the classification boundary is defined by $\Gamma = \{\mathbf{x} : P(\mathbf{x}, \sigma_0) = 0\}$.

The method will generate a family of discriminant functions $\{\text{sgn}(P(\mathbf{x}, \sigma_0)) : \sigma_0 \geq 0\}$. According to the visual sensation and perception principle, there should be an interval within which we can always observe the image properly and clearly (this is the so called "visual validity principle"). That is, the discriminant functions can classify the data properly with the variation of the scale σ_0. The problem is to determine such an interval, or, more specifically, a suitable scale σ^*at which the $\text{sgn}(P(\mathbf{x}, \sigma^*))$can perfectly classify the image.

It should be noted that the expected scale σ^* will lie within an bounded interval $[\varepsilon, N]$. When $\sigma^* < \varepsilon$, the method can categorize every sample point without generalization; and when $\sigma^* > N$ (for a large constant N), $P(\mathbf{x}, \sigma)$becomes nearly a constant and the method fails to classify the image. If only a finite number of scales from which σ^* is selected, the well-known cross-validation approach can be applied. So the next step is to select a finite number of scales from $[\varepsilon, N]$. According to Weber's law in physiology, a person cannot recognize the difference between two images whose fraction for line length of the scale parameters is smaller than 0.029 (Coren et al. 1994). Thus, a reasonable discretization scheme can then be defined as $\Delta_\sigma = 0.029$.

With the above discretization scheme, we can obtain a finite number of scales $\{\sigma_i : i = 1, 2, ..., M\}$ where $M = (N - \varepsilon)/\Delta_\sigma$, or, correspondingly, a finite number of discriminant functions

$$\{\text{sgn}(g(\mathbf{x}, \sigma_i))\} : i = 1,\ 2,\ \ldots, M. \tag{4.48}$$

Applying any cross-validation approach to $\{\sigma_i : i = 1,\ 2,\ \ldots,\ M\}$ can then give the expected scale σ^*. Figure 4.21 depicts such a process.

Meng and Xu (2006) and Meng et al. (2008) have developed a learning theory for the vision based method corresponding to the statistical learning theory. It shows why the best compromise of generalization and approximation can be achieved at the scale σ^*; why σ^* has to be in a bounded interval $[\varepsilon,\ N]$; and how such an interval can be specified. It also investigates the convergence to the optimal classification discriminant function when the training samples tend to infinity.

4.6.3 Experimental Results

4.6.3.1 Benchmark Problems

To demonstrate the feasibility and high efficiencyof the Vision-based Classification method, 11 groups of IDA benchmark problems (cf. http://ida.first.gmd.de/raetsch/data/benchmarks.htm) have been used to test against the support vector machine with Gaussian kernel (Xu et al. 2006). The dimension and size of the training datasets and test datasets related to the problems are listed in Table 4.29. Its performance on classifying the datasets is shown in Table 4.30. In the simulations, the five-fold cross validation is used to select the scale in vision-based method and the spread parameter in the support vector machine.

We can observe that both methods are very successfully in classifying the data sets and predicting new data. However, as far as the training time is concerned, the vision-based method significantly outperforms the support vector machine. Without

Table 4.29 The statistics of 11 benchmark problems used in simulations

Problems	Input dim	Size of training set	Size of training set
Banana	2	400	4,900
Broast-cancer	9	200	77
Diabotis	8	468	300
Flare-sola	9	666	400
German	20	700	300
Heart	13	170	100
Image	18	1,300	1,010
Thyroid	5	140	75
Titanic	3	150	2,051
Twonorm	20	400	7,000
Waveform	21	400	4,000

Table 4.30 Performance of the vision-based classification method

Problems	Training time		Prediction error	
	SVM	VBC	SVM (%)	VBC (%)
Banana	4501.41	7.83	11.53 ± 0.66	10.81 ± 0.51
Broast-cancer	773.46	6.63	26.04 ± 4.74	24.82 ± 4.07
Diabotis	7830.93	13.68	23.79 ± 1.80	25.84 ± 1.81
Flare-sola	20419.71	24.23	32.43 ± 1.82	35.01 ± 1.72
German	24397.02	41	23.61 ± 2.07	25.27 ± 2.39
Heart	538.11	4.3	15.95 ± 3.26	17.22 ± 3.51
Image	1,346,476	129.7	2.96 ± 0.60	3.62 ± 0.63
Thyroid	368.43	3.13	4.80 ± 2.19	4.35 ± 2.34
Titanic	403.59	3.53	22.42 ± 1.02	22.31 ± 1.00
Twonorm	449.84	15.63	2.96 ± 0.23	2.67 ± 0.39
Waveform	4586.25	18.75	9.88 ± 0.44	10.64 ± 0.98
Mean	12,825	24.40	16.01 ± 1.71	16.60 ± 1.76

increasing the misclassification rate (i.e., loss of generalization capability), the vision-based method only costs 0.2% of the computational effort of the support vector machine. That is, it saves approximately 500 times of the computation cost of the support vector machine. This shows the very high efficiency of the method.

4.6.3.2 Spiral Classification Problem

To show the capability of the vision-based method in solving classification problems involving irregular discriminant function, it is applied to a simulated image in which a spiral discriminant function is discovered at $\sigma^* = 0.05$ (Fig. 4.21).

Remark 4.5. The scale space theory provides a useful way of modeling the blurring process of images. It is, however, a linear model and isotropic. To improve classification performance and to handle more complicated classification tasks, we should further explore the possibility of using nonlinear and anisotropic models of the blurring process.

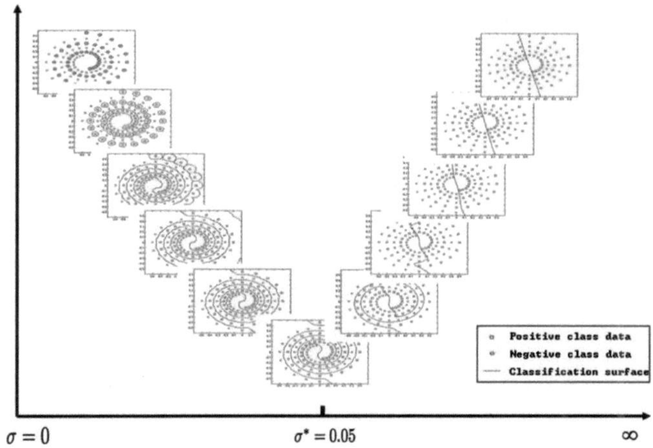

Fig. 4.21 Simulation result of a spiral classification problem. (The optimal discriminant function is spiral and it is found at σ')

4.7 A Remark on the Choice of Classifiers

Facing a large number of classifiers, one might wish to compare which is the one that is best for classification. Similar to the evaluation of clustering algorithms, it is difficult if not impossible to make such judgment. As discussed above, classifiers are constructed under different assumptions and on different bases. A classifier may work best for one problem but may not work as well for another problem. One needs to have a thorough examination of a classification problem and the data involved in order to choose the classifier that is most appropriate for the task. Notwithstanding that there are some objective guidelines for assessing the performance of a classifier, accuracy and robustness are perhaps the most common criteria for assessment (Hand 1986; Knoke 1986; Fukunaga and Hayes 1989). We not only want a classifier to be accurate, but we also want it to be robust to a certain degree of data non-conformity. Generalization, reflected by the classification error rate (total error of misclassification), is another basis one might want to employ to assess classifiers. Ideally, classifiers will not under- or over-fit in order to generalize. In classification, we often need to be able to explain how classes are separated. Under this requirement, interpretability or comprehensibility is another quality one might require of a classifier. A classifier might be accurate but relatively incomprehensible. Training time, computational complexity, scalability, flexibility to data types, ability to handle missing values, ability to select the optimal set of differentiating features, and requirement of prior knowledge are other criteria one might want to employ to evaluate classifiers. It is thus impossible to have a classifier that is best in everything. According to the objectives and situations under which classifications are performed, we would like our classifiers to be specific and yet as all rounded as possible.

Chapter 5
Discovery of Spatial Relationships in Spatial Data

5.1 On Mining Spatial Relationships in Spatial Data

Study of relationships in space has been the core of geographical research. In the simplest case, we might be interested in their characterization by some simple indicators. Sometimes we might be interested in knowing how things co-vary in space. From the perspective of data mining, it is the discovery of spatial associations in data. Often time, we are interested in relationships in which the variation of one phenomenon can be explained by the variations of the other phenomena. In terms of data mining, we are looking for some kinds of causal relationships that might be expressed in functional forms. Statistics in general and spatial statistics in particular have been commonly employed in such studies (Cliff and Ord 1972; Anselin 1988; Cressie 1993).

Regardless of what relationships that are of interest, the geographers' main concern is whether they are local or global. In the characterization of a spatial phenomenon, for example, is it appropriate to use an overall mean to describe the central tendency of a distribution in space? Will it be too over-sweeping an indicator so that it hides the distinct local variations that would be more telling otherwise? The task of data mining is thus to discover whether significant local variations are embedded in a general distribution, and if yes, we need to unravel the appropriate parameters and/or functional form for their description. In the identification of spatial associations, we often wonder if spatial autocorrelations are local or global. Again, it is essential to have a means to unravel such associative relationships. To discover causal relationship in space, the local vs. global issue rests on whether the effect of an explanatory variable on the dependent variable can be summarized by a global parameter, or whether it is localized with different effects at different points in space. In a word, the basic issue is on the discovery of spatial non-stationarity from data.

The inappropriateness of using global estimates to represent local relationships has long been a concern of not only the geographers, but also the statisticians and other social scientists. Simpsons's (1951) study of the local effect on interaction in

Y. Leung, *Knowledge Discovery in Spatial Data*, Advances in Spatial Science, 223
DOI 10.1007/978-3-642-02664-5_5, © Springer-Verlag Berlin Heidelberg 2010

contingency table, Linneman's (1996) examination of international trade flows, Cox's (1969) and Johnston's (1973) local analyzes in voting behavior are early examples. Over the years, researchers, particularly geographers, have developed methods for local and global analyzes. The geographical analysis machine (Openshaw et al. 1987), a limited version of the "scan statistics" (Kulldorf et al. 1997), for example, is catered for the study of point patterns with local variations that might not be appropriately captured by the global statistics described by Dacey (1960), Tinkler (1971), and Boots and Getis (1988). Differing from the concept advanced by Cliff and Ord (1972) Which gives a global statistic to describe spatial association, Getis and Ord (1992), Anselin (1995, 1998), Ord and Getis (1995, 2001) propose some local statistics to depict local variations in the study of spatial autocorrelation. It has been demonstrated that local clusters that cannot be detected by the global statistic can be identified by the local statistics. Leung et al. (2003d) make the analysis more rigorous by generalizing the local statistics into quadratic forms.

Besides the development of local statistics for the description of spatial dependency, the local and global issue has also surfaced in the study of spatial relationships within the framework of regression analysis. Similar to the study of spatial association, a key issue in the analysis of causal relationship is to discover whether a cause-effect relation is non-stationary in space. Specifically, we are interested in finding out whether the spatial effect is local or global. Within the context of regression, if the parameters of a regression model are functions of the locations on which the observations are made, then local patterns exist and the spatial relationship is non-stationary. The relationship can then be represented by the varying-parameter regression model (Cleveland 1979). In spatial terminology, the relationship is said to be captured by the geographically weighed regression (Brunsdon et al. 1996). Thus, the data mining task is to determine whether the underlying structure is global or local in terms of some statistics. For complex systems, however, spatial non-stationarity is not restricted to only the variation of parameters of a universal model. Spatial data manifesting such systems may contain several populations embedded in a mixture distribution. In other words, the functional form representing the relationship varies over space. Local relationships take on different functional expressions, and our task is to unravel all of them in a spatial database. It is particularly important to develop robust data mining methods in a highly noisy environment (Leung et al. 2001a).

In this chapter, the discovery of spatial associations is first discussed in Sect. 5.2. The emphasis is placed on the employment of various measures for the mining of global and local associations in space with rigorous statistical test. Discovery of non-stationarity of spatial relationship is then discussed in Sect. 5.3. Local variations are unraveled by detecting the significant variations of the parameters of a regression model in space. The general framework is the parameter-varying regression with geographically weighted regression as a special case. Spatial autocorrelation in geographically weighted regression is further discussed in Sect. 5.4. A more general model of geographically weighted regression is briefly discussed in Sect. 5.5. In Sect. 5.6, spatial non-stationarity is extended to situations in which relationships take on different forms in space. The regression-class mixture decomposition method is employed to mine local variations of spatial relationships captured by different functional forms.

5.2 Discovery of Local Patterns of Spatial Association

5.2.1 On the Measure of Local Variations of Spatial Associations

Many geographical problems can only be adequately analyzed by taking into account the relative locations of observations our failure in taking necessary steps to account for spatial association in spatial data sets often lead to misleading conclusions (see, for example Anselin and Griffith 1988; Arbia 1989). The well known statistics for the identification of global patterns of spatial association are Moran's I (Moran 1950) and Geary's c (Geary 1954). They are used as an overall measure of spatial dependency about the whole data set. The properties of these two statistics and their null distributions have been intensively studied over the years (see, for example Cliff and Ord 1981; Anselin 1988; Tiefelsdorf and Boots 1995; Hepple 1998; Tiefelsdorf 1998, 2000; Leung et al. 2003d).

However, with the increasingly large geo-referenced data sets obtained from complex spatial systems, stationarity of dependency over space may be an unrealistic presumption. Thus, there has been a surge of interest in discovering local patterns of spatial association based on the local forms of statistics in recent years. The local forms of statistics mainly focus on exceptions to the general patterns represented by conventional global forms, and the search of local areas exhibiting spatial heterogeneities with significant local departures from randomness.

The commonly used statistics for detecting local patterns of spatial association are Ord and Getis G_i or G_i^* statistic (Ord and Getis 1995) and Anselin's LISAs (Anselin 1995), including local Moran's I_i and local Geary's c_i. As defined in Anselin (1995), a LISA must indicate the extent of spatial clustering of observations around a reference location, and it must obey the additivity requirement for any coding scheme of the spatial link matrix. That is, the sum of values of a LISA at all locations must be proportional to a global indicator of spatial association. With its additivity, a LISA can also be used as a diagnosis of local instability in measures of global spatial association in the presence of significant global association. However, the G_i or G_i^* statistic, while being a statistic for local spatial association, is not a LISA in the sense of the additivity requirement because its individual components are not related to a global statistic of spatial association (Anselin 1995). In addition to the fundamental works by Anselin (1995), Getis and Ord (1992) as well as Ord and Getis (1995), the properties of these local statistics have been extensively studied and applied to many real-world and simulated spatial data sets (see, for example, Bao and Henry 1996; Sokal et al. 1998; Tiefelsdorf and Boots 1997; Fotheringham and Brunsdon 1999; Unwin 1996; Wilhelm and Steck 1998).

One of the important issues in the studies of local spatial associations is to find out the null distributions of these local statistics because only when their null distributions are made available can the other challenging subjects be addressed (Tiefelsdorf 2000). In this aspect, Tiefelsdorf and associates have defined the local Moran's I_i as a ratio of quadratic forms. By means of this definition and under either the assumption of spatial independence or a conditional on a global spatial process, they have investigated the unconditional and conditional exact distribution of I_i and its moments with the statistical theory for ratios of quadratic forms (Boots and Tiefelsdorf 2000; Tiefelsdorf 1998, 2000; Tiefelsdorf and Boots 1997). Unfortunately, the null distributions of other local statistics have not been examined along this line of reasoning. Furthermore, normal approximation and randomized permutation are still the common approaches for deriving the p-values of the local statistics. Some GIS modules for spatial statistical analysis also employ the normal approximation to compute the null distribution of I_i (Boots and Tiefelsdorf 2000). Nevertheless, there are problems with these two methods.

For the local statistics I_i, c_i, and G_i or G_i^*, the underlying spatial structure or spatial contiguity is typically star-shaped. Cliff and Ord (1981, Chap. 2) have shown that the null distributions of global Moran's I and Geary's c with star-shaped spatial structures deviate markedly from the normal distribution. A series of experiments performed by Anselin (1995), Boots and Tiefelsdorf (2000) and Sokal et al. (1998) have also demonstrated that the normal approximation to the distribution of the local Moran's I_i is inappropriate because of the excessive kurtosis of the distribution of I_i. Although asymptotic normality is a reasonable assumption to the null distribution of G_i orG_i^*, a misleading significance level may be obtained if the number of neighbors at a specific location is too small and the weights for describing the contiguities are too uneven (Ord and Getis 1995).

Although randomized permutation approach seems to provide a reliable basis for inference for both the LISAs and the G_i or G_i^* (Anselin 1995), this approach may suffer from resampling error and very large sample sizes needed for resampling are rather expensive for the purpose of routine significance test (Costanzo et al. 1983). Furthermore, in the significance tests of spatial association with these local statistics, empirical distribution functions are calculated by resampling from the observations under the assumption of equi-probability of selection across the space. If the spatial units are not uniformly defined, the assumption of equi-probability of selection may not hold and the derived test values may be biased (Bao and Henry 1996). In the regression context, if spatial association among the residuals is to be tested, then the randomized permutation approach is inappropriate since regression residuals are correlated (Anselin and Rey 1991).

Given the above shortcomings in performing the significance tests for local spatial association by normal approximation and randomized permutation, it is especially useful to develop the exact or some more accurate approximate methods for testing local spatial association. The idea is to develop the exact and approximate p-values of the aforementioned local statistics for testing local spatial clusters when global autocorrelation is not significant. Such a structure discovery process addresses essentially the following statistical test issues:

1. Is a reference location surrounded by a cluster of high or low values? Or
2. Is the observed value at this location positively (similarly) or negatively (dissimilarly) associated with the surrounding observations?

To offer a more formal approach in line with classical statistical framework, Leung et al. (2003d) have developed an exact method for computing the p-values of the local Moran's I_i, local Geary's c_i and the modified Ord and Getis G statistics based on the distributional theory of quadratic forms in normal variables. Furthermore, an approximate method, called three-moment χ^2 approximation, with explicit calculation formulae, has also been proposed to achieve a computational cost lower than the exact method. Their study not only provides exact tests for local patterns of spatial association, but also put the tests of several local statistics within a unified statistical framework.

5.2.2 Local Statistics and their Expressions as a Ratio of Quadratic Forms

I first introduce in this section the local Moran's I_i and Geary's c_i of Anselin's LISAs (Anselin 1995) as well as G_i and G_i^* of Ord and Getis G statistics (Ord and Getis 1995), and express them as ratios of quadratic forms in observations. By taking the square of G_i and G_i^* in particular, the analysis of G_i and G_i^* can be brought within the common framework of ratios of quadratic forms.

Let $\mathbf{x} = (x_1, x_2, \cdots, x_n)^T$ be the vector of observations on random variable X at n locations and let $\mathbf{W} = (w_{ij})_{n\times n}$ be a symmetric spatial link matrix which is defined by the underlying spatial structure of the geographical units where the observations are made. The simplest form of $\boldsymbol{W}$ can be such a matrix with elements taking the value one if the corresponding units i and j come in contact and zero otherwise. It should be noted that the link matrix can also incorporate information on distances, flows and other types of linkages.

5.2.2.1 Local Moran's I_i

For a reference location i, the local Moran's I_i in its standardized form is (Anselin 1995)

$$I_i = \frac{(x_i - \bar{x}) \sum_{j=1}^{n} w_{ij} (x_j - \bar{x})}{\frac{1}{n} \sum_{j=1}^{n} (x_j - \bar{x})^2}, \tag{5.1}$$

where $\bar{x} = \frac{1}{n} \sum_{j=1}^{n} x_j$, $(w_{i1}, w_{i2}, \cdots, w_{in})$ is the ith row of the symmetric spatial link matrix $\boldsymbol{W}$ and $w_{ii} = 0$ by convention. A large positive value of I_i indicates spatial clustering of similar values (either high or low) around location i, and a large negative value indicates a clustering of dissimilar values, that is, a location with high value is surrounded by neighbors with low values and vice versa.

We actually can express I_i as a ratio of quadratic forms as follows (Leung et al. 2003d):

$$I_i = \frac{\mathbf{x}^T \mathbf{B}\mathbf{W}(\mathbf{I}_i) B\mathbf{x}}{\frac{1}{\mathrm{n}} \mathbf{x}^T B\mathbf{x}} \tag{5.2}$$

where

$$(x_1 - \bar{x}, \cdots, x_n - \bar{x})^T = \left(\mathbf{I} - \frac{1}{n}\mathbf{1}\mathbf{1}^T\right)\mathbf{x} = \mathbf{B}\mathbf{x}, \tag{5.3}$$

in which $\mathbf{I}$ is the identity matrix of order n, $\mathbf{B} = \mathbf{I} - \frac{1}{n}\mathbf{1}\mathbf{1}^T$ is an idempotent and symmetric matrix, $\mathbf{1} = (1, 1, \cdots, 1)^T$, and $\mathbf{W}(I_i)$ is the $n \times n$ symmetric star-shaped matrix defined as:

$$\mathbf{W}(I_i) = \frac{1}{2}\begin{pmatrix} 0 & \cdots & 0 & w_{1i} & 0 & \cdots & 0 \\ \vdots & \ddots & \vdots & \vdots & \vdots & \ddots & \vdots \\ 0 & \cdots & 0 & w_{i-1,i} & 0 & \cdots & 0 \\ w_{i1} & \cdots & w_{i,i-1} & 0 & w_{i,i+1} & \cdots & w_{in} \\ 0 & \cdots & 0 & w_{i+1,i} & 0 & \cdots & 0 \\ \vdots & \ddots & \vdots & \vdots & \vdots & \ddots & \vdots \\ 0 & \cdots & 0 & w_{ni} & 0 & \cdots & 0 \end{pmatrix} \tag{5.4}$$

Since $\sum_{i=1}^{n} \mathbf{W}(I_i) = \mathbf{W}$, we have

$$\sum_{i=1}^{n} I_i = \frac{\mathbf{x}^T \mathbf{B}\mathbf{W}\mathbf{B}\mathbf{x}}{\frac{1}{\mathrm{n}} \mathbf{x}^T \mathbf{B}\mathbf{x}} = s\mathbf{I}, \tag{5.5}$$

where $s = \sum_{i=1}^{n} \sum_{j=1}^{n} w_{ij}$, and $\mathbf{I}$ is the global Moran's statistic (Cliff and Ord 1981, p. 47). This means that, when we take $\mathbf{W}(I_i)$ as a local link matrix, the additivity requirement is fulfilled by I_i.

5.2.2.2 Local Geary's c_i

The local Geary's c_i at a reference location i is defined by Anselin (1995) as

$$c_i = \frac{\sum_{j-1}^{n} w_{ij} \left(x_i - x_j\right)^2}{\frac{1}{n} \sum_{j=1}^{n} \left(x_j - \bar{x}\right)^2}, \tag{5.6}$$

where $w_{ij} = 0$. A small value of c_i suggests a positive spatial association (similarity) of observation i with its surrounding observations, while a large value of c_i suggests a negative association (dissimilarity) of observation i with its surrounding observations.

Based on Leung et al. (2003d), c_i can again be expressed as a ratio of quadratic forms as:

$$c_i = \frac{\mathbf{x}^T\mathbf{B}\mathbf{W}(c_i)\mathbf{B}\mathbf{x}}{\frac{1}{\mathrm{n}}\mathbf{x}^T\mathbf{B}\mathbf{x}}, \tag{5.7}$$

where $\mathbf{W}(c_i) = \mathbf{D}(i) - 2\mathbf{W}(I_i)$ is symmetric, and

$\mathbf{D}(i) = \mathrm{diag}\left(w_{i1}, \cdots, w_{i,i-1}, w_{i+}, w_{i,i+1} \cdots, w_{in}\right)$ is a diagonal matrix with the i th element in its main diagonal being $w_{i+} = \sum_{j=1}^{n} w_{ij}$.

According to the symmetry of $\mathbf{W}$ and $w_{ii} = 0$ for all i, it is easy to prove that

$$\sum_{i=1}^{n} \mathbf{W}(c_i) = \sum_{i=1}^{n} \mathbf{D}(i) - 2\sum_{i=1}^{n} \mathbf{W}(I_i) = 2(\mathbf{D} - \mathbf{W}), \tag{5.8}$$

where $\mathbf{D} = \mathrm{diag}(w_{1+}, w_{2+}, \cdots, w_{n+})$. From Cliff and Ord (1981, p. 167) as well as Leung et al. (2003), the global Geary's c can be expressed as

$$c = \frac{n-1}{s}\frac{\mathbf{x}^T\mathbf{B}(\mathbf{D} - \mathbf{W})\mathbf{B}\mathbf{x}}{\mathbf{x}^{\mathrm{T}}\mathbf{B}\mathbf{x}}. \tag{5.9}$$

Therefore

$$\sum_{i=1}^{n} c_i = \frac{\mathbf{x}^T\mathbf{B}\left[\sum_{i-1}^{n} \mathbf{W}(c_i)\right]\mathbf{B}\mathbf{x}}{\frac{1}{\mathrm{n}}\mathbf{x}^T\mathbf{B}\mathbf{x}} = \frac{2ns}{n-1}c. \tag{5.10}$$

That is, the additivity requirement is still fulfilled by c_i with the expression in (5.7).

5.2.2.3 G Statistics Expressed as Ratios of Quadratic Forms

Ord and Getis G_i and G_i^* statistics in their original forms (Getis and Ord 1992) are, respectively,

$$G_i = \frac{\sum_{j \neq 1} w_{ij}x_j}{\sum_{j \neq 1} x_j} \tag{5.11}$$

and

$$G_i^* = \frac{\sum_{j=1}^{n} w_{ij} x_j}{\sum_{j=1}^{n} x_j}. \tag{5.12}$$

For simplicity, d in $w_{ij}(d)$ (the weight for the link of location j and a given location i, with j being within distance d from i) is omitted here. The statistics G_i and G_i^* in (5.11) and (5.12) require that the underlying variable X has a natural origin and is positive (Getis and Ord 1992). In order to overcome this restriction, Ord and Getis (1995) have standardized them as

$$G_i = \frac{\sum_{j \neq i} w_{ij} \left[x_j - \bar{x}(i)\right]}{\left\{\frac{1}{n-1} \sum_{j \neq 1} \left[x_j - \bar{x}(i)\right]^2\right\}^{\frac{1}{2}}}, \tag{5.13}$$

and

$$G_i^* = \frac{\sum_{j=i}^{n} w_{ij} \left(x_j - \bar{x}\right)}{\left[\frac{1}{n} \sum_{j=1}^{n} \left(x_j - \bar{x}\right)^2\right]^{\frac{1}{2}}}, \tag{5.14}$$

where $\bar{x}(i) = \frac{1}{n-1} \sum_{j \neq i} x_j$. Here, the scale factor in each statistic is omitted because it does not affect the p-value to be derived. A large positive value of G_i or G_i^* indicates a spatial clustering of observations of high values while a large negative value indicates a spatial clustering of observations of low values. However, unlike the LISAs, these two local statistics are not related to a global one and therefore the additivity requirement is not satisfied. In order to put G_i and G_i^* into the framework of ratios of quadratic forms, Leung et al. (2003d) take the square of G_i and G_i^* and obtain the modified G statistics, respectively, as follows:

$$\tilde{G}_i = (G_i)^2 = \frac{\left\{\sum_{j \neq i} w_{ij} \left[x_j - \bar{x}(i)\right]\right\}^2}{\frac{1}{n-1} \sum_{j \neq i} \left[x_j - \bar{x}(i)\right]^2}, \tag{5.15}$$

and

$$\tilde{G}_i^* = \left(G_i^*\right)^2 = \frac{\left[\sum_{j=1}^{n} w_{ij} \left(x_j - \bar{x}\right)\right]^2}{\frac{1}{n} \sum_{j=1}^{n} \left(x_j - \bar{x}\right)^2}. \tag{5.16}$$

A large value of the transformed statistic G_i or G_i^* indicates a spatial clustering of observations of high values or low values. With this modification, G_i and G_i^* can then be expressed as a ratio of quadratic forms and their null distributions can be obtained by the distributional theory of quadratic forms. Statistically, it is equivalent to the use of G_i or G_i^* and the modified one for exploring local spatial association except that a spatial clustering of high values or low values cannot be identified by the extreme values of the modified statistic $\tilde{G}_i$ or $\tilde{G}_i^*$. However, the loss of directional association can be compensated by reexamining the values of the observations at location i and its neighbors after a significant value of $\tilde{G}_i$or $\tilde{G}_i^*$ is obtained at location i. Since $\tilde{G}_i$and $\tilde{G}_i^*$ can be expressed as a ratio of quadratic forms in a similar way, we henceforth only need to discuss the statistic $\tilde{G}_i^*$.

It should be noted that the numerator of $\tilde{G}_i^*$ in (5.16) can be written as

$$\begin{aligned}\left[\sum_{j=1}^{n} w_{ij}\left(x_j - \bar{x}\right)\right]^2 &= (x_1 - \bar{x}, \cdots, x_n - \bar{x})\mathbf{w}(i)\mathbf{w}^T(i)(x_1 - \bar{x}, \cdots, x_n - \bar{x})^T \\ &= x^T\mathbf{B}\mathbf{w}(i)\mathbf{w}^T(i)\mathbf{B}\mathbf{x} \end{aligned} \tag{5.17}$$

Therefore, we obtain

$$\tilde{G}_i^* = \frac{\mathbf{x}^T\mathbf{B}\mathbf{W}\left(\tilde{G}_i^*\right)\mathbf{B}\mathbf{x}}{\frac{1}{\mathrm{n}}\mathbf{x}^T\mathbf{B}\mathbf{x}}, \tag{5.18}$$

where

$$\mathbf{W}\left(G_i^*\right) = \mathbf{w}(i)\mathbf{w}^T(i) \tag{5.19}$$

is a symmetric matrix.

5.2.2.4 The Null Distributions of I_i, c_i and G_i^* and Their p-values for Spatial Association Test

Based on the above measures, we can derive the p-values of these local statistics to test for local spatial clusters in the absence of global spatial autocorrelation. Assume that the underlying distribution for generating the observations is normal. Then under the null hypothesis:

H_0: no local spatial association is present.

The variables $x_1, x_2, \cdots,$ and x_n are independent and identically distributed as $N(\mu, \sigma^2)$, a normal distribution with mean μ and variance σ^2. Therefore, $\mathbf{x} \sim N(\mu\mathbf{1}, \sigma^2\mathbf{I})$. In this case, for a specific spatial structure that is stipulated by the spatial link matrix **W**, the null distributions of the aforementioned local statistics can be obtained via the distributional theory of quadratic forms in normal variables. Therefore, significance tests for local spatial association can be performed by computing the p-values of the local statistics. In the following discussion, the exact and approximate methods for deriving the p-values of the local statistics I_i, c_i and $\tilde{G}_i^*$ are introduced.

The Exact Method

Under the null hypothesis H_0, $\mathbf{x} \sim N(\mu \mathbf{1}, \sigma^2 \mathbf{I})$, we have $y = \frac{1}{\sigma}(\mathbf{x} - \mu \mathbf{1}) \sim N(0, \mathbf{I})$. Substituting $\mathbf{x} = \sigma \mathbf{y} + \mu \mathbf{1}$ into the expression of I_i in (5.2) and noting that $\mathbf{1}^T \mathbf{B} = \mathbf{1}^T \left(\mathbf{I} - \frac{1}{n}\mathbf{1}\mathbf{1}^T\right) = \mathbf{0}$ and $\mathbf{B}\mathbf{1} = \left(\mathbf{I} - \frac{1}{n}\mathbf{1}\mathbf{1}^T\right)\mathbf{1} = \mathbf{0}$, we have, by omitting the scale factor $1/n$,

$$I_i = \frac{\mathbf{y}^T \mathbf{B} \mathbf{W}(I_i) \mathbf{B} \mathbf{y}}{\mathbf{y}^T \mathbf{B} \mathbf{y}}. \tag{5.20}$$

Similar expressions for c_i and $\tilde{G}_i^*$ can be obtained by replacing $\mathbf{W}(I_i)$ with $\mathbf{W}(c_i)$ and $\mathbf{W}\left(\tilde{G}_i^*\right)$ respectively.

For any real number r, the value of the null distribution function of I_i at r can be expressed as

$$P_{H_0}(I_i \le r) = P\left\{\mathbf{y}^T \mathbf{B}[\mathbf{W}(I_i) - r\mathbf{I}]\, \mathbf{B} \mathbf{y} \le 0\right\}. \tag{5.21}$$

Since $\mathbf{B}[\mathbf{W}(I_i) - r\mathbf{I}]\,\mathbf{B}$ is a symmetric matrix with real elements and y is distributed as $N(0,1)$, the Imhof's results on the distribution of quadratic forms (Hepple 1998; Imhof 1961; Leung et al. 2003d; Tiefelsdorf and Boots 1995) can be used to obtain the null distribution of I_i. That is,

$$P_{H_0}(I_i \le r) = \frac{1}{2} - \frac{1}{\pi}\int_0^{\infty} \frac{\sin[\theta(t)]}{t\rho(t)}\, dt, \tag{5.22}$$

where

$$\theta(t) = \frac{1}{2}\sum_{k=1}^{m} [h_k \arctan(\lambda_k t)], \tag{5.23}$$

$$\rho(t) = \prod_{k=1}^{m} \left(1 + \lambda_k^2 t^2\right)^{\frac{1}{4}h_k}, \tag{5.24}$$

With $\lambda_1, \lambda_2, \cdots, \lambda_m$ being the distinct nonzero eigenvalues of the matrix $\mathbf{B}[\mathbf{W}(I_i) - r\mathbf{I}]\,\mathbf{B}$, and $h_1, h_2, \cdots, h_m$ being their respective orders of multiplicity.

The same formulae for computing the null distributions of c_i and $\tilde{G}_i^*$ can be obtained by replacing $\lambda_1, \lambda_2, \cdots, \lambda_m$ and $h_1, h_2, \cdots, h_m$ with the eigenvalues and their orders of multiplicity of the matrices $\mathbf{B}[\mathbf{W}(c_i) - r\mathbf{I}]\,\mathbf{B}$ and $\mathbf{B}\left[\mathbf{W}(\tilde{G}_i^*) - r\mathbf{I}\right]\mathbf{B}$ respectively.

As a special case of the above results, we can obtain the exact p-values of the statistics I_i, c_i and $\tilde{G}_i^*$ for the spatial association test. Let r_1, r_c and r_G be, respectively, the observed values of I_i, c_i and $\tilde{G}_i^*$ which can be computed from (5.1), (5.6) and (5.16), or from (5.2), (5.7) and (5.18), by omitting the scale factor $1/n$ in each expression. For I_i, the p-value for testing positive spatial autocorrelation (a spatial cluster of similar values) is $\mathrm{P}_{\mathrm{H}_0}(I_i \geq r_I)$, and the p-value for testing negative spatial autocorrelation (a spatial cluster of dissimilar values) is $\mathrm{P}_{\mathrm{H}_0}(I_i \leq r_I)$. For c_i, the p-value for testing positive spatial autocorrelation is $\mathrm{P}_{\mathrm{H}_0}(c_i \leq r_c)$ and the p-value for testing negative spatial autocorrelation is $\mathrm{P}_{\mathrm{H}_0}(c_i \geq r_c)$. For G_i^*, the p-value for testing a spatial clustering of observations of high or low values is $\mathrm{P}_{\mathrm{H}_0}\left(G_i^* \geq r_G\right)$. All these p-values can be calculated through the corresponding exact formulae in (5.22)–(5.24). The derivations of $\theta(t)$ and $\rho(t)$ in (5.22) for I_i, G_i and C_i given in Leung et al. (2003). For I_i, we have

$$\begin{aligned}\theta(t) =& \frac{1}{2}\{\arctan[\lambda_1(1) - (r)\,t] + \arctan[(\lambda_1(2) - r)\,t] \\ & -(n-3)\arctan(rt)\},\end{aligned} \tag{5.25}$$

$$\rho(t) = \left\{1 + [\lambda_1(1) - r]^2 t^2\right\}^{\frac{1}{4}} \left\{1 + [\lambda_1(2) - r]^2 t^2\right\}^{\frac{1}{4}} \left(1 + r^2 t^2\right)^{\frac{n-3}{4}}, \tag{5.26}$$

where $\lambda_I(1)$ and $\lambda_I(2)$ are the non-zero eigenvalues of the matrix $\mathbf{BW}(I_i)\,\mathbf{B}$. For G_i, we have

$$\theta(t) = \frac{1}{2}\{\arctan[(\lambda_G - r)t] - (n-2)\arctan(rt)\}, \tag{5.27}$$

$$\rho(t) = \left[1 + (\lambda_G - r)^2 t^2\right]^{\frac{1}{4}} \left[1 + r^2 t^2\right]^{\frac{n-2}{4}}. \tag{5.28}$$

For c_i, we have

$$\begin{aligned}\theta(t) =& \frac{1}{2}\{\arctan[(w_{i+} - 1 - r)t] + (w_{i+} - 1)\arctan[(1-r)t]\} \\ & -(n - w_{i+} - 1)\arctan(rt),\end{aligned} \tag{5.29}$$

$$\rho(t) = \left[1 + (w_{i+} + 1 - r)^2 t^2\right]^{\frac{w_{i+}-1}{4}} \left(1 + r^2 t^2\right)^{\frac{n - w_{i+} - 1}{4}}. \tag{5.30}$$

The Approximate Method

Computing numerically the eigenvalues of a $n \times n$ matrix and an integral on an infinite interval is in fact computationally expensive. Therefore, the above exact method for computing the p-values of the statistics is not very efficient in practice, especially when the sample size n of a data mining task is large. Some approximate methods may be useful in solving this problem. As pointed out above, the null distributions of LISAs cannot be effectively approximated by the normal distribution. Leung et al. (2003d) hence propose a higher-moments procedure, called three-moment χ^2 approximation, to compute the p-values of the local statistics for spatial association test and derive the explicit computation formulae which can significantly reduce the computational overhead.

The main idea of the three-moment χ^2 approximation is to approximate the distribution of a quadratic form in normal variables by that of a linear function of a χ^2 variable with appropriate degrees of freedom, say $a + b_{\chi_d^2}$. The coefficients a and b of the linear function and the degrees of freedom d are chosen in such a way that the first three moments of $a + b_{\chi_d^2}$ are made to match those of the quadratic form. This method was originally proposed by Pearson (1959) to approximate the distribution of a noncentral χ^2 variable. Imhof (1961) has extended this method to approximate the distribution of a general quadratic form in normal variables.

For local Moran's I_i, we have

$$\begin{aligned} P_{H_0}(I_i \le r) &= P\{\mathbf{y}^T\mathbf{B}[\mathbf{W}(I_i) - r\mathbf{I}]\mathbf{B}\mathbf{y} \le 0\} \\ &\approx \begin{cases} P\{\chi_d^2 \le d - \frac{1}{b}tr[\mathbf{B}[\mathbf{W}(I_i) - r\mathbf{I}]\mathbf{B}]\}, & if \quad tr\{\mathbf{B}[\mathbf{W}(I_i) - r\mathbf{I}]\mathbf{B}\}^3 > 0, \\ P\{\chi_d^2 \ge d - \frac{1}{b}tr[\mathbf{B}[\mathbf{W}(I_i) - r\mathbf{I}]\mathbf{B}]\}, & if \quad tr\{\mathbf{B}[\mathbf{W}(I_i) - r\mathbf{I}]\mathbf{B}\}^3 < 0, \end{cases} \end{aligned} \tag{5.31}$$

where

$$b = \frac{tr\{\mathbf{B}[\mathbf{W}(I_i) - r\mathbf{I}]\mathbf{B}\}^3}{tr\{\mathbf{B}[\mathbf{W}(I_i) - r\mathbf{I}]\mathbf{B}\}^2}, \tag{5.32}$$

$$d = \frac{\left\{tr[\mathbf{B}[\mathbf{W}(I_i) - r\mathbf{I}]\mathbf{B}]^2\right\}^3}{\left\{tr[\mathbf{B}[\mathbf{W}(I_i) - r\mathbf{I}]\mathbf{B}]^3\right\}^2}. \tag{5.33}$$

Therefore, the approximate p-value of I_i for testing local positive or negative spatial autocorrelation can be computed via (5.31) if the observed value $r\mathbf{I}$ is obtained.

For local Geary's c_i, the probability $P_{H_0}(c_i \le r)$ can be computed by the same formulae as those in (5.31)–(5.33) except that the matrix $\mathbf{B}[\mathbf{W}(I_i) - r\mathbf{I}]\mathbf{B}$ is replaced by $\mathbf{B}[\mathbf{W}(c_i) - r\mathbf{I}]\mathbf{B}$. For the modified statistic $\tilde{G}_i^*$, the probability $P_{H_0}(G_i^* \le r)$ can still be calculated by replacing the matrix $\mathbf{B}[\mathbf{W}(I_i) - r\mathbf{I}]\mathbf{B}$ in (5.31), (5.32) and (5.33) with $\mathbf{B}\left[\mathbf{W}\left(G_i^*\right) - r\mathbf{I}\right]$.

When the underlying variable for the generating data is normally distributed and the null hypothesis of "no local spatial association" is true, each of the local statistics I_i, c_i and $\tilde{G}_i^*$ can then be expressed as a ratio of quadratic forms in standard normal variables. Therefore, a well known result saying that " a ratio of quadratic forms in normal variables with the matrix in its denominator being idempotent is distributed independently of its denominator" (see for example Cliff and Ord 1981, p. 43 as well as Stuart and Ord 1994, pp. 529–530 for the proof) can be employed to obtain the exact moments of I_i, c_i and $\tilde{G}_i^*$. According to this result, we have from (5.20) that for any positive integer k,

$$\mathrm{E}\left(I_i^k\right) = \frac{\mathrm{E}[\mathbf{y}^T\mathbf{BW}(I_i)\mathbf{By}]^k}{\mathrm{E}(\mathbf{y}^T\mathbf{By})^k}. \tag{5.34}$$

Similar to the derivation in Tiefelsdorf (2000, pp. 100–102), for example, we can obtain in particular

$$E(I_i) = \frac{1}{n-1} tr[\mathbf{BW}(I_i)\mathbf{B}], \tag{5.35}$$

$$Var(I_i) = \frac{1}{(n-1)^2(n+1)}\left\{ (n-1)\, tr\, [\mathbf{BW}(I_i)\mathbf{B}]^2 - [tr\,(\mathbf{BW}(I_i)\mathbf{B})\,]^2\right\}. \tag{5.36}$$

Leung et al. (2003d) show that the normal approximation of the null distribution of I_i can be expressed as

$$\mathrm{P}_{\mathrm{H}_0}(I_i \le r) \approx \Phi\left(\frac{r - \mathrm{E}(I_i)}{\sqrt{\mathrm{Var}(I_i)}}\right), \tag{5.37}$$

where $\Phi(x)$ is the distribution function of $N(0,1)$. And, we can obtain similar normal approximation formulae as those in (5.37) for the null distributions of c_i and G_i^* respectively. Simulations conducted by Leung et al. (2003d) demonstrate that this approximation approach performs generally better than normal approximation and very accurate in some instances.

It should be emphasized that both the exact and approximate p-values of I_i, c_i and G_i^* are obtained under the assumptions that global spatial autocorrelation is insignificant and that the underlying distribution for generating observations is normal. The first assumption means that the results can only be used in significance test for local spatial clusters that the global statistics fail to detect. This is one of the two important purposes that the LISAs intend to serve (Anselin 1995). In practice, a test for the non-existence of a global spatial autocorrelation should first be performed. If global autocorrelation is not significant, results obtained by Leung et al. (2003d) can then be used to assess the significance of local spatial clusters.

5.3 Dicovery of Spatial Non-Stationarity Based on the Geographically Weighted Regression Model

5.3.1 *On Modeling Spatial Non-Stationarity within the Parameter-Varying Regression Framework*

In spatial analysis, ordinary linear regression (OLR) model has been one of the most useful statistical means to identify the nature of relationships among variables. In this technique, a variable$y$, called the dependent variable, is modeled as a linear function of a set of independent variables $x_i, x_2, \cdots, x_p$. Based on n observations $(y_i; x_{i1}, x_{i2}, \cdots, x_{ip})$, $i = 1, 2, \cdots, n$, taken from a study region, the model can be expressed as

$$y_i = \beta_o + \sum_{k=1}^{p} \beta_x x_{ik} + \varepsilon_i, \tag{5.38}$$

where $\beta_0, \beta_1, \cdots, \beta_p$ are parameters and $\varepsilon_1, \varepsilon_2, \cdots, \varepsilon_n$ are error terms which are generally assumed to be independent normally distributed random variables with zero means and constant variance σ^2. In this model, each of the parameters can be thought of as the "slopes" between the dependent variable and one of the independent variables. The least squares estimate of the parameter vector can be written as

$$\hat{\boldsymbol{\beta}} = \left(\hat{\beta}_0 \hat{\beta}_1 \cdots \hat{\beta}_p\right)^T = \left(\mathbf{X}^T \mathbf{X}\right)^{-1} \mathbf{X}^T \mathbf{Y} \tag{5.39}$$

where

$$\mathbf{X} = \begin{pmatrix} 1 & x_{11} & \cdots & x_{1p} \\ 1 & x_{21} & \cdots & x_{2p} \\ \cdots & \cdots & \cdots & \cdots \\ 1 & x_{n1} & \cdots & x_{np} \end{pmatrix}, \quad \mathbf{Y} = \begin{pmatrix} y_1 \\ y_2 \\ \vdots \\ y_n \end{pmatrix}. \tag{5.40}$$

Statistical properties of these estimates have been well studied and various hypothesis tests have also been established.

Although the OLR model has been used extensively in the study of spatial relationships, it cannot incorporate spatial non-stationarity in space since the relationships between the dependent variable and the independent variables, manifested by the slopes (parameters), are assumed to be global across the study area. However, in many real-life situations, there is ample of evidence indicating the lack of uniformity in the effects of space. Local variations of relationships over space commonly exist in spatial data sets and the assumption of stationarity or structural stability over space may be unrealistic (see for example, Anselin 1988; Fotheringham et al. 1996; Fotheringham 1997). It is shown that, as stated in Brunsdon et al (1996), (1) relationships can vary significantly over space and a "global" estimate of the relationships may obscure interesting geographical phenomena; (2) variation over space can be sufficiently complex that it invalidates simple trend-fitting exercises. So when analyzing spatial data, particularly in data mining, we should take into account this kind of spatial non-stationarity.

Over the years, some approaches have been proposed to incorporate spatial structural instability or spatial drift into the models. For example, Anselin (1988, 1990) has investigated regression models with spatial structural change. Casetti (1972, 1986), Jones and Casetti (1992), Fotheringham and Pitts (1995) have studied spatial variations by the expansion method. Basing on the locally weighted regression method, Cleveland (1979), Cleveland and Devlin (1988), Casetti (1982), Foster and Gorr (1986), Gorr and Olligschlaeger (1994), Brunsdon et al. (1996, 1997), Fotheringham et al (1997a,b) have examined the following varying-parameter regression model:

$$y_i = \beta_{i0} + \sum_{k=1}^{p} \beta_{ik} x_{ik} + \varepsilon_i. \tag{5.41}$$

Unlike the OLR model in (5.38), this model allows the parameters to vary in space. However, this model in its unconstrained form is not implementable because the number of parameters increases with the number of observations, i.e., the curse of dimensionality. Hence, strategies for limiting the number of degrees of freedom used to represent variation of the parameters over space should be developed when the parameters are estimated.

There are several methods for estimating the parameters. For example, the method of spatial adaptive filter (Foster and Gorr 1986; Gorr and Olligschlaeger 1994) uses generalized damped negative feedback to estimate spatially-varying parameters of the model in (5.41). However, this approach incorporates spatial relationships in a rather ad hoc manner and produces parameter estimates that cannot be tested statistically. Locally weighted regression method and kernel regression method (Cleveland 1979; Casetti 1982; Cleveland and Devlin 1988; Cleveland et al. 1988; Brunsdon 1995; Wand and Jones 1995) focus mainly on the fit of the dependent variable rather than on spatially varying parameters. Furthermore, the weighting system depends on the location in the "attribute space" (Openshaw 1993) of the independent variables. Along this line of thinking, Brunsdon et al. (1996, 1997), Fotheringham et al. (1997a,b, 2002) suggest a so-called geographically weighted regression (GWR) technique. The mathematical representation of the GWR model is actually the same as the varying-parameter regression model in (5.41). In the following subsection, I will outline the GWR model and the basic issues involved in using it as a means to unravel local variations in spatial relationships.

5.3.2 *Geographically Weighted Regression and the Local–Global Issue About Spatial Non-Stationarity*

In the GWR Model, the parameters are assumed to be functions of the locations on which the observations are obtained. That is,

$$y_i = \beta_{i0} + \sum_{k=1}^{p} \beta_{ik} x_{ik} + \varepsilon_i, \quad i \in C = \{1, 2, \cdots n\}, \tag{5.42}$$

where C is the index set of locations of n observations and β_{ik} is the value of the kth parameter at location i.

The parameters in the GWR model are estimated by the weighted least squares approach. The weighting matrix is taken as a diagonal matrix where each element in its diagonal is assumed to be a function of the location of observation. Suppose that the weighting matrix at location i is $\mathbf{W}(i)$. Then the parameter vector at location i is estimated as

$$\hat{\boldsymbol{\beta}}(i) = \left(\mathbf{X}^T \mathbf{W}(i) \mathbf{X}\right)^{-1} \mathbf{X}^T \mathbf{W}(i) \mathbf{Y}, \tag{5.43}$$

where $\mathbf{W}(i) = diag(w_1(i), w_2(i), \cdots, w_n(i))$ and **X, Y** are the same matrices as in Eq. (4.4). Here we assume that the inverse of the matrix $\mathbf{X}^T\mathbf{W}(i)\mathbf{X}$ exists.

According to the principle of the weighted least squares method, the generated estimators at location i in (5.43) are obtained by solving the following optimization problem. That is, determine the parameters $\beta_0, \beta_1, \cdots, \beta_p$ at each location i so that

$$\sum_{j=1}^{n} w_j(i) \left(y_j - \beta_0 - \beta_1 x_{j1} - \cdots - \beta_p x_{jp}\right)^2 \tag{5.44}$$

is minimized. Given appropriate weights $w_j(i)$ which are a function of the locations at which the observations are made, different emphases can be given to different observations for generating the estimated parameters at location i.

5.3.2.1 Possible Choices of the Weighting Matrix

The role of the weighting matrix is to place different emphases on different observations in generating the estimated parameters. In spatial analysis, observations close to a location i are generally assumed to exert more influence on the parameter estimates at location i than those farther away. When the parameters at location i are estimated, more emphases should be placed on the observations which are close to location i. A simple but natural choice of the weighting matrix at location i is to exclude those observations that are farther than some distance d from location i. This is equivalent to setting a zero weight on observation j if the distance from i to j is greater than d. If the distance from i to j is expressed as d_{ij}, the elements of the weighting matrix at location i can be chosen as

$$w_j(i) = \begin{cases} 1, & \text{if } d_{ij} \leq d \\ 0, & \text{if } d_{ij} > d \end{cases}, \quad j = 1, 2, \cdots, n. \tag{5.45}$$

The above weighting function suffers from the problem of discontinuity over the study area. One way to overcome this problem is to specify $w_j(i)$ as a continuous and monotone decreasing function of d_{ij}. One obvious choice can be

$$w_j(i) = \exp\left(-\theta d_{ij}^2\right), \quad j = 1, 2, \cdots, n, \tag{5.46}$$

so that if i is a point at which observation is made, the weight assigned to that observation will be unity and the weights of the others will decrease according to a Gaussian curve as d_{ij} increases. Here, θ is a non-negative constant depicting the way the Gaussian weights vary with distance. Given d_{ij}, the larger is θ, the less emphasis is placed on the observation at location j. The problem in (5.46) amounts to assigning weights to all locations of the study area.

A compromise between the above two weighting functions can be reached by setting the weights to be zero outside a radius d and to decrease monotonically to zero inside the radius as d_{ij} increases. For example, we can take the elements of the weighting matrix as a bi-square function, i.e.,

$$w_j(i) = \begin{cases} \left(1 - \frac{d_{ij}^2}{d^2}\right)^2, & \text{if } d_{ij} \leq d \\ 0, & \text{if } d_{ij} > d \end{cases}, \quad j = 1, 2, \ldots, n. \tag{5.47}$$

The weighting function in (5.46) is the most common choice in practice.

Compared with other methods, the GWR technique appears to be a relatively simple but useful geographically-oriented method to explore spatial non-stationarity. Based on the GWR model, not only can variation of the parameters be explored, but significance of the variation can also be tested. Unfortunately, at present, only Monte Carlo simulation has been used to perform tests on the validity of the model. In this technique, under the null hypothesis that the global linear regression model holds, any permutation of the observations $\left(y_i; x_{i1}, x_{i2}, \cdots, x_{ip}\right)$, $i = 1, 2, \cdots, n$, among the geographical sampling points are equally likely to occur. The observed values of the statistics proposed can then be compared with these randomization distributions and the significant tests can be performed accordingly. The computational overhead of this method is however considerable, especially for a large data set. Also, since the validity of these randomization distributions is limited to the given data set, this in turn restricts the generality of the proposed statistics. The ideal way to test the model is to construct appropriate statistics and to perform the tests in a conventional statistical manner.

To test whether relationships unraveled from spatial data are local or global, the following two questions are the most important and should be rigorously tested within the conventional hypothesis testing framework:

1. Does a GWR model describe the data significantly better than an OLR model?

That is, on the whole, do the parameters in the GWR model vary significantly over the study region?

2. Does each set of parameters $\beta_{ik}, i = 1, 2, \cdots, n$, exhibit significant variation over the study region?

That is, the effect of which independent variable has significant local variation?

For the first question, it is, in fact, a goodness-of-fit test for a GWR model. It is equivalent to test whether or not $\theta = 0$ if we use (5.46) as the weighting function. In the second case, for any fixed k, the deviation of $\beta_{ik}, i = 1, 2, \cdots, n$, can be used to evaluate the variation of the slope of the kth independent variable. Since it is very difficult to find the null distribution of the estimated parameter, say θ in (5.46), in the weighting matrix, a Monte-Carlo technique has been employed to perform the tests (Brunsdon et al. 1996; Fotheringham et al. 1997a). However, as pointed out above, the computational overhead of the method is considerable. Furthermore, the validity of the reference distributions obtained by the randomized permutation is limited to the given data set, and it in turn may restrict the generality of the corresponding statistics.

5.3.2.2 Goodness-of-Fit Test of the Independent Variables

Based on the notion of residual sum of squares and the following assumptions, some statistics are constructed in Leung et al (2000b):

Assumption 5.1. *The error terms $\varepsilon_1, \varepsilon_2, \cdots, \varepsilon_n$ are independently and identically distributed as a normal distribution with zero mean and constant variance σ^2.*

Assumption 5.2. *Let $\hat{y}_i$ be the fitted value of y_i at location i. For all $i = 1, 2, \cdots, n$, $\hat{y}_i$ is an unbiased estimate of $E(y_i)$. That is, $E(\hat{y}_i) = E(y_i)$ for all i.*

Assumption 5.1 is in fact the conventional assumption in theoretical analysis of regression. Assumption 5.2 is in general not exactly true for local linear fitting except that the exact global linear relationship between the dependent variable and the independent variables exist (see Wand and Jones 1995, pp. 120–121 for the univariate case). However, the local-regression methodology is mainly oriented towards the search for low-bias estimates (Cleveland et al. 1988). In this sense, the bias of the fitted value could be neglected. So, Assumption 5.2 is a realistic one in the GWR model since this technique still belongs to the local-regression methodology.

1. The residual sum of squares and its approximated distribution

Let $\mathbf{x}_i^{\mathrm{T}} = (1\, x_{i1} \cdots x_{ip})$ be the ith row of $\mathbf{X}, i = 1, 2, \cdots, n$, and $\hat{\boldsymbol{\beta}}(i)$ the estimated parameter vector at location i. Then the fitted value of y_i is

$$\hat{y}_i = \mathbf{x}_i^T \hat{\boldsymbol{\beta}}(i) = \mathbf{x}_i^T \left(\mathbf{X}^T \mathbf{W}(i) \mathbf{X}\right)^{-1} \mathbf{X}^T \mathbf{W}(i) \mathbf{Y}. \tag{5.48}$$

Let $\hat{\mathbf{Y}} = (\hat{y}_1\hat{y}_2\cdots\hat{y}_n)^T$ be the vector of the fitted values and $\hat{\varepsilon} = (\hat{\varepsilon}_1\hat{\varepsilon}_2\cdots\hat{\varepsilon}_n)^T$ the vector of the residuals. Then

$$\hat{\mathbf{Y}} = \mathbf{LY}, \tag{5.49}$$

$$\hat{\varepsilon} = \mathbf{Y} - \hat{\mathbf{Y}} = (\mathbf{I} - \mathbf{L})\mathbf{Y}, \tag{5.50}$$

where

$$\mathbf{L} = \begin{pmatrix} \mathbf{X}_1^T\left(\mathbf{X}^T\mathbf{W}(1)\mathbf{X}\right)^{-1}\mathbf{X}^T\mathbf{W}(1) \\ \mathbf{X}_2^T\left(\mathbf{X}^T\mathbf{W}(2)\mathbf{X}\right)^{-1}\mathbf{X}^T\mathbf{W}(2) \\ \vdots \\ \mathbf{X}_n^T\left(\mathbf{X}^T\mathbf{W}(n)\mathbf{X}\right)^{-1}\mathbf{X}^T\mathbf{W}(n) \end{pmatrix} \tag{5.51}$$

Denote the residual sum of squares by RSS_g. Then

$$RSS_g = \sum_{i=1}^{n} \hat{\varepsilon}_i^2 = \hat{\boldsymbol{\varepsilon}}^T\hat{\boldsymbol{\varepsilon}} = \mathbf{Y}^T(\mathbf{I} - \mathbf{L})^T(\mathbf{I} - \mathbf{L})\mathbf{Y}. \tag{5.52}$$

This quantity measures the goodness-of-fit of a GWR model for the given data and can be used to estimate σ^2, the common variance of the error terms $\varepsilon_i,\quad i = 1, 2, \cdots, n$.

2. Goodness-of-Fit Test

Using the residual sum of squares and its approximated distribution, we can test whether a GWR model describes a given data set significantly better than an OLR model. If a GWR model is used to fit the data, under Assumption 5.2, Leung et al. (2000b) show that the residual sum of squares can be expressed as (5.52) and the distribution of $\delta_1 RSS_g/\delta_2\sigma^2$ can be approximated by a chi-square distribution with δ_1^2/δ_2 degrees of freedom, where $\delta_1 = tr\left[(\mathbf{I} - \mathbf{L})^T(\mathbf{I} - \mathbf{L})\right]$, $\delta_2 = tr\left[(\mathbf{I} - \mathbf{L})^T(\mathbf{I} - \mathbf{L})\right]^2$, and σ^2 is the common variance of the error terms whose unbiased estimate is RSS_g/δ_1.

If an OLR model is used to fit the data, the residual sum of squares is $RSS_o = \mathbf{Y}^T(\mathbf{I} - \mathbf{Q})\mathbf{Y}$, where $\mathbf{Q} = \mathbf{X}\left(\mathbf{X}^T\mathbf{X}\right)^{-1}\mathbf{X}^T$ and $\mathbf{I} - \mathbf{Q}$ is idempotent. So, RSS_o/σ^2 is exactly distributed as a chi-square distribution with $n - p - 1$ degrees of freedom (Neter et al. 1989; Hocking 1996).

If the null hypothesis, H_o : there is no significant difference between OLR and GWR models for the given data, is true, then the quantity RSS_g/RSS_o is close to one. Otherwise, it tends to be small. Let

$$F = \frac{RSS_g/\delta_1}{RSS_o/(n - p - 1)}. \tag{5.53}$$

Then a small value of F supports the alternative hypothesis that the GWR model has a better goodness-of-fit. On the other hand, the distribution of F may reasonably be approximated by an F- distribution with δ_1^2/δ_2 degrees of freedom in the numerator and $n-p-1$ degrees of freedom in the denominator. Given a significance level α, we denote by $F_{1-\alpha}\left(\delta_1^2/\delta_2, n-p-1\right)$ the upper $100(1-\alpha)$ percentage point. If $F < F_{1-\alpha}\left(\delta_1^2/\delta_2, n-p-1\right)$, we reject the null hypothesis and conclude that the GWR model describes the data significantly better than the OLR model. Otherwise, we will say that the GWR model cannot improve the fitness significantly compared with the OLR model. Testing the goodness-of-fit via the analysis of variance method and a stepwise procedure for selecting the independent variables are also given in Leung et al. (2000b).

3. Test for Variation of each set of Parameters

After a final model is selected, we can further test whether or not each set of parameters in the model varies significantly across the study region. For example, if the set of parameters $\{\beta_{ik};\ i=1,2,\cdots,n\}$ of x_k (if $k=0$, the parameters examined correspond to the intercept terms) is tested not to vary significantly over the region, we can treat the coefficient of x_k to be constant and conclude that the slope between x_k and the dependent variable is uniform over the area when the other variables are taken to be fixed. Statistically, it is equivalent to testing the hypotheses

$$
\begin{aligned}
&H_0: \beta_{1k} = \beta_{2k} = \cdots = \beta_{nk} \ \textit{for a given } k,\\
&H_1: \textit{not all } \beta_{ik}, i = 1,2,\cdots,n, \ \textit{are equal}.
\end{aligned}
$$

First, we must construct an appropriate statistic which can reflect the spatial variation of the given set of parameters. A practical and yet natural choice is the sample variance of the estimated values of $\beta_{ik}, i=1,2,\cdots,n$. We denote by V_k^2 the sample variance of the n estimated values, $\hat{\beta}_{ik}, i=1,2,\cdots,n$, for the kth parameter. Then

$$
V_k^2 = \frac{1}{n}\sum_{i=1}^{n}\left(\hat{\beta}_{ik} - \frac{1}{n}\sum_{i=1}^{n}\hat{\beta}_{ik}\right)^2, \tag{5.54}
$$

where $\hat{\beta}_{ik}$ $(i=1,2,\cdots,n)$ are obtained by (5.43).

The next stage is to determine the sampling distribution of V_k^2 under the null hypothesis H_o. Let $\hat{\boldsymbol{\beta}}_k = \left(\hat{\beta}_{1k}\hat{\beta}_{2k}\cdots\hat{\beta}_{nk}\right)^T$ and $\mathbf{J}$ be a $n\times n$ matrix with unity for each of its elements. Then V_k^2 can be expressed as

$$
V_k^2 = \frac{1}{n}\hat{\boldsymbol{\beta}}_k^T\left(\mathbf{I} - \frac{1}{n}\mathbf{J}\right)\hat{\boldsymbol{\beta}}_k. \tag{5.55}
$$

Under the null hypothesis that all of the $\beta_{ik},\ i=1,2,\cdots,n$, are equal, we may assume that the means of the corresponding estimated parameters are equal, i.e.,

$$
E\left(\hat{\beta}_{ik}\right) = E\left(\hat{\beta}_{2k}\right) = \cdots = E\left(\hat{\beta}_{nk}\right) = \mu_k \tag{5.56}
$$

Thus,

$$E\left(\hat{\boldsymbol{\beta}}_k\right) = \mu_k \mathbf{1}\,, \tag{5.57}$$

where $\mathbf{1}$ is a column vector with unity for each element. From (5.57) and the fact that $\mathbf{1}^T\left(\mathbf{I} - \frac{1}{n}\mathbf{J}\right) = \mathbf{0}$ and $\left(\mathbf{I} - \frac{1}{n}\mathbf{J}\right)\mathbf{1} = \mathbf{0}$, we can further express V_k^2 as

$$V_k^2 = \frac{1}{n}\left[\hat{\boldsymbol{\beta}}_k - E\left(\hat{\boldsymbol{\beta}}_k\right)\right]^T \left(\mathbf{I} - \frac{1}{n}\mathbf{J}\right)\left[\hat{\boldsymbol{\beta}}_k - E\left(\hat{\boldsymbol{\beta}}_k\right)\right]. \tag{5.58}$$

Furthermore, let e_k be a column vector with unity for the $(k+1)$th element and zero for other elements. Then

$$\hat{\beta}_{ik} = \mathbf{e}_k^T \hat{\boldsymbol{\beta}}(i) = \mathbf{e}_k^T \left(\mathbf{X}^T \mathbf{W}(i)\mathbf{X}\right)^{-1} \mathbf{X}^T \mathbf{W}(i)\mathbf{Y} \tag{5.59}$$

and

$$\hat{\boldsymbol{\beta}}_k = \left(\hat{\beta}_{1k}\hat{\beta}_{2k}\cdots\hat{\beta}_{nk}\right)' T = \mathbf{B}\mathbf{Y}\,, \tag{5.60}$$

where

$$\mathbf{B} = \begin{pmatrix} \mathbf{e}_k^T \left(\mathbf{X}^T \mathbf{W}(1)\mathbf{X}\right)^{-1} \mathbf{X}^T \mathbf{W}(1) \\ \mathbf{e}_k^T \left(\mathbf{X}^T \mathbf{W}(2)\mathbf{X}\right)^{-1} \mathbf{X}^T \mathbf{W}(2) \\ \vdots \\ \mathbf{e}_k^T \left(\mathbf{X}^T \mathbf{W}(n)\mathbf{X}\right)^{-1} \mathbf{X}^T \mathbf{W}(n) \end{pmatrix}. \tag{5.61}$$

Substituting (5.60) into (5.58), we obtain

$$\begin{aligned} V_k^2 &= \frac{1}{n}(\mathbf{Y} - E(\mathbf{Y}))^T \mathbf{B}^T \left(\mathbf{I} - \frac{1}{n}\mathbf{J}\right)\mathbf{B}(\mathbf{Y} - E(\mathbf{Y})) \\ &= \boldsymbol{\varepsilon}^T \left(\frac{1}{n}\mathbf{B}^T \left(\mathbf{I} - \frac{1}{n}\mathbf{J}\right)\mathbf{B}\right)\boldsymbol{\varepsilon}, \end{aligned} \tag{5.62}$$

where $\boldsymbol{\varepsilon} \sim N(\mathbf{0},\ \sigma_2\mathbf{I})$ and $\frac{1}{n}\mathbf{B}^T\left(\mathbf{I} - \frac{1}{n}\mathbf{J}\right)\mathbf{B}$ is positive semidefinite.

Similar to the method employed above, the distribution of $\gamma_1 V_k^2/\gamma_2\sigma^2$ can be approximated by a chi-square distribution with γ_1^2/γ_2 degrees of freedom, where

$$\gamma_i = tr\left(\frac{1}{n}\mathbf{B}^T\left(\mathbf{I} - \frac{1}{n}\mathbf{J}\right)\mathbf{B}\right)^i, \quad i = 1, 2. \tag{5.63}$$

Since σ^2 is unknown, we cannot use $\gamma_1 V_k^2/\gamma_2\sigma^2$ as a test statistic directly. However, we know that the distribution of $\delta_1^2\hat{\sigma}^2/\delta_2\sigma^2$ can be approximated by a chi-square distribution with δ_1^2/δ_2 degrees of freedom, where $\hat{\sigma}^2$ is an unbiased estimator of σ^2, and $\delta_i = tr\left((\mathbf{I}-\mathbf{L})^T(\mathbf{I}-\mathbf{L})\right)^i, i = 1, 2$. So, for the statistic

$$F_3(k) = \frac{V_k^2/\gamma_1}{\hat{\sigma}^2}, \tag{5.64}$$

under the assumption in (5.56), its distribution can be approximated by a F-distribution with γ_1^2/γ_2 degrees of freedom in the numerator and δ_1^2/δ_2 degrees of freedom in the denominator. Therefore, we can take F_3 as a test statistic. The large value of F_3 supports the alternative hypothesis H_1. For a given significance level α, find the upper 100α percentage point $F_\alpha\left(\gamma_1^2/\gamma_2, \delta_1^2/\delta_2\right)$. If $F_3 \geq F_\alpha \left(\gamma_1^2/\gamma_2, \delta_1^2/\delta_2\right)$, reject H_0, accept H_0 otherwise.

The simulation results in Leung et al. (2000b) have shown that the test power of the proposed statistics is rather high and their p-values are rather robust to the variation of the parameter in the weighting matrix.

5.3.3 Local Variations of Regional Industrialization in Jiangsu Province, P.R. China

The technique of GWR is employed to explore in Huang and Leung (2002) the relationships between the level of industrialization (the share of industrial output in the total output of industry and agriculture) and various factors over the study area. There are many aspects, such as social, economic, human, geographical, historical and financial factors, that are related to the process of industrialization. The determinant factors of regional industrialization include the share of urban labor in total population (UL), GDP per capita (GP), fixed capital investment per unit of GDP (IG), and the share of township and village enterprises output in gross output value of industry and agriculture (TVGIA). UL is an indicator of the level of urbanization. GP represents the level of economic development. UL and GP set up the context of industrialization in an area. On the other hand, IG and TVGIS are considered factors directly related to the process of industrialization.

Before investigating possible spatial variations in the determinants of industrialization across Jiangsu Province, the global regression equation representing the average relationships of 75 spatial units between the level of industrialization and various factors is obtained as follows:

$$\begin{aligned}
\mathrm{Y} = {} & 41.211 + 0.440\ \mathrm{UL} + 0.0008066\ \mathrm{GP} + 0.381\ \mathrm{IG} + 0.391\ \mathrm{TVGIA} \\
& (14.353) \quad (4.190) \qquad (3.302) \qquad\qquad (4.268) \qquad\quad (7.598) \\
\mathrm{R} = {} & 0.913\ \mathrm{R}^2 = 0.834\ \text{Adjusted}\ \mathrm{R}^2 = 0.824 \\
& \text{Significance level} = 0.001
\end{aligned} \tag{5.65}$$

The numbers in brackets are t-statistics of the estimated parameters. The R-squared value of the above model is 0.834, which means that the equation explains 83.4% of the variance of the level of industrialization in 1995.

To consider the spatial variation of relationships between the level of industrialization and various determinants, the GWR model is applied. To estimate parameters β^*_{ik}, $i = 1, 2, \ldots, n$; $k = 1, 2, \ldots, p$, the study adopts the commonly used Gaussian function

$$W_{ij} = exp\left(-\theta\ d_{ij}{}^2\right),\ i,j = 1,2,\ldots n \tag{5.66}$$

to calculate weight W_{ij} in the weighting matrix. Here, d_{ij} is the geometric distance between the central points of locations i and j. However, β is a nonnegative parameter and different θ will result in different weights. Thus, the estimated parameters of GWR are not unique. The best θ is chosen by the following procedure:

Assume that there are many different possible values of θ. Then, for each θ, the weighting matrix W_i, i = 1, 2, . . ., n, is obtained from using (5.66). Consequently, many weighting matrices can likewise be obtained. A weighted OLS calibration is then used to obtain many sets of β^*_i, i = 1, 2, . . ., n in (5.29). It should be noted that the observations at location i are not included in the estimation of its parameters. Thus, many different values of the estimated independent $Y_{\neq i}{}^*(\theta)$, fitted value of Y_i, can be estimated at this stage, and therefore the scores of the residuals sum of squares, $\sum_i [Y_i - Y_{\neq i}{}^*(\theta)]^2$, can also be calculated. Finally, the best value of θ is selected by minimizing the score of residuals sum of squares.

Applying the above procedure to the analysis of industrialization in Jiangsu province, the best value of θ is obtained. Figure 5.1 shows the CV score against the parameter θ. Thus, the minimum score of the CV value is obtained when θ equals 0.9. That is,

$$\sum_{i=1}^{n}\left[Y_i - Y^*_{\neq i}(0.9)\right]^2 = \min\sum_{i=1}^{n}\left[Y_i - Y^*_{\neq i}(\theta)\right]^2. \tag{5.67}$$

Thus, the weighting matrix $\mathbf{W}_i$, i = 1, 2, . . ., n, is estimated, where $W_{ij} = exp\left(-0.9d_{ij}{}^2\right)$.

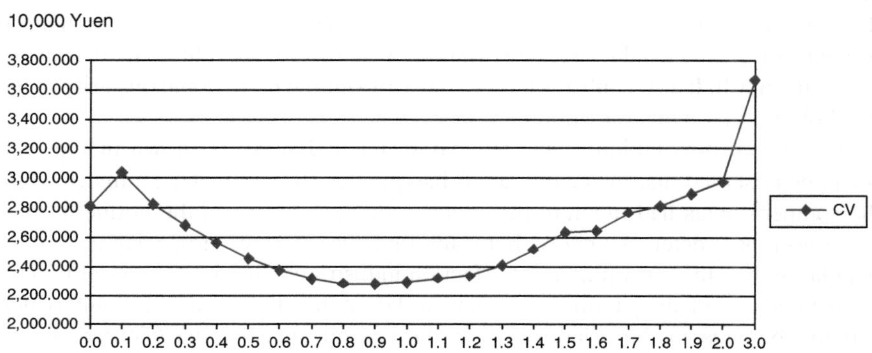

Fig. 5.1 The CV score against the parameter θ

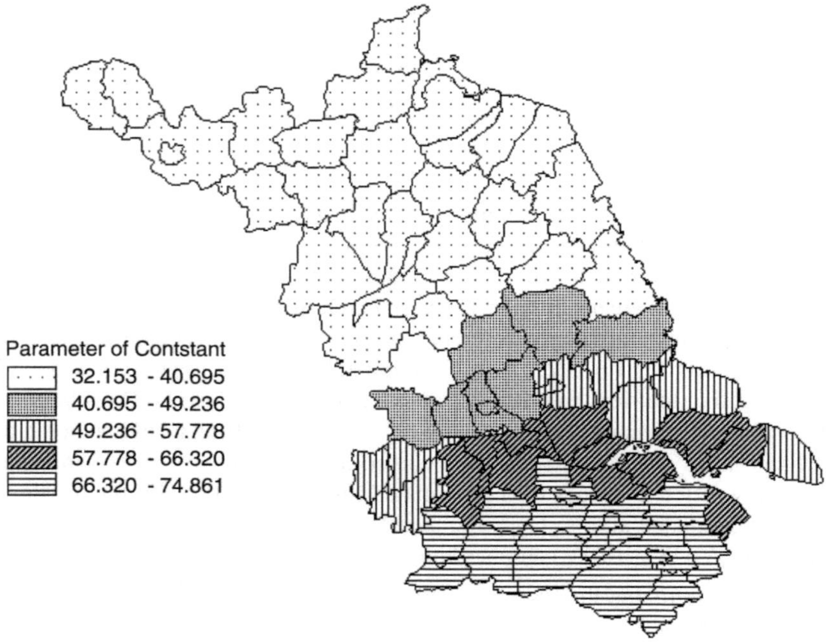

Fig. 5.2 Spatial distribution of the regression constant in Jiangsu

Spatial distributions of the parameter estimates are shown in Figs. 5.2–5.7. Based on the spatial distributions of the parameter estimates, there appears to be significant local variations in the relationships between various factors and industrial development across Jiangsu province. Figure 5.2 shows the spatial distribution of the intercept terms in Jiangsu province in 1995. In principle, the intercept term measures the fundamental level of industrialization excluding the effects of all factors on regional industrialization across Jiangsu province. It is henceforth referred to as "the basic level of regional industrialization." There is a clear spatial variation with higher constant parameters in the southern region and lower ones in the northern region. Thus the basic level of regional industrialization in Jiangsu province displayed a ladder-step distribution which varies from high in the south to low in the north. It also confirms the existence of significant regional disparity in the level of regional industrialization.

The spatial distribution of the UL parameter in Jiangsu is shown in Fig. 5.3. It can be observed that the central areas had greater UL parameter estimates while the southern areas had medium parameter estimates, where as the northern areas had lower parameter estimates. It means that the share of urban labor in total population had the most important effect on industrialization in the central region. On the other hand, the parameter estimate of UL in the global model is 0.440 which actually belongs to the relationship in the central areas of the GWR analysis. Therefore, the relationship of the global model was essentially similar to those of the local models in the central region. This is possibly due to the fact that the

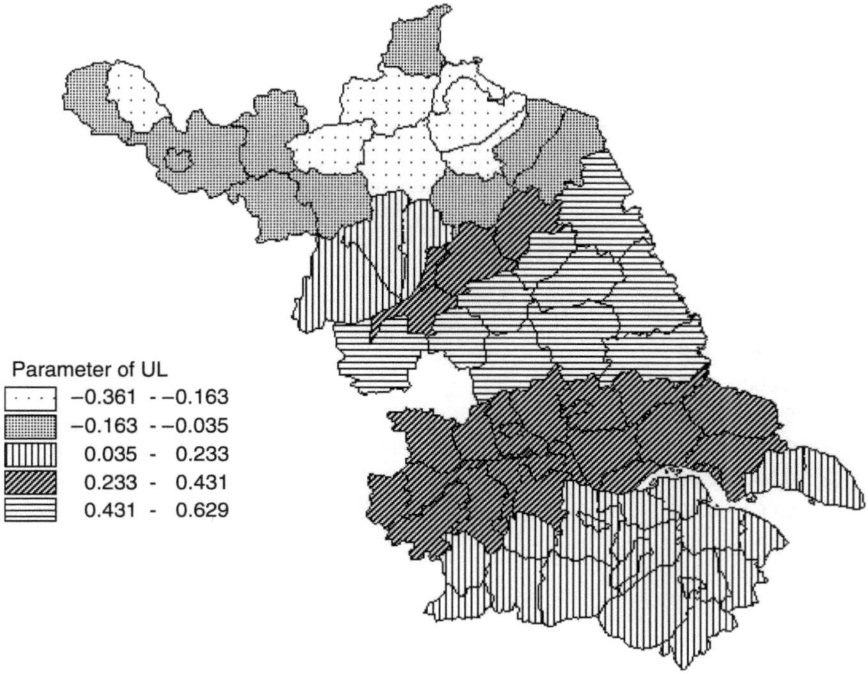

Fig. 5.3 Spatial distribution of the UL parameter in Jiangsu

condition of industrialization in the central region lies between that of the southern and northern regions.

The spatial variation in the GP parameter in Fig. 5.4 depicts the differing effect of GDP per capita on the level of industrialization across Jiangsu in 1995. Similar effect of GDP per capita on regional industrialization was found in most areas but some areas in the northern region exhibited a certain extent of spatial variation in 1995. It means that GDP per capita played a more important role in some northern areas than in other areas.

The spatial distribution of the IG parameter in Fig. 5.5 shows a trend differing from those of the constant and the GP parameters. The fixed capital investment per unit of GDP had the smallest effect on regional industrialization in the southern areas. On the contrary, it exerted the greatest effect on the development of regional industrialization in the central and northern areas. It means that capital investment per unit of GDP was more important in the central and northern regions than in the southern region. It also indicates that the development of regional industrialization in the southern region did not rely very much on the amount of capital investment. It should be observed that the IG parameter in the global model is 0.381. Clearly, the global model represents an average relationship across the study areas.

The spatial distribution of the TVGIA parameter in Fig. 5.6 is very similar to that of the UL parameter in Fig. 5.3. The TVGIA factor had greater effect on regional

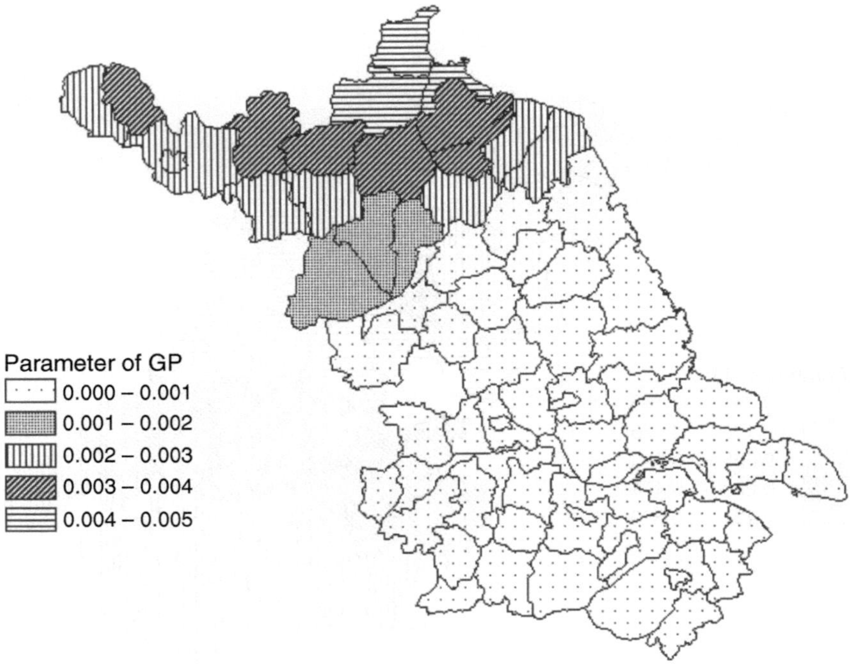

Fig. 5.4 Spatial distribution of the GP parameter in Jiangsu

industrialization in some central and northern areas. It is apparent that TVEs were more important to industrialization in the central and northern areas. The parameter estimate of TVGIA in the global model is 0.391 which is located in the second last group with larger UL parameter in Fig. 5.6. Thus, the global model mainly represents some central and northern areas belonging to the last group of Fig. 5.6.

Another important spatial distribution obtained from the GWR analysis is the spatial variation in the goodness-of-fit statistic, R-square, shown in Fig. 5.7. It shows that the R-square value varies from 0.665 to 0.963. As previously analyzed, the global model explains 83.4% of the variance of the level of industrialization which is between the minimum and the maximum values of R-square. Therefore, some local models have a better fit than the global model, while the others are not. It can be observed that the northern areas usually have higher R-square values. It can then be inferred that the relationships between the selected factors and the level of regional industrialization are much better captured by the regression model in the northern region. However, the development of regional industrialization in the southern and the central regions may be affected by other factors or areas outside Jiangsu province. It is very reasonable to suggest that the economic development of Shanghai plays a very important role in the regional industrialization of the southern or the central areas in Jiangsu since they are close in terms of geographical location. But, the analysis of GWR did not consider the external effect coming from

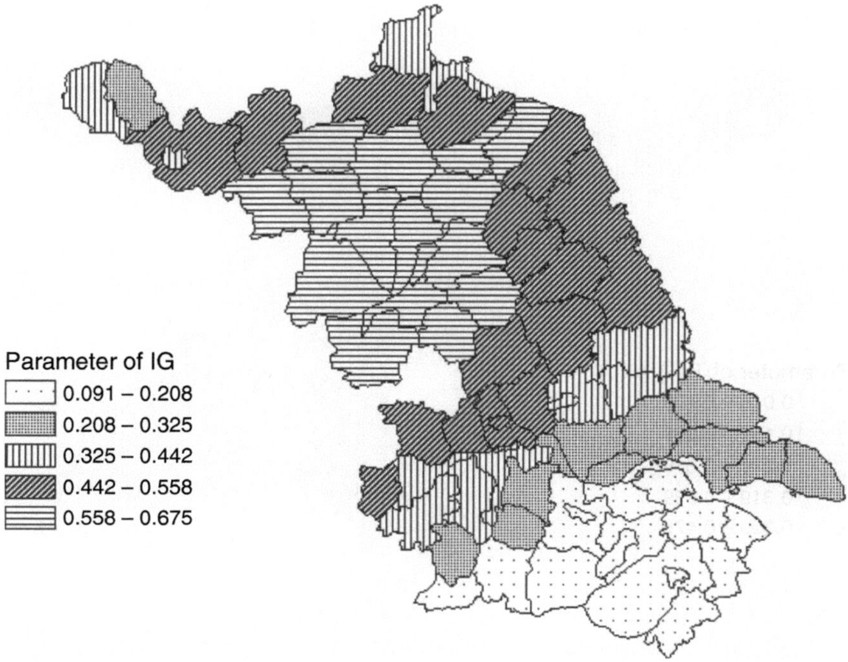

Fig. 5.5 Spatial distribution of the IG parameter in Jiangsu

areas outside Jiangsu. It may be the reason for the smaller R-square value in the central and the southern regions. Such relationships between Shanghai and Jiangsu are not considered since no consistent data are available.

The parameter estimates of various factors affecting regional industrialization in Jiangsu province show different spatial variations, indicating possible spatial nonstationarity. Thus, the GWR technique appears to be a useful method to unravel spatial nonstationarity. However, from the statistical viewpoint, two critical questions still remain. One is whether the GWR model describes the relationship significantly better than the OLR model. The other is whether each set of parameter estimates b_{ij}^*, $i = 1, 2,\ldots, n$; $j = 1, 2,\ldots, P$ exhibit significant spatial variation over the study areas (Leung et al. 2000).

From the result of Table 5.1, it is clear that at the significance level of 0.0081, the GWR model performs better than the OLR model in the analysis of regional industrialization of Jiangsu province. Thus, the relationships of regional industrialization and the factors affecting it exhibit significant spatial nonstationarity over the county-level areas in Jiangsu province.

In terms of the spatial variation of the estimated parameters, the test result shows that the constant parameter and the GP parameter have robust spatial nonstationarity over the whole study area. Statistically, the other three factors, UL, IG and TVGIA, did not have significant spatial variation. Therefore, spatial variation of the

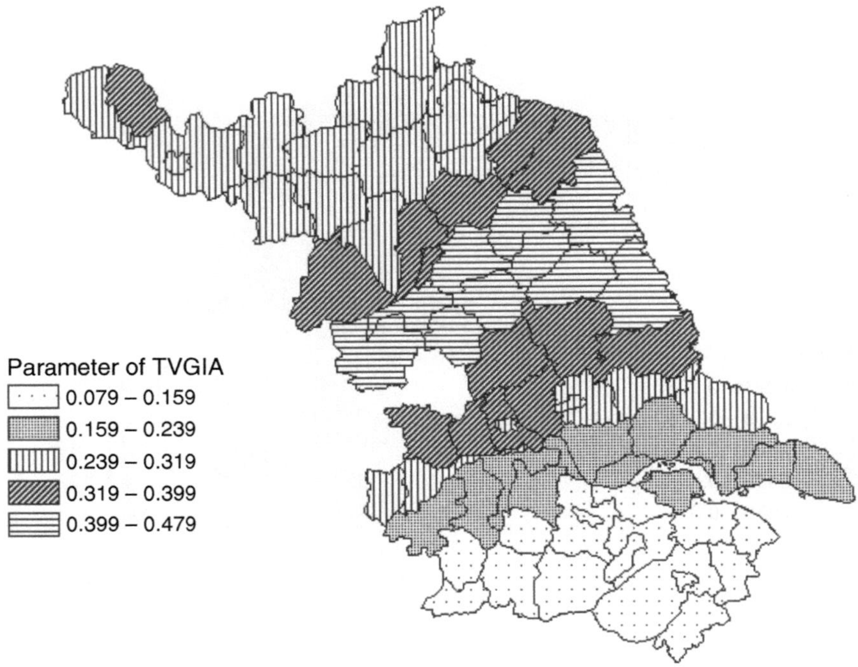

Fig. 5.6 Spatial distribution of the TVGIA parameter in Jiangsu

effect of economic factors on regional industrialization is mainly represented by the factors of the basic level of industrialization and GDP per capita among county-level areas in Jiangsu.

In the GWR analysis, it is assumed that spatial relationships between two areas show the distance-decay effect. However, with the advancement of information technology, friction of distance may be weakened. Nevertheless, in developing countries such as China, distance decay still plays a crucial role in the interaction between areas. Therefore, in the study of regional economic development in China, the GWR technique appears to be an effective tool to explore variations among different localities.

5.3.4 Discovering Spatial Pattern of Influence of Extreme Temperatures on Mean Temperatures in China

It has been recognized that the increase in global mean temperature has close relationship with temperature extremes. Extensive studies have been carried out on the extreme temperature events in different regions of the world (Beniston and Stephenson 2004; Bonsal et al. 2001; DeGaetano 1996; DeGaetano and Allen 2002;

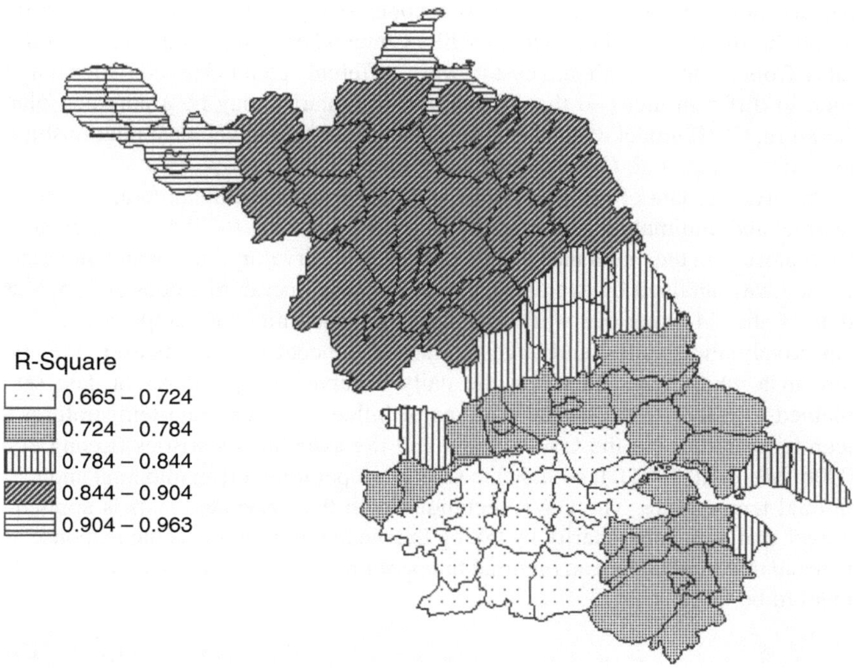

Fig. 5.7 Spatial distribution of the R-Square value in Jiangsu

Table 5.1 Test statistics of the GWR model

Statistics	Value	NDF	DDF	*p*-value
F_1	0.53931	58.15	70	0.0081
$F_3(0)$	10.954	3.96	58.15	1.179×10^{-6}
$F_3(1)$	0.923	3.16	58.15	0.44
$F_3(2)$	2.726	2.31	58.15	0.0066
$F_3(3)$	1.567	4.39	58.15	0.19
$F_3(4)$	1.694	4.71	58.15	0.15

Note NDF and DDF represent the degrees of freedom of the numerator and denominator of the corresponding F-distributions, respectively

Heino et al. 1999; Prieto et al. 2004; Robeson 2004) in general and China (Gong et al. 2004; Qian and Lin 2004; Yan et al. 2001; Zhai and Pan 2003; Zhai et al. 1999) in particular. For China as a whole, the frequency of extremely low temperature exhibits a significant decreasing trends while that of extremely high temperature a slightly decreasing or insignificant trend, which may be a main cause of the increase of mean temperature.

In the study of extreme temperatures, concentration has been placed on the temporal trends of extreme temperatures. While spatial characteristics have generally been analyzed on a station-by-station basis (Beniston and Stephenson 2004; Bonsal et al. 2001; Gong et al. 2004; Prieto et al. 2004; Qian and Lin 2004), such analysis, however, does not take into account spatial autocorrelation of the data among the stations. For large territory like China where temperature varies considerably from north to south and east to west, different spatial characteristics may be found in different areas so that spatial non-stationarity may be a common place. Therefore, GWR model would be a useful technique to unravel local relationships if they exist. Wang et al. (2005) give such a study.

The original data of the study consist of daily observed mean temperature and maximal and minimal temperatures of 40 years from 1961 to 2000 collected at 110 observatories on the mainland of China. At each observatory, the mean temperature in a day was obtained by averaging the observed temperature values at 2, 5, 8 and 20 h of the 24-h period, while the maximal and minimal temperatures were, respectively, the smallest and largest values of the continuously measured temperature in a whole day. Based on the daily observed temperatures, a data set is obtained to discover the spatial patterns of influence of extreme temperatures on mean temperature via the GWR model and the associated statistics (Leung et al. 2000b; Mei et al. 2004). It contains the mean temperature, mean maximal and mean minimal temperature. The GWR technique with the associated tests is applied to unravel spatial nonstationarity by taking the mean temperature as the response and the mean maximal and mean minimal temperature as the explanatory variables. The model to be fitted is

$$y_i = \beta_0(u_i, v_i) + \beta_1(u_i, v_i)x_{i1} + \beta_2(u_i, v_i)x_{i2} + \varepsilon_i, \quad i = 1, 2, \ldots, 110, \qquad (5.68)$$

where $(y_i; x_{i1}, x_{i2})$, $i = 1, 2, \ldots, 110$, are the observations of mean temperature and mean maximal and mean minimal temperatures at the 110 observatories located at longitude u_i and latitude v_i. Based on the Gaussian kernel function, the distance between any two observatories is computed according to the longitudes and latitudes of the observatories to formulate the weight. The optimal bandwidth value is selected by the cross-validation approach.

For the data set, the bandwidth value selected is $h_o = 0.42$ (kilometer $\times\ 10^3$) and the p-values for testing the significant variation of the three coefficients are, respectively, $p_o = 0.0004239,\ \ p_1 = 0.0007347\ \ \textit{and}\ \ p_2 = 0.0000159$, which shows that variation of each coefficient across the mainland of China is very significant.

Based on Fig. 5.8, the contribution rate of mean maximal temperature to mean temperature over 40 years varies rather significantly over the mainland of China. In the northwestern region where the latitude is great than about 45°, it is discovered that the rates (largest) range from about 0.6 to 1.182 from north to south. That is, the sharpest increase in mean temperature with the increase of mean maximal temperature is discovered in the coldest area of China. On the other hand, the smallest contribution rates which vary from about 0.2 to 0.4 are detected around Bohai Bay, the southwestern region and the northern part of Xingjiang province. The remaining part of Mainland China, northwest to southeast, shows a roughly homogenous contribution rates arranging from about 0.6 to 0.8. It is interesting to observe that the contribution rates of mean maximal temperature to mean temperature appear in apparent regional clusters.

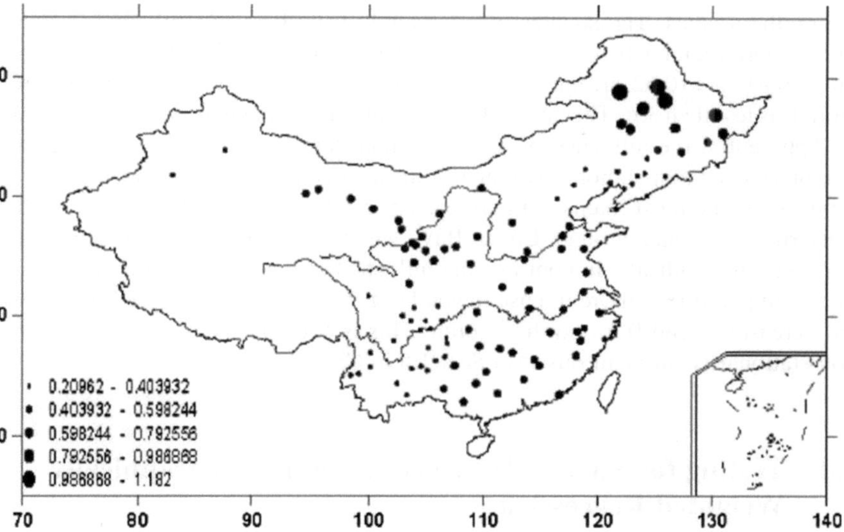

Fig. 5.8 Spatial distribution of the estimates for the coefficient $\beta_1(u_i, v_i)$ of mean maximal temperature over 40 years

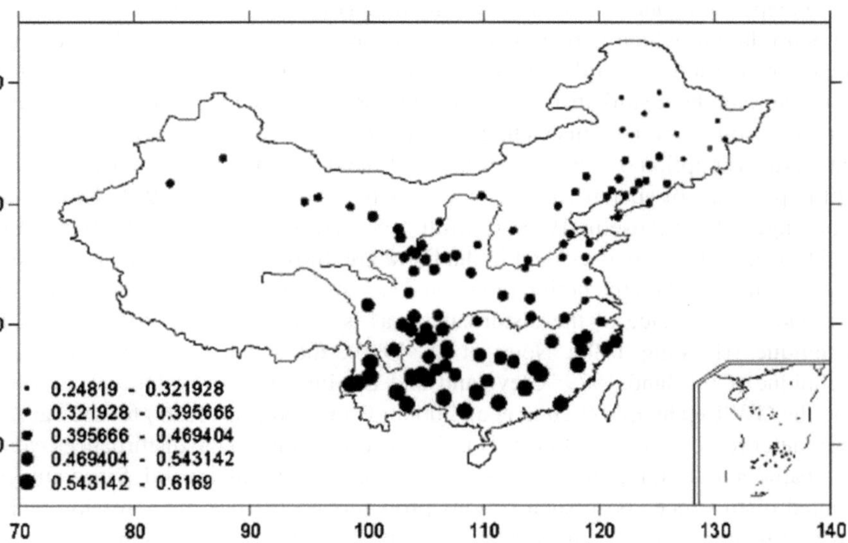

Fig. 5.9 Spatial Distribution of the estimates for the coefficient $\beta_2(u_i, v_i)$ of mean minimal temperature over 40 years

From Fig. 5.9, the contribution rates of mean minimal temperature to mean temperature over 40 years reveal a significant increasing trend from north to south over the mainland of China. Specifically, when mean minimal temperature increases a unit, the increase of mean temperature is greater in the southern areas than in the northern areas. The smallest rates, roughly from 0.25 to 0.39, are observed in the northern region where latitude is greater than about 44°. The largest rates, arranging from 0.47 to 0.62, are unraveled mainly on the south of the Yangzi river where the latitude is less than 30°. The rates in the remaining areas range from about 0.32 to 0.47.

Apparently, the influence of mean maximal temperature on mean temperature exhibits spatial non-stationarity that appears as several obvious spatial clusters. The influence is the most intense in northeastern region and the least intense in southwestern region and around Bohai Bay, while the influence is moderate from northwest to southeast. In contrast, the influence of mean minimal temperature on mean temperature is more intense in southern than in the northern region, showing an increasing trend from north to south. This is actually the answer to the spatial non-stationarity problem raised in Sect 1.5 in Chap 1.

5.4 Testing for Spatial Autocorrelation in Geographically Weighted Regression

It should be observed that one of the important assumptions for the GWR technique to be applied to the varying-parameter model in (5.41) is that the disturbance terms are independent and identically distributed. However, the existence of spatial autocorrelation, which is one of the main characteristics of spatial data sets, may invalidate certain standard methodological results. For example, spatial autocorrelation among the disturbance terms in the OLR model can lead to inefficient least-squares estimators and misleading statistical inference results. Furthermore, the standard assumption of constant variance of the disturbance terms may fail to hold in the presence of spatial autocorrelation (Cliff and Ord 1973, 1981; Krämer and Donninger 1987; Anselin 1988; Griffith 1988; Anselin and Griffith 1988; Cordy and Griffith 1993). As is evident in the literature, most statistical tests in regression analysis are based on the notion of residual sum of squares, more specifically on the estimator of variance of the disturbances, as is adopted in the well known OLR technique (Hocking 1996; Neter et al. 1996), the locally weighted regression technique (Cleveland 1979; Cleveland and Devlin 1988; Cleveland et al. 1988), and the GWR technique (Leung et al. 2000b; Brunsdon et al. 1999) for the varying-parameter regression model in (5.41). Heteroscedasticity in the disturbances caused by spatial autocorrelation thus makes such testing methods invalid. Since autocorrelated disturbances pose such serious problems on the use of regression techniques, it is then extremely important to be able to test for their presence.

For the OLR technique, this problem has long been investigated. Substantial effort has been devoted to the tests for spatial autocorrelation in the OLR model. Two basic types of test methods are commonly used in the literature. One is the generalized form of Moran's I_0 (Moran 1950), in order not to confuse it with the notation of the identity matrix I the Moran's statistic is denoted by I_0 instead of the conventional I in this discussion, or Geary's c (Geary 1954) to the OLR residuals suggested by Cliff and Ord (1972, 1973, 1981). The other is the likelihood-function-based methods such as the Lagrange multiplier form of test (Burridge 1980) or the likelihood ratio test (Griffith 1988; Anselin 1988). Both types rely upon the asymptotic distribution of the statistics under the null hypothesis of no spatial autocorrelation. Recently, based on the theoretical results by Imhof (1961) and the algebraic results by Koerts and Abrahamse (1968), Tiefelsdorf and Boots (1995, with corrections 1996), as well as Hepple (1998) have independently derived the exact distributions of Moran's I_0 and Geary's C for the OLR residuals under the null hypothesis of no spatial autocorrelation among the normally distributed disturbances. Based on the test statistics of Moran's I_0 and Geary's c, Leung et al. (2000c) first extend the exact test method developed by Tiefelsdorf and Boots (1995), and Hepple (1998) for the OLR residuals to the GWR case.

A statistical procedure is developed by Leung et al. (2000c) to test for spatial autocorrelation among the residuals of the GWR model. They focus on the test of spatial autocorrelation among the disturbance terms $\varepsilon_1, \varepsilon_2, \cdots, \varepsilon_n$ of the model in (5.41) when the GWR technique is employed to calibrate it. Similar to the case of the OLR model, the null hypothesis for testing spatial autocorrelation in the varying-parameter model can still be formulated as:

H_0: There is no spatial autocorrelation among the disturbances, or alternatively

$$Var(\boldsymbol{\varepsilon}) = E(\boldsymbol{\varepsilon}\boldsymbol{\varepsilon}^T) = \sigma^2\mathbf{I}$$

where $\boldsymbol{\varepsilon} = (\varepsilon \quad \varepsilon_2 \quad \cdots \quad \varepsilon_n)^T$ is the disturbance vector.

The alternative hypothesis is that there exists (positive or negative) spatial autocorrelation among the disturbances with respect to a specific spatial weight matrix $\mathbf{W}$ which is defined by the underlying spatial structure such as the spatial contiguity or adjacency between the geographical units where observations are made. The simplest form of $\mathbf{W}$ can be the one that assigns 1 to two units that come in contact and 0 otherwise. It can also incorporate information on distances, flows, and other types of linkages.

Since the disturbance vector $\boldsymbol{\varepsilon} = (\varepsilon \quad \varepsilon_2 \quad \cdots \quad \varepsilon_n)^T$ is not observable, the autocorrelation among the residuals is tested instead, i. e., the errors which result by comparing each local GWR estimate of each y with the actual value. When the model in (5.41) is calibrated by the GWR technique, we obtain the results from (5.48) to (5.52).

Spatial autocorrelation based on Moran's I_0 and Geary's c

For the residuals $\hat{\boldsymbol{\varepsilon}} = (\hat{\varepsilon}_1 \quad \hat{\varepsilon}_2 \quad \cdots \quad \hat{\varepsilon}_n)^T$ in (5.49) and (5.50), and a specific spatial weight matrix $\mathbf{W} = (w_{ij})$, Moran's I_0 takes the form of

$$I_0 = \frac{n}{s} \frac{\sum_{i=1}^{n}\sum_{j=1}^{n} w_{ij}\hat{\varepsilon}_i\hat{\varepsilon}_j}{\sum_{i=1}^{n}\hat{\varepsilon}_i^2} = \frac{n}{s}\frac{\hat{\boldsymbol{\varepsilon}}^T\mathbf{W}\hat{\boldsymbol{\varepsilon}}}{\hat{\boldsymbol{\varepsilon}}^T\hat{\boldsymbol{\varepsilon}}}, \tag{5.69}$$

where $s = \sum_{i=1}^{n} \sum_{j=1}^{n} w_{ij}$. The spatial weight matrix is commonly used in its row-standardized form. That is, the row elements are normalized (summed to 1) and this may make W asymmetric. Nevertheless, if $\mathbf{W}$ is asymmetric, we can construct from it a new symmetric spatial weight matrix as

$$\mathbf{W}^* = \left(w_{ij}^*\right) = \frac{1}{2}\left(\mathbf{W} + \mathbf{W}^T\right). \tag{5.70}$$

Since $\hat{\boldsymbol{\varepsilon}}^T \mathbf{W}^T \hat{\boldsymbol{\varepsilon}}^T = \hat{\boldsymbol{\varepsilon}}^T \mathbf{W} \hat{\boldsymbol{\varepsilon}}^T$, we have

$$\frac{\hat{\boldsymbol{\varepsilon}}^T \mathbf{W}^* \hat{\boldsymbol{\varepsilon}}^T}{\hat{\boldsymbol{\varepsilon}}^T \hat{\boldsymbol{\varepsilon}}} = \frac{\hat{\boldsymbol{\varepsilon}}^T \mathbf{W} \hat{\boldsymbol{\varepsilon}}}{\hat{\boldsymbol{\varepsilon}}^T \hat{\boldsymbol{\varepsilon}}}. \tag{5.71}$$

Thus, without loss of generality, we can assume that $\mathbf{W}$ is symmetric. Also, the term n/s in (5.69) is purely a scaling factor which can be omitted from the test statistic without affecting the p-value of the statistic. Hence, we can write Moran's I_0 as

$$I_0 = \frac{\hat{\boldsymbol{\varepsilon}}^T \mathbf{W} \hat{\boldsymbol{\varepsilon}}}{\hat{\boldsymbol{\varepsilon}}^T \hat{\boldsymbol{\varepsilon}}}, \tag{5.72}$$

where $\mathbf{W}$ is a specific symmetric spatial weight matrix of order n.

It is known that a large value of I_0 supports the alternative hypothesis that there exists positive autocorrelation among the residuals and a large negative value of I_0 supports the alternative hypothesis that there exists negative autocorrelation among the residuals. For these two alternatives, the p-values of I_0 are, respectively: $p = P\{I_0 \geq r\}$, and $p = P\{I_0 \leq r\}$, where r is the observed value of I_0. It should be noted that the above two alternatives belong to the one-tailed test. For spatial autocorrelation which corresponds to a two-tailed test, considering the complexity of the distribution of I_0, we may simply take the p-value as $2P\{I_0 \geq r\}$ if $P\{I_0 \geq r\} \leq 1/2$ or $2(1 - P\{I_0 \geq r\})$ if $P\{I_0 \geq r\} > 1/2$.

Thus, for a given significance level α, if $p \geq \alpha$, one fails to reject the null hypothesis H_0 and concludes that there is no spatial autocorrelation among the residuals. If $p < \alpha$, one, depending on the assumed alternative hypothesis, rejects H_0 and concludes that there exists positive or negative autocorrelation among the residuals. Leung et al. (2000c) show how the p-values can be computed via the Imhof result (Imhof 1961).

Similarly, for the residual vector $\hat{\boldsymbol{\varepsilon}} = (\hat{\varepsilon}_1 \quad \hat{\varepsilon}_2 \quad \cdots \quad \hat{\varepsilon}_n)^T$ and a specific spatial weight matrix $\mathbf{W} = \left(w_{ij}\right)$, Geary's C is obtained as

$$c = \frac{(n-1)}{2s} \frac{\sum_{i=1}^{n} \sum_{j=1}^{n} w_{ij} \left(\hat{\varepsilon}_i - \hat{\varepsilon}_j\right)^2}{\sum_{i=1}^{n} \hat{\varepsilon}_i^2}. \tag{5.73}$$

With respect to a given spatial weight matrix **W**, a small value of c supports the alternative hypothesis that there exists positive spatial autocorrelation among the residuals and a large value of c supports the one saying that there exists negative spatial autocorrelation. For simplicity, we still use r to represent the observed value of c. The p-values of c for testing H_0 against the above two alternatives are, respectively, $P\{c \leq r\}$ and $P\{c \geq r\}$. They can again computed by the Imhof method.

To circumvent the computation overhead of the resulting Imhof method, particularly for large sample, the three-moment χ^2 approximation to the null distributions of the testing statistics is derived in Leung et al. (2000c). Based on their simulation runs on the Imhof and approximation tests, the following observations are made:

1. The statistics of Moran's I_0 and Greay's c formed by the GWR residuals are quite powerful in exploring spatial autocorrelation among the disturbances of the varying-parameter model, especially for exploring positive autocorrelation. This also implies that in deriving the p-values of the test statistics, it is reasonable to assume that the fitted value of y_i is an unbiased estimate of the $E(y_i)$ for all i. However, the test statistics are not so sensitive to moderate negative autocorrelation. Some improvement on the proposed testing methods will be necessary in order to overcome this shortcoming.
2. The three-moment χ^2 approximation to the p-values of I_0 and c is very accurate. Compared with the computational overhead in obtaining the p-values in the Imhof method, this approximation method is very time-saving, especially for cases with large sample size.
3. The p-values of I_0 and c are quite robust to the variation of the parameter θ in the weighting function for calibrating the model. This makes the testing methods applicable in practice since θ could still be predetermined by the cross-validation procedure without considering spatial autocorrelation. Although there is some loss in the significance of spatial autocorrelation, the testing methods still give useful indications which are sufficient to achieve certain practical purposes, especially for exploring positive autocorrelation.

For both the Imhof method and the three-moment χ^2 approximation method proposed in Leung et al. (2000c), the assumption that the disturbance terms are normally distributed plays an important role in deriving the p-values of I_0 and c. Although it is a common assumption in regression analysis, this condition is not easy to satisfy in practice. Therefore, it will be useful to investigate the null distributions of the test statistics for the GWR model under some more general conditions. Moreover, some improvements on the proposed methods are still needed to make them more powerful in order to test for moderate negative autocorrelation.

It should be noted that the measures of spatial autocorrelation in Leung et al. (2000c), both Moran's I_0 and Geary'sc, are global statistics and therefore, as shown in the simulations, global association among the GWR residuals can be efficiently tested by the proposed methods. They may be insensitive to local spatial autocorrelation. A more practical situation may be to use some local statistics to test more general association among the GWR residuals. The LISA method, i.e., local indicators of spatial association (Anselin 1995) seems to be a promising method to achieve this purpose. Though it will be more difficult to develop formal statistical testing methods such as those proposed in this paper, it deserves to be investigated in further research.

5.5 A Note on the Extentions of the GWR Model

As a further refinement of the basic GWR model, the mixed GWR model, which is a combination of the ordinary linear regression model and the spatially varying coefficient model, was firstly proposed by Brunsdon et al. (1999) to model the situation in which the impact of some explanatory variables on the response is spatially homogeneous and that of the remaining explanatory variables varies over space.

A spatially varying coefficient regression model that the GWR technique calibrates is of the form

$$y_i = \sum_{j=1}^{p} \beta_j(u_i, v_i)x_{ij} + \varepsilon_i, \quad i = 1, 2, \cdots, n, \tag{5.74}$$

where $(y_i; x_{i1}, \cdots, x_{ip})$ are observations of the response y and explanatory variables $x_1, x_2, \cdots, x_p$ at location (u_i, v_i), and $\varepsilon_1, \varepsilon_2, \cdots, \varepsilon_n$ are independent random errors with mean zero and common variance σ^2. Generally, one takes $x_1 \equiv 1$ to accommodate a spatially varying intercept in the model. The GWR technique (Brunsdon et al. 1996; Fotheringham et al. 2002) calibrates the model in (5.1) with the locally weighted least-squares procedure in which the weights in each focal spatial point are generated by a given kernel function and the distance between this focal point and each of the observational locations (u_i, v_i), $i = 1, 2, \cdots, n$. A mixed GWR model (Brunsdon et al. 1999; Fotheringham et al. 2002) takes some of the coefficients $\beta_j(u, v)$ $(j = 1, 2, \cdots, p)$ to be constant and, after properly adjusting the order of the explanatory variables, is of the form

$$y_i = \sum_{j=1}^{q} \beta_j x_{ij} + \sum_{j=q+1}^{p} \beta_j(u_i, v_i)x_{ij} + \varepsilon_i, \quad i = 1, 2, \cdots, n. \tag{5.75}$$

By first smoothing the spatially varying coefficients $\beta_j(u, v)$ $(j = q + 1, \cdots, p)$ with the GWR technique and then estimating the constant coefficients $\beta_j (j = 1, \cdots, q)$ with the ordinary least-squares method, a two-step calibration procedure has been proposed by Fotheringham et al. (2002).

As an extension of the mixed GWR model, it is of interest and practical use to consider another kind of regression models that combines a geographical expansion model with a spatially varying coefficient model. That is, some regression coefficients in a spatially varying coefficient model are assumed to be globally vertain parametric functions of spatial coordinates. Leung et al. (2008b) coin this model the semi-parametric spatially varying coefficient model for the reason that some regression coefficients are parametric functions of spatial coordinates and the others are nonparametric.

Motivated by the geographical expansion method (Casetti 1982, 1997; Jones and Casetti 1992). We can assume that some coefficients in the spatially varying coefficient model in (5.74) are certain parametric functions of spatial coordinates, say $\beta_j(u, v; \theta_{j1}, \cdots, \theta_{jl_j})(j = 1, \cdots, q)$, and the semi-parametric spatially varying coefficient model can be defined as

$$y_i = \sum_{j=1}^{q} \beta_j(u_i, v_i; \theta_{j1}, \cdots, \theta_{jl_j})x_{ij} + \sum_{j=q+1}^{p} \beta_j(u_i, v_i)x_{ij} + \varepsilon_i, \quad i = 1, 2, \cdots, n. \tag{5.76}$$

For simplicity in estimation and sufficiency in application, each of the parametric coefficients $\beta_j(u_i, v_i; \theta_{j1}, \cdots, \theta_{jl_j})(j = 1, \cdots, q)$ is taken to be a linear combination of some known functions of spatial coordinates (u, v), that is,

$$\beta_j(u, v; \theta_{j1}, \cdots, \theta_{jl_j}) = \sum_{k=1}^{l_j} \theta_{jk} g_{jk}(u, v). \tag{5.77}$$

Here for each $(j = 1, 2, \cdots, q)$, $g_{j1}(u, v), g_{j2}(u, v), \cdots, g_{jl_j}(u, v)$ are known linearly independent functions.

The semi-parametric spatially varying coefficient model so constructed includes several commonly used spatial regression models as its special cases. The followings are typical cases:

1. When $q = 0$, the model in (5.76) is the spatially varying coefficient model that the GWR technique calibrates.
2. When $q =$ p, the model in (5.76) becomes a kind of geographical expansion models. In particular, when all of the $\beta_j(u, v; \theta_{j1}, \cdots, \theta_{jl_j})(j = 1, \cdots, p)$ are polynomial functions of spatial coordinates u and v, the resulting models become the most commonly used expansion models in geographical research.
3. Let $l_1 = l_2 \cdots = l_q = 1$ and $g_{j1}(u, v) \equiv 1$ for each $j = 1, 2, \cdots, q$. Then the semi-parametric spatially varying coefficient model becomes the mixed GWR model. Furthermore, of $q = p$, the model degenerates into an ordinary linear regression model. Based on the local linear fitting procedure in Wang et al. (2008) and the OLS method, Leung et al. (2008b) derive a two-step setimation procedure for the model, with its effectiveness supported by some simulation studies.

5.6 Discovery of Spatial Non-Stationarity Based on the Regression-Class Mixture Decomposition Method

5.6.1 *On Mixture Modeling of Spatial Non-Stationarity in a Noisy Environment*

In the study of spatial relationship, we generally assume that a single regression model can be applied to a large or complicated spatial data set manifestating certain spatial structure or pattern. Though parameter-varying regression in general and GWR in particular intend to study spatial non-stationarity, they still assume a single model for the whole data set. Local variations are captured by the varying parameters. Unfortunately, conventional regression analysis is usually not appropriate for the study of very large data sets, especially those with noise contamination for the follow reasons:

1. Regression analysis handles a data set as a whole. Even with the computer hardware available today, there are no effective means, such as processors and storage, for manipulating and analyzing a large amount of data.
2. More importantly, it might be unrealistic to assume that a single model can fit a large data set. It is highly likely that we need multiple models to fit a large data set. That is, spatial patterns hidden in a data set may take on different forms that cannot be accurately represented by a single model.
3. Classical regression analysis is based on stringent model assumptions. However, the real world, a large data set in particular, does not behave in accordance with these assumptions. In a noisy environment, it is very common that inliers (patterns) are out-numbered by outliers so that many robust methods fail.

To overcome the above difficulties, we may want to view a complicated data set as a mixture of many populations. If we view each spatial pattern described by a regression model as a population, then the data set is a mixture of a finite number of such populations. Spatial knowledge (patterns/relationships) discovery can then be treated as the identification of these models through mixture modeling.

Mixture modeling is the modeling of a statistical distribution by a mixture of distributions, known as components or classes. Finite mixture densities have served as important models for the analysis of complex phenomena in statistics (McLachland and Basford 1988). This model deals with the unsupervised discovery of clusters within data (McLachlan 1992). In particular, mixtures of normal populations are most frequently studied and applied in practice. In estimating mixture parameters, the maximum likelihood (ML) method, the maximum likelihood estimator (MLE) in particular, has become the most extensively adopted approach (Redner and Walker 1984). Although the use of the expectation maximization (EM) algorithm greatly reduces the computational difficulty for the MLE of mixture models, the EM algorithm still has drawbacks. The slow convergence of the generated sequence of iterates in some applications is a typical example. Other methods such as the method of moments and the moment generating function (MGF) method generally involve the problem of simultaneously estimating all of the mixture parameters. It is clearly a very difficult task of estimation in large data sets. Therefore, the development of an efficient method to unravel patterns in mixtures is important.

In addition to the efficiency of an estimation method, another important feature that needs to be addressed is robustness. To be useful in practice, a method needs to be very robust, especially for large data sets. It means that the performance of a method should not be significantly affected by small deviations from the assumed model and it should not deteriorate drastically due to noise and outliers. Discussions on and comparison with several popular clustering methods from the point of view of robustness are summarized in Dave and Krishnapuram (1997). Obviously, robustness in spatial knowledge discovery is also necessary. Some attempts have been made in recent years (Hsu and Knoblock 1995; John and Langley 1995) and the problem needs to be further studied.

To have an efficient and robust method for the mining of regression classes in large data sets, especially under contamination with noise, Leung et al. (2001a) introduce a new concept named "regression-class" which is defined by a regression model. The concept is different from the existing conceptualization of class (cluster) based on commonsense or a certain distance measure. As a generalization of classes, a regression class contains more useful information. Their model assumes that there is a finite number of this kind of regression classes in a large data set. Instead of considering the whole data set, sampling is used to identify the corresponding regression classes. A novel framework, formulated in a recursive paradigm, for mining multiple regression classes in a data set is constructed. Based on a highly robust model-fitting (MF) estimator and an effective Gaussian mixture decomposition algorithm (GMDD) in computer vision (Zhuang et al. 1992, 1996), the proposed method, coined regression-class mixture decomposition (RCMD), only involves the parameters of a regression class at each time of the mining process. Thus, it greatly reduces the difficulty of parametric estimation and achieves a high degree of robustness. The method is suitable for small, medium, and large data sets and has many promising applications in a variety of disciplines including computer vision, pattern recognition, and economics.

It is necessary to point out that identifying some regression classes is different from the conventional classification problem, which is concerned with modeling the conditional distribution of a response/dependent variable Y given a set of carriers/independent variables X. It also differs from other models, such as piecewise regression and regression tree, in which different subsets of X follow different regression models. The RCMD method not only can solve the identity problem of regression classes, but may also be extended to other models such as piecewise regression. It can be employed to discover local variations taking different functional forms.

5.6.2 The Notion of a Regression Class

Intuitively, a regression class ("reg-class" in abbreviation) is equated with a regression model (Leung et al. 2001a). To state it formally, for a fixed integer i, a reg-class G_i is defined by the following regression model with random carriers

$$G_i: \quad Y = f_i(\mathbf{X}, \boldsymbol{\beta}_i) + e_i, \tag{5.78}$$

where $Y \in \mathbf{R}$ is the response variable; the explanatory variable that consists of carriers or regressors $\mathbf{X} \in \mathbf{R}^p$ is a random (column)vector with a probability density function (p.d.f.) $p(\bullet)$, the error term e_i is a random variable with a p.d.f. $\psi\ (u;\sigma_i)$ having a parameter σ_i, $E(e_i) = 0$, and $\mathbf{X}$ and e_i are independent. Here, $f_i(\cdot,\cdot) : \mathbf{R}^p \times \mathbf{R} \to \mathbf{R}$ is a known regression function, and $\boldsymbol{\beta}_i \in \mathbf{R}^{q_i}$ is an unknown regression parameter (column) vector. Although the dimension of $\boldsymbol{\beta}_i$ and q_i may be different for different G_i, we usually take $q_i = p$ for simplicity. Henceforth, we assume that e_i is distributed according to a normal distribution, i.e.,

$$\psi(u;\sigma_i) = \frac{1}{\sigma_i}\phi(\frac{u}{\sigma_i}), \tag{5.79}$$

where $\phi\ (\cdot)$ is the standard normal p.d.f.

For convenience of discussion, let

$$r_i(\mathbf{x}, y; \boldsymbol{\beta}_i) \equiv y - f_i(\mathbf{x}, \boldsymbol{\beta}_i). \tag{5.80}$$

Definition 5.1. *A random vector $(\mathbf{X}, Y)$ belongs to a regression class G_i (denoted as $(X,Y) \in G_i$) if it is distributed according to the regression model G_i.*

Thus, under Definition 5.1, a random vector $(\mathbf{X}, Y) \in G_i$ implies that $(\mathbf{X}, Y)$ has a p.d.f.

$$p_i(\mathbf{x}, y; \theta_i) = p(\mathbf{x})\psi(r_i(\mathbf{x}, y; \boldsymbol{\beta}_i); \sigma_i), \ \boldsymbol{\theta}_i = (\boldsymbol{\beta}_i^T, \sigma_i)^T. \tag{5.81}$$

For practical purpose, the following definition associated with Definition 4.1 may be used:

Definition 5.2. *A data point $(\mathbf{X}, Y)$ belongs to a regression class G_i (denoted as $(\mathbf{x}, y) \in G_i$) if it satisfies $p_i(\mathbf{x}, y; \boldsymbol{\theta}_i) \geq b_i$, i.e.,*

$$G_i \equiv G_i(\boldsymbol{\theta}_i) \equiv \{(\mathbf{x}, y) : p_i(\mathbf{x}, y; \boldsymbol{\theta}_i) \geq b_i\}, \tag{5.82}$$

where the constant $b_i > 0$ is determined by $P[p_i(\mathbf{X}, Y; \boldsymbol{\theta}_i) \geq b_i, (\mathbf{X}, Y) \in G_i] = a$, a is a probability threshold specified a priori and approaches to one.

Assume that there are m reg-classes $G_1, G_2, \ldots, G_m$ in a data set under study and that m is known in advance (m can actually be determined at the end of the mining process when all plausible reg-classes have been identified). The objective of knowledge discovery in mixture spatial distriution is to find all m reg-classes, to identify the parameter vectors and to make predication or interpretation by the models. To lower computation cost, we need to randomly sample from a data set to search for the reg-classes. Assumed that $\{(\mathbf{x}_1, y_1), ..., (\mathbf{x}_n, y_n)\}$ are the observed values of a random sample of size n taken from a data set. Thus they can be considered as realized values of n independently and identically distributed (i.i.d.) random vectors with a common mixture distribution population

$$p(\mathbf{x}, y; \boldsymbol{\theta}) = \sum_{i=1}^{m} \pi_i p_i(\mathbf{x}, y; \boldsymbol{\theta}_i), \tag{5.83}$$

i.e, they consist of random observations from m reg-classes with prior probabilities $\pi_1, ..., \pi_m(\pi_1 + ... + \pi_m = 1,\ \pi_i \geq 0{,}1 \leq i \leq m),\ \boldsymbol{\theta}^T = (\boldsymbol{\theta}_1^T, ..., \boldsymbol{\theta}_m^T)$.

5.6.3 *The Discovery of Regression Classes under Noise Contamination*

In a noisy data set, regression classes are distribution amidst a large number of outliers. Thus, how to unreal reg-classes under noise contamination becomes a challenge in the discovery of relevant relationships in the overall data set. Leung et al. (2001a) scrutinize the problem under two situations.

The case in which $\pi_1, ..., \pi_m$ are known

In this case all unknown parameters consist of the aggregate vector $\boldsymbol{\theta} = (\boldsymbol{\theta}_1^T, ..., \boldsymbol{\theta}_m^T)^T$. If the vector $\boldsymbol{\theta}_0^T = (\boldsymbol{\theta}_1^{0^T}, ..., \boldsymbol{\theta}_m^{0^T})$ of true parameters is known *a priori*, and the outliers are absent ($\varepsilon_i \equiv 0$, $1 \leq i \leq m$), then the posterior probability that $(\mathbf{x}_j, y_j)$ belongs to G_i is given by

$$\tau_i(\mathbf{x}_j, y_j; \boldsymbol{\theta}_i^0) = \frac{\pi_i p_i(\mathbf{x}_j, y_j; \boldsymbol{\theta}_i^0)}{\sum_{k=1}^{m} \pi_k p_k(\mathbf{x}_j, y_j; \boldsymbol{\theta}_k^0)},\ 1 \leq i \leq m. \tag{5.84}$$

A partitioning of the sample $Z = \{(\mathbf{x}_1, y_1), ..., (\mathbf{x}_n, y_n)\}$ into m reg-classes can be made by assigning each $(\mathbf{x}_j, y_j)$ to the population to which it has the highest estimated posterior probability of belonging to G_i if

$$\tau_i(\mathbf{x}_j, y_j; \boldsymbol{\theta}_i^0) > \tau_k(\mathbf{x}_j, y_j; \boldsymbol{\theta}_k^0),\ 1 \leq k \leq m,\ k \neq i. \tag{5.85}$$

This is just the Bayesian decision rule:

$$d = d(\mathbf{x}, y; \boldsymbol{\theta}_0) = \arg \max_{1 \leq i \leq m} [\pi_i p_i(\mathbf{x}, y; \boldsymbol{\theta}_i^0)],\ \mathbf{x} \in \mathbf{R}^p, y \in \mathbf{R}, 1 \leq d \leq m, \tag{5.86}$$

which classifies the sample Z and "new" observation with minimal error probability. As $\boldsymbol{\theta}_0$ is unknown, the so-called "plug-in" decision rule is often used:

$$d = d(\mathbf{x}, y; \hat{\boldsymbol{\theta}}_0) = \arg\max_{1 \le i \le m} [\pi_i p_i(\mathbf{x}, y; \hat{\boldsymbol{\theta}}_i^0)], \tag{5.87}$$

where $\hat{\boldsymbol{\theta}}_0$ is the MLE of $\boldsymbol{\theta}_0$ constructed by the sample Z from the mixture population,

$$\hat{\boldsymbol{\theta}}_0 = \arg\max_{\boldsymbol{\theta} \in \boldsymbol{\Theta}} l(\boldsymbol{\theta}), \tag{5.88}$$

$$l(\boldsymbol{\theta}) = \ln \prod_{j=1}^{n} p_0(\mathbf{x}_j, y_j; \boldsymbol{\theta}) = \sum_{j=1}^{n} \ln p(\mathbf{x}_j, y_j; \boldsymbol{\theta}), \tag{5.89}$$

where Θ is a parameter space.

For the case in which $p_i(\mathbf{x}, y; \boldsymbol{\theta}_i)$ is contaminated, i.e., the ε_i-contaminated neighborhood is:

$$\mathrm{B}(\varepsilon_i) = \{p_i^{\varepsilon}(\mathbf{x}, y; \boldsymbol{\theta}_i) : p_i^{\varepsilon}(\mathbf{x}, y; \boldsymbol{\theta}_i) = (1 - \varepsilon_i) p_i(\mathbf{x}, y; \boldsymbol{\theta}_i) + \varepsilon_i h_i(\mathbf{x}, y)\}, \tag{5.90}$$

where $h_i(\mathbf{x}, y)$ is any p.d.f. of outliers in G_i, and ε_i is an unknown fraction of an outlier present in G_i.

The effect of outliers on the MLE $\hat{\boldsymbol{\theta}}_0$ under ε-contaminated models can now be studied. Under this situation, Z is the random sample from the mixture p.d.f.:

$$p_{\varepsilon}(\mathbf{x}, y; \boldsymbol{\theta}_0) = \sum_{i=1}^{m} \pi_i p_i^{\varepsilon}(\mathbf{x}, y; \boldsymbol{\theta}_i^0). \tag{5.91}$$

Let ∇_{θ}^k be the operator of the k-th order differentiation with respect to $\boldsymbol{\theta}$, $\mathbf{0}$ be a zero matrix with all elements being zero and $\mathbf{1}$ be a matrix with all elements being 1. Denote

$$\begin{aligned} I_{\varepsilon}(\boldsymbol{\theta}; \boldsymbol{\theta}_0) &= -E_{\varepsilon}[\ln p_0(\mathbf{X}, Y; \boldsymbol{\theta})] \\ &= -\iint_{R^{p+1}} \ln p_0(\mathbf{x}, y; \boldsymbol{\theta}) p_{\varepsilon}(\mathbf{x}, y; \boldsymbol{\theta}_0) d\mathbf{x} dy, \end{aligned} \tag{5.92}$$

$$B_i(\boldsymbol{\theta}) = \iint_{R^{p+1}} [h_i(\mathbf{x}, y) - p_i(\mathbf{x}, y; \boldsymbol{\theta}_i^0)] \ln p_0(\mathbf{x}, y; \boldsymbol{\theta}) d\mathbf{x} dy, \tag{5.93}$$

$$J_{\varepsilon}(\boldsymbol{\theta}_0) = -\iint_{R^{p+1}} p_{\varepsilon}(\mathbf{x}, y; \boldsymbol{\theta}_0) \nabla_{\theta}^2 \ln p_0(\mathbf{x}, y; \boldsymbol{\theta})|_{\theta = \theta_0} d\mathbf{x} dy. \tag{5.94}$$

It can be observed that $I_0(\boldsymbol{\theta}_0; \boldsymbol{\theta}_0)$ is the Shannon entropy for the hypothetical mixture $p_0(\mathbf{x}, y; \boldsymbol{\theta})$. Furthermore, $J_0(\boldsymbol{\theta}_0)$ is the Fisher information matrix

$$J_\varepsilon(\boldsymbol{\theta}_0) = \iint_{R^{p+1}} p_0(\mathbf{x}, y; \boldsymbol{\theta}_0)\nabla_\theta \ln p_0(\mathbf{x}, y; \boldsymbol{\theta})[\nabla_\theta \ln p_0(\mathbf{x}, y; \boldsymbol{\theta})]^T|_{\theta=\theta_0} d\mathbf{x}dy, \quad (5.95)$$

and in regularity conditions

$$\nabla_\theta I_0(\boldsymbol{\theta}; \boldsymbol{\theta}_0)|_{\theta=\theta_0} = \mathbf{0},\ \nabla_\theta^2 I_\varepsilon(\boldsymbol{\theta}; \boldsymbol{\theta}_0)|_{\theta=\theta_0} = J_\varepsilon(\boldsymbol{\theta}_0). \quad (5.96)$$

Theorem 5.1. *If the family of p.d.f. $p(\mathbf{x}, y; \boldsymbol{\theta})$ satisfies the regularity condition (Kendall 1987), the function $I_0(\boldsymbol{\theta}; \boldsymbol{\theta}_0)$, $B_i(\boldsymbol{\theta})$ are thrice differentiable with respect to $\boldsymbol{\theta} \in \Theta$, and the point $\boldsymbol{\theta}_\varepsilon = \arg\min_{\theta\in\Theta} I_\varepsilon(\boldsymbol{\theta}; \boldsymbol{\theta}_0)$ is unique, then the MLE $\hat{\boldsymbol{\theta}}$ under ε-contamination is almost surely convergent (as), i.e.,*

$$\hat{\boldsymbol{\theta}} \xrightarrow{a.s.} \boldsymbol{\theta}_\varepsilon \ (n \to \infty) \quad (5.97)$$

and $\boldsymbol{\theta}_\varepsilon \in \Theta$ satisfies the asymptotic expansion:

$$\boldsymbol{\theta}_\varepsilon = \boldsymbol{\theta}_0 + [J_\varepsilon(\boldsymbol{\theta}_0)]^{-1} \sum_{i=1}^{m} \varepsilon_i \pi_i \nabla_\theta \mathrm{B}_i(\boldsymbol{\theta}_0) + \mathrm{O}\left(\|\boldsymbol{\theta}_\varepsilon - \boldsymbol{\theta}_0\|^2\right)\mathbf{1}. \quad (5.98)$$

(see Leung et al. (2001) for the proof)

Remark 5.1. It can be observed from Theorem 4.1 that in the presence of outliers in the sample, the estimator $\hat{\boldsymbol{\theta}}$ can become inconsistent. It should be noted that $|\nabla_\theta B_i(\boldsymbol{\theta})|$ depends on the contaminating density $h_i(\mathbf{x}, y)$, $1 \le i \le m$, and may have sufficiently large value.

From Theorem 5.1, we have the following result:

Corollary 5.1. *In the setting of Theorem 4.1, $\hat{\boldsymbol{\theta}}$ has an influence function*

$$IF(\mathbf{x}, y; \hat{\boldsymbol{\theta}}) = [J_0(\boldsymbol{\theta}_0)]^{-1} \nabla_\theta \ln p_0(\mathbf{x}, y; \boldsymbol{\theta})|_{\theta=\theta_0}.$$

(See Leung et al. (2001a) for the proof)

Remark 5.2. The influence function (IF) is an important concept in robust statistics. It can provide the richest quantitative information on robustness by describing the (approximate and standardized) effect of an additional observation in any point $(\mathbf{x}, y)$ on the estimator $\hat{\boldsymbol{\theta}}$. Roughly speaking, the IF measures the effect of infinitesimal perturbations on the estimator.

The case in which $\pi_1, ..., \pi_m$ are unknown

Here we adopt the method in McLachlan and Basford (1988). Let $\boldsymbol{\pi} = (\pi_1, ..., \pi_m)^T$, $\boldsymbol{\varphi} = (\boldsymbol{\pi}^T, \boldsymbol{\theta}^T)^T$, and

$$l(\boldsymbol{\varphi}) = \ln \prod_{j=1}^{n} p_{\varepsilon}(\mathbf{x}_j, y_j; \boldsymbol{\theta}) = \sum_{j=1}^{n} \ln[\sum_{i=1}^{m} \pi_i p_i^{\varepsilon}(\mathbf{x}_j, y_j; \boldsymbol{\theta}_i)], \tag{5.99}$$

$$\tau_i(\mathbf{x}_j, y_j; \boldsymbol{\varphi}) = \frac{\pi_i p_i^{\varepsilon}(\mathbf{x}_j, y_j; \boldsymbol{\theta}_i)}{\sum_{k=1}^{m} \pi_k p_k^{\varepsilon}(\mathbf{x}_j, y_j; \boldsymbol{\theta}_k)}, \ 1 \le i \le m. \tag{5.100}$$

It should be noted that $\pi_m = 1 - \sum_{i=1}^{m-1} \pi_i$. Therefore, for $1 \le k \le m-1$, the MLE of π_k, $\hat{\pi}_k$, satisfies

$$\nabla_{\pi_k} l(\boldsymbol{\varphi}) = \sum_{j=1}^{n} [\frac{\tau_k(\mathbf{x}_j, y_j; \boldsymbol{\theta}_k)}{\pi_k} - \frac{\tau_m(\mathbf{x}_j, y_j; \boldsymbol{\theta}_m)}{\pi_m}] = \mathbf{0}. \tag{5.101}$$

By simple computation, the likelihood equation for $\boldsymbol{\varphi}$, $\nabla_{\phi} l(\boldsymbol{\varphi}) = \mathbf{0}$, can thus be rewritten as

$$\nabla_{\theta_k} l(\boldsymbol{\varphi}) \Big|_{\theta_k = \hat{\theta}_k} = \sum_{j=1}^{n} \tau_k(\mathbf{x}_j, y_j; \hat{\varphi}) \nabla_{\theta} \ln p_k^{\varepsilon}(\mathbf{x}_j, y_j; \boldsymbol{\theta}_k) \Big|_{\theta_k = \hat{\theta}_k} = \mathbf{0}, \tag{5.102}$$

$$\hat{\pi}_k = \sum_{j=1}^{n} \tau_k(\mathbf{x}_j, y_j; \hat{\varphi}) \Big/ n, \ 1 \le k \le m. \tag{5.103}$$

There is a difficulty with the mixtures in that if $p_i(\mathbf{x}, y; \boldsymbol{\theta}_i)$ and $p_j(\mathbf{x}, y; \boldsymbol{\theta}_j)$ belong to the same parametric family, then $p(\mathbf{x}, y; \boldsymbol{\varphi})$ will have the same value when the cluster labels i and j are interchanged in $\boldsymbol{\varphi}$. That is, although this class of mixtures may be identifiable, $\boldsymbol{\varphi}$ is not. However, this lack of identifiability of $\boldsymbol{\varphi}$ due to the interchanging of cluster labels is of no concern in practice, as it can easily be overcome by the imposition of an appropriate constraint on $\boldsymbol{\varphi}$ (McLachlan and Basford 1988).

However, it may be very difficult to get $\hat{\boldsymbol{\theta}}_0$ because too many parameters are involved. As a matter of fact, the ML method for directly estimating the parameters of mixture densities actually has many practical implementation difficulties (Zhuang et al. 1996). For example, (1) when there are a large number of clusters in the mixture, the total number of parameters to be estimated can be very large in proportion to the available data samples; and (2) there may be singularities in the log-likelihood function, since the likelihood needs not be bounded from above (Vapnik 1995).

One of the main aims of robust statistics is to develop robust methods which can resist the effect of outliers in data sets. However, almost all of the robust methods tolerate only less than 50% of outliers. When there are multiple reg-classes in a data set, they cannot identify these classes because it is very common that the proportion of outliers with respect to a single class is more than 50%. Recently, several more robust methods have been developed for computer vision. For example, MINPRAN (Stewart 1995) is perhaps the first technique that reliably tolerates more than

50% of outliers without assuming a known bound for inliers. The method assumes that the outliers are randomly distributed within the dynamic range of the sensor, and the noise (outlier) distribution is known. When the outliers are non-uniform, adjustment of MINPRAN to suit other kinds of distributions has also been proposed. However, the assumptions of MINPRAN restrict its generality in practice.

Another highly robust estimator is the MF estimator (Zhuang et al. 1992), which is developed for a simple regression problem without carriers. It does not need assumptions such as those in MINPRAN. Indeed, no requirement is imposed on the distribution of outliers. So, it seems to be more applicable to a complex data set. Extended on the ideas of the MF estimator and GMDD, Leung et al. (2001a) derived the RCMD estimator to unreal regression classes.

5.6.4 The Regression-Class Mixture Decomposition (RCMD) Method for knowledge Discovery in Mixed Distribution

Since a mixture density is observed as a composition of simple structured densities or data structures, with respect to a particular density or structure, all other densities or structures can be readily classified as part of the outlier category in the sense that these other observations obey different statistics. Thus, a mixture density can be viewed as a contaminated density with respect to each cluster in the mixture. When all of the observations for a single density are grouped together, the remaining observations (clusters and true outliers) can then be considered to form an unknown outlier density. According to this idea, the mixture p.d.f. in (5.91) with respect to G_i can be rewritten as

$$\begin{aligned} p_\varepsilon(\mathbf{x}, y; \boldsymbol{\theta}) &= \pi_i(1-\varepsilon_i)p_i(\mathbf{x}, y; \boldsymbol{\theta}_i) + \pi_i\varepsilon_i h_i(\mathbf{x}, y) + \sum_{j \neq i}^{m} \pi_j p_j^{\varepsilon}(\mathbf{x}, y; \boldsymbol{\theta}_j) \\ &\equiv \pi_i(1-\varepsilon_i)p_i(\mathbf{x}, y; \boldsymbol{\theta}_i) + [1-\pi_i(1-\varepsilon_i)]g_i(x, y). \end{aligned} \tag{5.104}$$

Ideally, a sample point $(\mathbf{x}_k, y_k)$ from the above mixture p.d.f. is classified as an inlier if it is realized from $p_i(\mathbf{x}, y; \boldsymbol{\theta}_i)$ or as an outlier coming from the p.d.f. $g_i(\mathbf{x}, y)$ otherwise.

The given data set $Z = \{(\mathbf{x}_1, y_1), ..., (\mathbf{x}_n, y_n)\}$ is now generated by the mixture p.d.f. $p_\varepsilon(\mathbf{x}, y; \boldsymbol{\theta})$, i.e., it comes from $p_i(\mathbf{x}, y; \boldsymbol{\theta}_i)$ with probability $\pi_i(1-\varepsilon_i)$ together with an unknown outlier $g_i(x, y)$ with probability $[1-\pi_i(1-\varepsilon_i)]$.

Let D_i be the subset of all inliers with respect to G_i and $\bar{D}_i$ be its complement. From the Bayesian classification rule, we have

$$D_i = \left\{(\mathbf{x}_j, y_j) : p_i(\mathbf{x}_j, y_j; \boldsymbol{\theta}_i) > \frac{1-\pi_i+\pi_i\varepsilon_i}{\pi_i(1-\varepsilon_i)} g_i(\mathbf{x}_j, y_j)\right\}, \bar{D}_i = Z - D_i. \tag{5.105}$$

Define

$$d_i^0 = \min\{p_i(\mathbf{x}_j, y_j; \boldsymbol{\theta}_i) : \quad (\mathbf{x}_j, y_j) \in D_i\},$$

$$d_i^1 = \max\{p_i(\mathbf{x}_j, y_j; \boldsymbol{\theta}_i) : \quad (\mathbf{x}_j, y_j) \in \bar{D}_i\}.$$

Ideally the likelihood of any inlier being generated by $p_i(\mathbf{x}, y; \boldsymbol{\theta}_i)$is greater than the likelihood of any outlier being generated by $g_i(\mathbf{x}, y)$. Thus, we may assume that $d_i^0 > d_i^1$. Therefore, the Bayesian classification becomes

$$D_i = \{(\mathbf{x}_j, y_j) : p_i(\mathbf{x}_j, y_j; \boldsymbol{\theta}_i) > \frac{1 - \pi_i + \pi_i \varepsilon_i}{\pi_i(1 - \varepsilon_i)} \delta_i\}, \tag{5.106}$$

where we can choose $\delta_i \in [\pi_i(1\text{-}\varepsilon_i)d_i^1/(1\text{-}\pi_i\text{+}\pi_i\varepsilon_i),\ \pi_i(1\text{-}\varepsilon_i)d_i^0/(1\text{-}\pi_i\text{+}\pi_i\varepsilon_i)]$. So, if we assume that $g_i(\mathbf{x}_1, y_1) = ... = g_i(\mathbf{x}_n, y_n) = \delta_i$, then we will get equivalent results. Using this assumption, (5.100) becomes

$$p_\varepsilon(\mathbf{x}, y; \boldsymbol{\theta}) = \pi_i(1 - \varepsilon_i)p_i(\mathbf{x}, y; \boldsymbol{\theta}_i) + (1 - \pi_i + \pi_i\varepsilon_i)\delta_i. \tag{5.107}$$

The log-likelihood function of observing Z corresponding to (5.89) under ε-contamination becomes

$$l(\boldsymbol{\theta}_i) = n \ln[\pi_i(1 - \varepsilon_i)] + \sum_{j=1}^{n} \ln[p_i(\mathbf{x}_j, y_j; \boldsymbol{\theta}_i) + \frac{1 - \pi_i + \pi_i\varepsilon_i}{\pi_i(1 - \varepsilon_i)} \delta_i]. \tag{5.108}$$

Thus, in order to estimate $\boldsymbol{\theta}_i$ from Z, we need to maximize $l(\boldsymbol{\theta}_i)$ with each δ_i subject to $\sigma_i > 0$. Since the maximization of $l(\boldsymbol{\theta}_i)$ at δ_i with respect to $\boldsymbol{\theta}_i$ is equivalent to maximizing the G_i model-fitting function

$$l_i(\boldsymbol{\theta}_i; t_i) \equiv \sum_{j=1}^{n} \ln[p_i(\mathbf{x}_j, y_j; \boldsymbol{\theta}_i) + t_i] \tag{5.109}$$

at t_i with respect to $\boldsymbol{\theta}_i$, provided that $t_i = (1 - \pi_i + \pi_i\varepsilon_i)\delta_i/[\pi_i(1 - \varepsilon_i)]$, then we can discuss the problem of maximizing $l(\boldsymbol{\theta}_i)$ subject to $\sigma_i > 0$. Similar to Zhuang et al. (1996), we henceforth shall refer to each "t_i" (≥ 0) as a partial model. Since each t_i corresponds to a value δ_i of unknown outlier distribution $g_i(\mathbf{x}, y)$, we only use the partial information about the model without the knowledge of the whole shape of $g_i(\mathbf{x}, y)$.

Leung et al. (2001a) introduce a new concept as follows:

Definition 5.3. *For a reg-class G_i and the data set $Z = \{(\mathbf{x}_1, y_1), ..., (\mathbf{x}_n, y_n)\}$, the t-level set of G_i is defined as*

$$G_i(\boldsymbol{\theta}_i; t) = \{(\mathbf{x}_j, y_j) : p_i(\mathbf{x}_j, y_j; \boldsymbol{\theta}_i) > t\}, \tag{5.110}$$

the t-level support set of an estimator $\hat{\boldsymbol{\theta}}_i$ for $\boldsymbol{\theta}_i$ is defined as $G_i(\hat{\boldsymbol{\theta}}_i; t)$.

According to this concept, $G_i(\boldsymbol{\theta}_i; t)$ is the subset of all inliers with respect to G_i at a partial model t. Maximizing (5.109) may be approximately interpreted as maximizing the "likelihood" over the t-level set of G_{i}. It should be noted that the capacity of $G_i(\boldsymbol{\theta}_i; t)$ will decrease as a partial model level t increases. Moreover, the t-level support set of an estimator $\hat{\boldsymbol{\theta}}_i$ reflects the extent to which the data set supports this estimator at partial model level t.

Definition 5.4. *The RCMD estimator of the parametric vector θ_i for a reg-class G_i is defined by*

$$\hat{\boldsymbol{\theta}}_i^t = \arg\max_{\theta_{\mathrm{i}}} l_i(\boldsymbol{\theta}_i; t_i),\ \boldsymbol{\theta}_i = (\boldsymbol{\beta}_i^T, \sigma_i)^T,\ \sigma_i > 0.$$

When $m = 1$ and the random carriers disappear in (5.78), the RCMD estimator becomes a univariate MF estimator. In particular, when X is distributed uniformly (i.e., $p(\mathbf{x}) \equiv constant$ in some domain) and $e_i \sim N(0, \sigma_i^2)$, the maximization of $l_i(\boldsymbol{\theta}_i; t_i)$ is equivalent to maximizing

$$\bar{l}_i(\boldsymbol{\theta}_i; \bar{t}_i) \equiv \sum_{j=1}^{n} \ln\left[\psi[y_j - f_i(\mathbf{x}_j, \boldsymbol{\beta}_i); \sigma_i] + \bar{t}_i\right], \tag{5.111}$$

where $\bar{t}_i = t_i/c$. For simplicity, we still denote $\bar{t}_i$ and $\bar{l}_i$ by t_i and l_i, respectively. That is, the above expression is rewritten as

$$l_i(\boldsymbol{\theta}_i; t_i) \equiv \sum_{j=1}^{n} \ln\left[\psi[y_j - f_i(\mathbf{x}_j, \boldsymbol{\beta}_i); \sigma_i] + t_i\right]. \tag{5.112}$$

In this case, the corresponding expressions in (5.110) and (5.82) become, respectively,

$$G_i(\boldsymbol{\theta}_i; t_i) = \{(\mathbf{x}_j, y_j) : \psi[r_i(\mathbf{x}_j, y_j; \boldsymbol{\beta}_i); \sigma_i] > t_i\}, \tag{5.113}$$

$$G_i(\boldsymbol{\theta}_i) = \{(\mathbf{x}, y) : |r_i(\mathbf{x}, y; \boldsymbol{\beta}_i)| \leq 3\sigma_i\}, \tag{5.114}$$

which is based on the 3 σ-criterion of the normal distribution (i.e., a in (5.82) is 0.9972).Leung et al. (2001a) shows the convergence of $\hat{\boldsymbol{\theta}}_i^k$.

The RCMD method can be summarized as follows:

At each selected partial model $t_i^{(s)}$, $s = 0,1,\ldots,S$, $l_i(\boldsymbol{\theta}_i; t_i^{(s)})$ is maximized with respect to $\boldsymbol{\beta}_i$ and σ_i by using an iterative algorithm beginning with a randomly chosen initial $\boldsymbol{\beta}_i^{(0)}$ or by using a genetic algorithm (GA). Having solved $\max_{\beta_1,\sigma_i} l_i\left(\boldsymbol{\theta}_i; t_i^{(s)}\right)$ for $\hat{\boldsymbol{\beta}}_i(t_i^{(s)})$ and $\hat{\sigma}_i(t_i^{(s)})$, the possible reg-class $G_i(\hat{\boldsymbol{\theta}}_i(t_i^{(s)}))$ is calculated and it is followed by the test of normality on $G_i(\hat{\boldsymbol{\theta}}_i(t_i^{(s)}))$. If the test statistic is not significant (usually at level $\alpha = 0.01$), then the hypothesis that the respective distribution is normal should be accepted and a valid reg-class, $G_i(\hat{\boldsymbol{\theta}}_i(t_i^{(s)}))$, has been determined, otherwise we proceed to the next partial model if the upper bound $t_i^{(S)}$ has not been reached. It may be said that the identity of each $G_i(\hat{\boldsymbol{\theta}}_i(t_i^{(s)}))$ is based on its t-level set.

Throughout, a valid reg-class is subtracted from the current data set after it has been detected and the next reg-class will be identified in the new size-reduced data set by the recursive process. Individual reg-classes continue to be estimated recursively until there are no more valid reg-classes, or the size of the new data set gets to be too small for estimation. Thus, the RCMD method can handle an arbitrary number of reg-class models with single reg-class extraction. That is, the parameters of each reg-class can be estimated progressively and the data points are partitioned into inliers and outliers with respect to this reg-class. The RCMD procedure is depicted in Fig. 5.10 and the iterature and GA-based algorithms are detailed in Leung et al. (2001a).

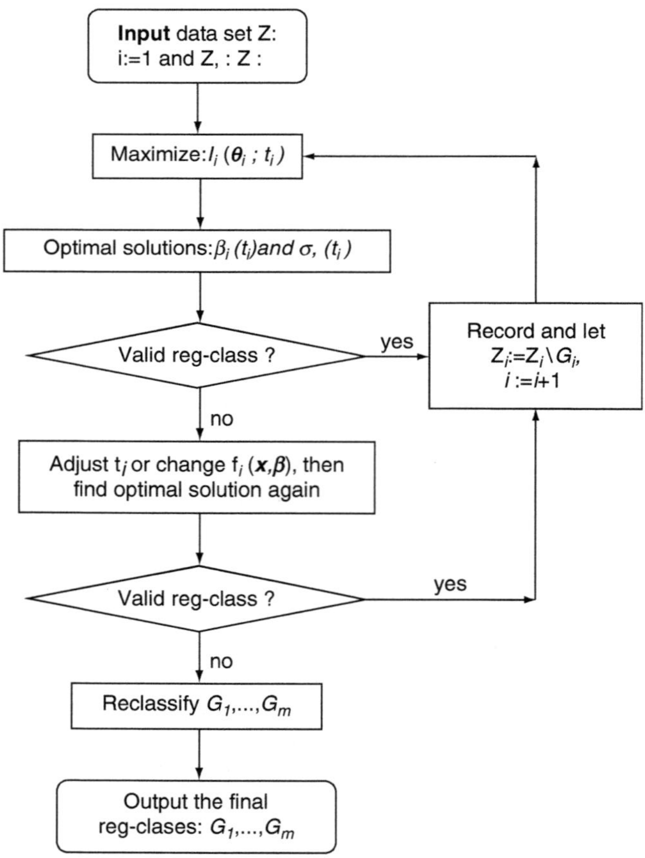

Fig. 5.10 Flowchart of the RCMD method

5.6.5 Numerical Results and Observations

The effectiveness of the RCMD method for data mining is demonstrated by some numerical simulations here.

Example 5.1. Assuming that there are nine points in a data set, where five points fit the regression model: $Y = \beta_1 X + e_1$, $e_1 \sim N(0, \sigma_1^2)$, $\beta_1 = 1, \sigma_1 = 0.1$, and the others fit the regression model: $Y = \beta_2 X + e_2$, $e_2 \sim N(0, \sigma_2^2)$, $\beta_2 = 0$, $\sigma_2 = 0.1$ (Fig. 5.11a). Now To unravel the two regression classes, we select $t_l = 0.1$, the objective function is the G_l model-fitting function

$$l_1(\theta_1; t_1) = \sum_{j=1}^{9} \ln \left[\frac{1}{\sqrt{2\pi}\sigma} \exp\left(-\frac{(y_j - x_j \beta)^2}{2\sigma^2}\right) + 0.1 \right],$$

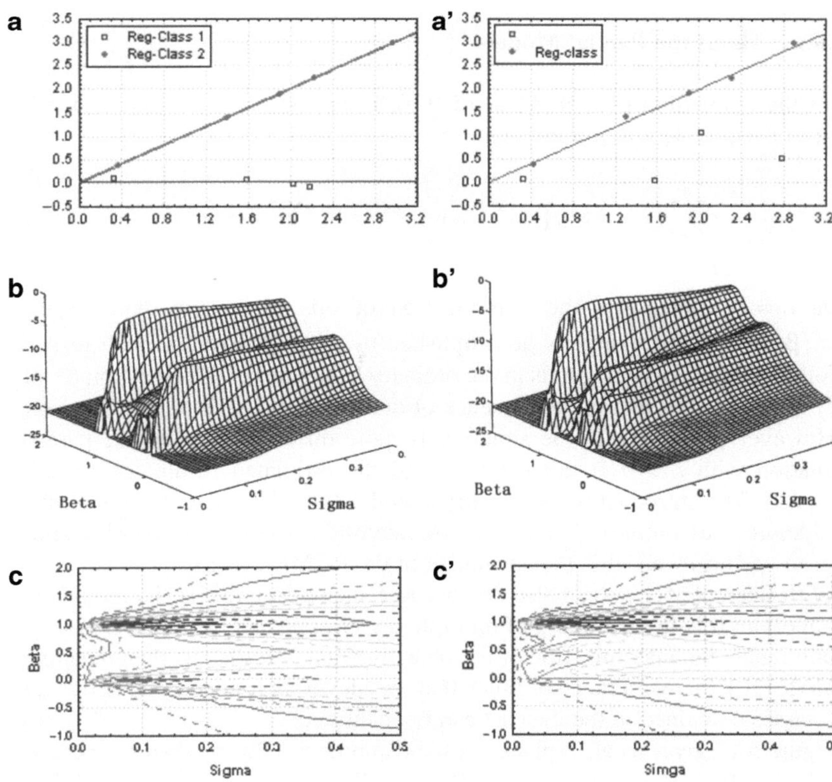

Fig. 5.11 Results obtained by the RCMD method for two reg-classes and one reg-class. (**a**) Scatterplot for two reg-classes. (**a'**) Scatterplot for one reg-class. (**b**) Objective function plot. (**b'**) Objective function plot. (**c**) Contour plot of objective function. (**c'**) Contour plot of objective function

which is depicted in Fig. 5.11b. It can be observed that this function have two obvious peaks, with each corresponding to the relevant reg-classes. Using the iterative algorithm or genetic algorithm, the two reg-classes are easily discovered. It is clearly shown in the contour plot of this function (Fig. 5.11c). For example, using the GA procedure, we can find: $\hat{\beta}_1 = 1.002$, $\hat{\sigma}_1 = 0.109$, and $l_{\max} = -2.167$. Using more exact maximization method, we obtain $\hat{\beta}_1 = 1.00231, \hat{\sigma}_1 = 0.109068$, and $l_{\max} = -2.016715$. The difference between the estimated values and the true parameters is in fact very small. On the other hand, if there is only one reg-class in this set (see Fig. 5.11 a'), our objective function is still very sensitive to this change. It can also find the only reg-class in the data set. As can be observed in the 3D and contour plots, there is only one peak which represents the reg-class (Fig. 5.11b', c').

5.6.6 Comments About the RCMD Method

5.6.6.1 About the Partial Models

From the expression of $l_i(\boldsymbol{\theta}_i; t_i)$ in (5.109), it can be observed that maximizing $l_i(\boldsymbol{\theta}_i; t_i)$ is equivalent to minimizing

$$\frac{1}{2\sigma_i^2}\sum_{j=1}^{n}[y_j - f_i(\mathbf{x}_j, \boldsymbol{\beta}_i)]^2 + n\ln(\sqrt{2\pi}\sigma_i) - \sum_{j=1}^{n}\ln p(\mathbf{x}_j), \tag{5.115}$$

when $t_i = 0$. Obviously, the minimization of this expression with respect to $\boldsymbol{\theta}_i = (\boldsymbol{\beta}_i^T, \sigma_i)^T$ can be directly accomplished by the minimization with respect to $\boldsymbol{\beta}_i$ followed by σ_i, which results in the ordinary least squares (OLS) estimates of $\boldsymbol{\beta}_i$. They are not robust and in the presence of outliers they give a poor estimation.

However, when $t_i > 0$, the situation is quite different. In fact, the parameter estimation with $t_i > 0$ is fairly robust and the estimated result can be greatly improved. The introduction to a partial model "$t_i > 0$" not only represents the consideration of outliers, but is also the simplification of this consideration in order to perform well. It is the advantage of the RCMD method.

With Example 5.1 we can also demonstrate such a fact: the partial model t plays an important role in the mining of multiple reg-classes, and if t is selected within a certain range, the maximization of the objective function $l(\boldsymbol{\theta}, t)$is then meaningful. From (5.110), there is a range of t such that the t-level set is nonempty. In this range, reg-classes contained in the data set can be identified.

Figure 5.12 gives us an explanation for Example 5.1. Even when t is very small (10^{-3}), the RCMD method is still effective. However, it becomes invalid when t equals zero. For the data in Example 5.1, when t changes from a very small positive number to approximately 5, the method remains valid. Once t exceeds five, the greater t is, the more difficult it becomes for the RCMD method to identify the reg-classes.

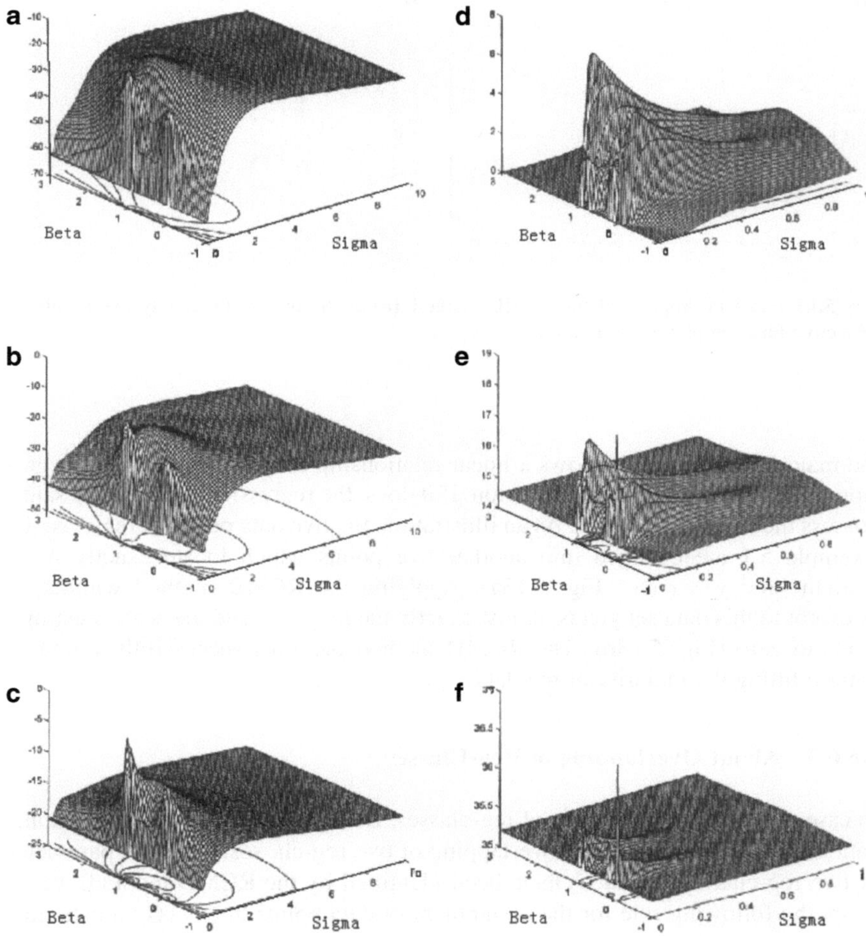

Fig. 5.12 Effect of partial model t on the mining of reg-classes. (**a**) $t = 0.001$. (**b**) $t = 0.01$. (**c**) $t = 0.1$. (**d**) $t = 1$. (**e**) $t = 5$. (**f**) $t = 50$

5.6.6.2 About Robustness

The RCMD estimator is asymptotically stable though it may be a biased estimator (see Theorem 2 in Leung et al. (2001a)). However, in practice it can be improved by other methods. As shown in the numerical examples in Leung et al. (2001a), the RCMD method also has a very high degree of robustness. It can resist more than 50% of outliers in a data set without assuming the type of distributions of the outliers. Besides, the method also possesses the *exact fit property* that many robust regression models possess. In robust regression, the exact fit property means that if

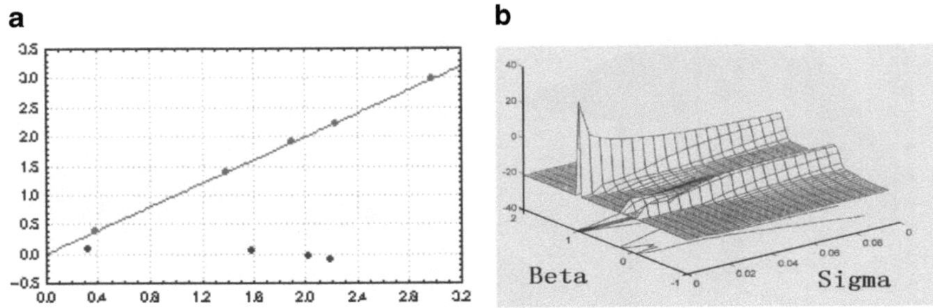

Fig. 5.13 Exact fit property of the RCMD method. (**a**) Scatterplot, with exactly five points. (**b**) Objective function plot located on the line: $y = x$

the majority of the data follows a linear relationship *exactly*, then a robust regression method should yield this equation. If it does, the regression technique is said to possess the exact fit property. As an illustration, the five data points in reg-class 1 in Example 5.1 are changed into another five points which locate exactly in the straight line: $y = x$ (see Fig. 5.13a). Applying the RCMD method without the intercept to this data set yields almost exactly the fit: $y = x$ and the scale σ estimate tends to zero (Fig. 5.13b). The RCMD method has thus successfully found the pattern fitting the majority of the data.

5.6.6.3 About Overlapping of Reg-Classes

In case there is an overlapping of reg-classes, Leung et al. (2001) propose another data classification rule for the overlapping of two reg-classes. Once the parameters of two reg-classes G_i and G_j have been identified by the RCMD method, we can adopt the following rule for the assignment of data points in $G_i \cap G_j$: a data point $(\mathbf{x}_k, y_k) \in G_i \cap G_j$ is assigned to G_i if

$$p_i(\mathbf{x}_k, y_k; \hat{\boldsymbol{\theta}}_i) > p_j(\mathbf{x}_k, y_k; \hat{\boldsymbol{\theta}}_j). \tag{5.116}$$

Combining (5.114) and (5.116), we can reclassify the data set into reg-classes. That is, although the points in the overlapping region are removed from the data set when the first reg-class has been detected, to which reg-class these points eventually belong will be determined only after all reg-classes have been found. Thus, based on the rule in (5.116), the final result in the partitioning of reg-classes is almost independent of the extraction order.

For substantiation, the RMCD method has been successfully applied to solve the problem of switching regression models, mixture of linear and non-linear structures, detection of curves, and mining of reg-classes in large data sets contaminated with noise (Leung et al. 2001).

The extension of the RCMD method for the mining of irregular geometric features in spatial database has been discussed in Chap. 5.2.

5.6.7 A Remote Sensing Application

To demonstrate the practicality of the RCMD algorithm, a real-life mining of line objects in remotely sensed data is also performed (Leung et al. 2001a). In their application, runways are identified in a remotely sensed image from LANDSAT Thematic Mapper (TM) data acquired over a suburb in Hangzhou, China. The region contains the runways and parking apron of a certain civilian aerodrome. The image consists of a finite rectangular 95×60 lattice of pixels (see Fig. 5.14a). To identify the runways, Band 5 is used as a feature variable. A feature subset of data, depicted in Fig. 5.14b, is first extracted by using a simple technique which selects a pixel point when its gray-level value is above a given threshold (e.g., 250). For the lattice coordinates of points in the subset, the RCMD method is then used to identify two runways, which can be viewed as two reg-classes. At $t = 0.05$ level, two line equations identified by the RCMD method are $y = 0.774x + 34.874$ and $y = 0.341x + 22.717$, respectively. The result shows an almost complete accordance with data points in Fig. 5.14b. In other words, line-type objects such as runways and highways in remotely sensed images can easily and accurately be detected. Compared with existing techniques such as the window method, the RCMD method can avoid the problem of selecting the appropriate window sizes and yet obtains the same results.

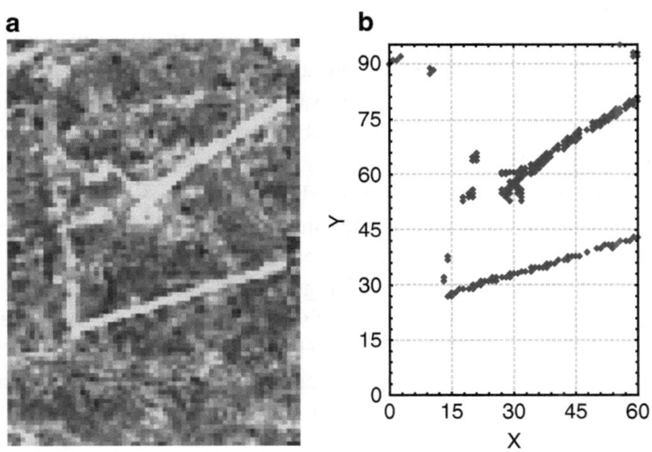

Fig. 5.14 Identification of line objects in remotely sensed data

5.6.8 An Overall View about the RCMD Method

It appears that RCMD is a promising method for a large variety of applications. As an effective means for data mining, the RCMD method has the following advantages:

1. The number of reg-classes does not need to be specified *a priori*.
2. The proportion of noise in the mixture can be large. Neither the number of outliers nor their distributions is part of the input. The method is thus very robust.
3. The computation is quite fast and effective, and can be implemented by parallel computing.
4. Mining is not limited to straight lines and planes as imposed by some previous methods. It can also extract many curves which can be linearized (such as polynomials) and can deal with high dimensional problems.
5. It estimates simultaneously the regression and scale parameters such as the MLE by using all of the information provided by the samples. Thus, the effect of the scale parameters on the regression parameters is considered. This is more effective than estimating separately the regression and scale parameters.

Though the RCMD method appears to be rather successful, at least by the simulation experiments, in the mining of reg-classes, there are problems which should be further investigated. As discussed in the literature, the singularity of the likelihood function for a mixture is an issue that needs to be investigated. Singularity means that the value of the likelihood function becomes infinite as the standard deviation of any one component approaches zero (Titterington et al. 1987). Since the RCMD method is based on the MLE, it is then natural to wonder whether or not singularity will occur in the objective function in (5.109). In light of the theory, the function $l_i(\theta_i; t_i)$ is not immune to singularities, but in practice this case rarely occurs. It should be observed that singularities occur only in the edge of the parametric spaces (search spaces). However, with good starting values, singularities are less likely to happen. The study in Caudill and Acharya (1998) indicates that the incidence of singularity decreases with the increase in sample size and the increase in the angle of separation of two linear reg-classes. Obviously, we need to further study this aspect within the RCMD framework, though many researchers think that the issue of singularity in MLE may have been overblown.

The second issue that deserves further study is the problem of sample size in the RCMD method. In RCMD, we analyze a very large data set by examining a sample taken from it. If a small fraction of reg-classes contains rare, but important, response variables, complications may arise. In this situation, retrospective sampling may need to be considered (O'hara Hines 1997). In general, how to select a suitable sample size in RCMD is a problem which needs theoretical and experimental investigations.

Chapter 6
Discovery of Structures and Processes in Temporal Data

6.1 A Note on the Discovery of Generating Structures or Processes of Time Series Data

Beyond any doubt, natural and man-made phenomena change over time and space. In our natural environment, temperature, rainfall, cloud cover, ice cover, water level of a lake, river channel morphology, surface temperature of the ocean, to name but a few examples, all exhibit dynamic changes over time. In terms of human activities, we have witnessed the change of birth rate, death rate, migration rate, population concentration, unemployment, and economic productivity throughout our history. In our interacting with the environment, we have experienced the time varying concentration of various pollutants, usage of natural resource, and global warming. For natural disasters, the occurrence of typhoon, flood, drought, earthquake, and sand storm are all dynamic in time. All of these changes might be seasonal, cyclical, randomly fluctuating, or trend oriented in a local or global scale.

To have a better understanding of and to improve our knowledge about these dynamic phenomena occurring in natural and human systems, we generally make a sequence of observations ordered by a time parameter within certain temporal domain. Time series are a special kind of realization of such variations. They measure changes of variables at points in time. The objectives of time series analysis are essentially the description, explanation, prediction, and perhaps control of the time varying processes. With respect to data mining and knowledge discovery, we are primarily interested in the unraveling of the generating structures or processes of time series data. Our aim is to discover and characterize the underlying dynamics, deterministic or stochastic, that generate the time varying phenomena manifested in chronologically recorded data.

Study of time series has been a long tradition in data analysis. Theoretical investigations and applications have been made in a large variety of fields such as statistics (Box and Jenkins 1976; Tong 1990; Wei 1990; Fuller 1996; Kantz and Schreider 2004), physics (Frisch 1995), economics (Granger 1980; Enders 1995), hydrology (Beran 1994), and geography (Bennett 1979). Models used to describe

Y. Leung, *Knowledge Discovery in Spatial Data*, Advances in Spatial Science,
DOI 10.1007/978-3-642-02664-5_6, © Springer-Verlag Berlin Heidelberg 2010

the generating processes of time series range from one extreme, the deterministic processes dictated by some physical laws, to the other extreme, the processes of complete randomness such as the random walk. Some time series exhibit strictly predictable deterministic trends that can be described by deterministic functions depicting exponential growth or cyclical fluctuations. In general, particularly in complex systems, time series are stochastic in nature.

Within a stable environment, time series are stationary. Roughly speaking, a linear system is stationary if all of its moments are fixed and constant over time. A dynamical system is stationary if the evolution operator remains unchanged over time. Specifically, a time series is weakly stationary if the mean and variance are constant over time and the autocovariance function depends only on the time lag. If the mean, variance and autocovariance structure are constant over time, a time series is strictly stationary. Over the years, methods have been developed for the analysis of stationary time series (Box and Jenkins 1976; Box et al. 1994).

In reality, natural and human processes are generally non-stationary. Model parameters often depend on time. Mean and variance, for example, are functions of time. Thus, time series are usually generated by some non-stationary processes which we need to identify. Of particular interest is the scaling behavior of the time series data in the local-global context. It has been observed in a large variety of processes that there are no characteristic scales of space or time by which the whole can be distinguished from the parts. It is a paradigm shift from models such as Markov chains or Poisson processes which seek a characteristic scale that plays a more important role in the analysis of a time-varying process. Random walk is perhaps the simplest stochastic model for the study of such scale-based non-stationary time series. It is an additive structure in which the present is the summation of unrelated events in the past. It is scale invariant because there is no characteristic scale indicating a cut-off in the development of the walk. The beauty of random walk is its simplicity. It is often not a realistic structure for generating time series in practice.

Many real-life time series actually exhibit long memory (Beran 1992; Rangarajan and Ding 2003). Simple systems have correlation functions decaying exponentially over time. However, complex natural and man-made structures and processes generally have long-range spatial and temporal correlations. Distant past in these process often continue to exert effect in a law-like manner. They are self-similar processes with scaling behavior that holds for all scales. That is, the process and its dilations are statistically self-similar. Processes with long-range dependency have covariance functions decaying very slowly to zero as a power law. In terms of memory, they are of longer range than the exponentially decaying correlations. Fractional Brownian motion (fBm), a generalization of random walk and Brownian motion, is a typical strongly correlated process with power-law behavior (Mandelbrot and Van Ness 1968). It is a non-stationary process with Gaussian stationary increments. Time series generated by fBm exhibits strong spatial and temporal correlation. Wavelet analysis (Daubechies 1992) is another powerful method for the study of scaling behavior in data. Self-similarity is captured by the wavelet coefficients.

Therefore, processes with long-range dependence have no characteristic scale of time. Instead of looking for characteristic scales in time series, we look for relations and mechanisms over a wide range of scales. The underlying mechanisms of long range dependence processes have similar statistical properties on different time scales. A stochastic process, such as fBm, is said to be self-similar if its statistics stay invariant with respect to a change in time scale. Long-range dependent processes are often referred to as fractal processes because their sample path displays self-similarity. That is, the exponent of their moments is restricted by a constant self-similarity parameter.

Due to the variations at small intervals, however, non-constant scaling parameter often exists in time series. Nonlinear processes may involve simultaneous appearance of periodic and chaotic behavior. Spatio-temporal intermittency due to parameter fluctuations around some critical values is a common place in non-stationary time series involving multiple scaling behaviors. Stochastic process with multiple scaling is often called an intermittent process. The heavy-tailed distribution is often a power-law type with slow decaying autocorrelation function. Multifractals is a typical multiscaling structure with irregularly varied sample paths (Mandelbrot 1999). In terms of data mining, we need to discover the multiplicative scheme that generates such multifractal processes.

The purpose of this chapter is not to discuss time series analysis in general. I only focus on the discovery of mechanisms that generate time series with scaling behavior. Special attention is paid to self-similar and intermittent processes with long range dependence. Our ability to handle non-stationary uniscaling and multiscaling behaviors is essential to the mining of useful structures in complex systems manifested by spatio-temporal data.

In the remaining part of this chapter, our discussion is concentrated on the analysis of time series with long range dependence. Wavelet analysis of signals/ functions at all scales and times is first examined in Sect. 6.2. The multifractal approach to the mining of intermittents, transient, noisy and aperiodic processes from time series data is then discussed in Sect. 6.3. The identification of intermittency in air quality is employed to substantiate the theoretical arguments. A formal characterization of such time series is given in Sect. 6.4. To account for spatial variability over time, the multifractal approach is further extended to discover spatial variability of rainfall intensity in Sect. 6.5. In Sect. 6.6, a methodology for the analysis of multifractaility and long-range dependence in remote sensing data is proposed for further study.

6.2 The Wavelet Approach to the Mining of Scaling Phenomena in Time Series Data

6.2.1 A Brief Note on Wavelet Transform

Since the 1980s, wavelet transform has been found to be instrumental in analyzing temporal signals with scaling behavior. It is particularly effective in the discovery of self-similar processes. Significant applications have been made in the study of a large variety of time series data including seismic signals, climatic data, river runoffs, atmospheric turbulence, DNA sequencing, and finance.

The wavelet transform is actually the convolution of the wavelet function with the signal so that it can be scaled and more revealingly examined under another representation in both frequency and time. By plotting the wavelet transform in terms of scale and location, we can build a picture of correlation between the wavelet and the signal under study.

6.2.2 *Basic Notions of Wavelet Analysis*

There are two broad classes of wavelet transforms: the continuous and discrete wavelet transforms. The continuous wavelet transform deals with time series defined over the entire real axis. The discrete wavelet transform, on the other hand, is constructed for time series that are observed over a range of discrete points in time (translations). They are briefly described in the following discussion.

6.2.2.1 The Continuous Wavelet Transform

In order to transform a function/signal into another form that unfolds it in time and scale, we need to manipulate a wavelet, a localized waveform, along the time axis with a scaling (dilation between the finest and the coarsest scales) process. Such a process of translating and scaling a function is called the wavelet transform. Though there are a large number of choices for a wavelet, its selection depends on the signal under scrutiny and the purpose of a particular application. The commonly employed wavelet in continuous wavelet transform is the so-called Mexican hat wavelet

$$\psi(t) = \left(1 - t^2\right)e^{-t^2/2}, \tag{6.1}$$

which is essentially the second derivative of the Gaussian distribution function $\frac{1}{\sigma\sqrt{2\pi}}e^{-t^2/\sigma^2}$ without the usual term $1/\sqrt{2\pi}$ and with $\sigma^2 = 1$ (Fig. 6.1). Actually, all derivatives of such Gaussian function can be employed as a wavelet to unravel a signal in terms of time and scale. The Mexican hat is the basic structure, generally called the mother wavelet, on which translation and scaling are performed.

The Haar wavelet (Harr 1910):

$$\psi(t) = \begin{cases} 1\ , & 0 \leq t < \frac{1}{2}\ , \\ -1\ , & \frac{1}{2} \leq t < 1\ , \\ 0\ , & \textit{otherwise}\ , \end{cases} \tag{6.2}$$

is perhaps the earliest proposal of a wavelet that can be employed to transform a signal by a step function (Fig. 6.2).

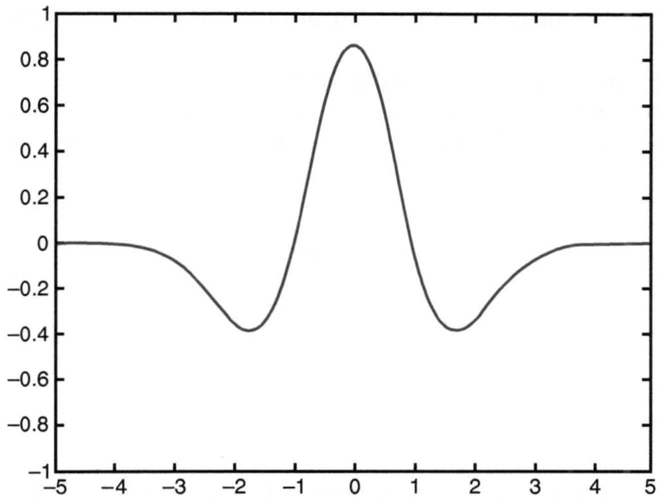

Fig. 6.1 The Maxican hat wavelet

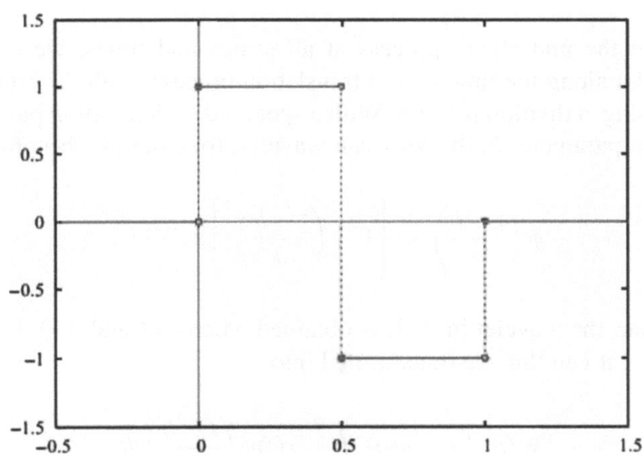

Fig. 6.2 The Haar wavelet

A wavelet by definition is any function $\psi(\cdot)$ whose integral is zero:

$$\int_{-\infty}^{\infty} \psi(t)dt = 0, \tag{6.3}$$

and it is square integrable:

$$\int_{-\infty}^{\infty} \psi^2(t)dt = 1. \tag{6.4}$$

Thus, it is essentially a small wave that grows and decays in a limited time period.

To make it practical for solving different problems, additional conditions can be imposed on a wavelet (Daubechies 1992). The admissibility condition is the common condition adopted in many studies. A wavelet is admissible if its Fourier transform

$$\Psi(f) = \int_{-\infty}^{\infty} \psi(t) e^{-i2\pi f t} dt \tag{6.5}$$

is such that

$$C_\psi = \int_0^{\infty} \frac{|\Psi(f)|^2}{f} df < \infty. \tag{6.6}$$

Plotting the squared magnitude of the Fourier transform against the frequency for the Mexican hat wavelet, for example, we obtain the energy spectrum

$$|\,\Psi(f)\,|^2 = 32\pi^5 f^4 e^{-4\pi^2 f^2}. \tag{6.7}$$

To unravel the underlying process at all scales and times, we can move the mother wavelet along the time axis, a translation process, with different stretching and compressing, a dilation process. With a specified scale/dilation parameter a and the translation parameter b, the Mexican wavelet, for example, becomes

$$\psi\left(\frac{t-b}{a}\right) = \left[1 - \left(\frac{t-b}{a}\right)^2\right] e^{-\frac{1}{2}\left(\frac{t-b}{a}\right)^2}. \tag{6.8}$$

In particular, the wavelet in (6.2) is obtained when a=1 and b=0. For a continuous signal x(t), it can thus be transformed into

$$\mathrm{W}(a, b) = g(a) \int_{-\infty}^{\infty} x(t) \psi\left(\frac{t-b}{a}\right) dt. \tag{6.9}$$

It should be noted that the complex conjugate is used in (6.9) when $\psi(t)$ is complex (e.g., the Morlett wavelet in (6.12)).

For the conservation of energy, the weighting function $g(a)$ is by convention adopted as $1/\sqrt{a}$, albeit other functional form can be customized for specific applications. Under this situation, the continuous wavelet transform in (6.9) becomes

$$\mathrm{W}(a, b) = \frac{1}{\sqrt{a}} \int_{-\infty}^{\infty} x(t) \psi\left(\frac{t-b}{a}\right) dt. \tag{6.10}$$

The convolution is thus the inner product of the wavelet and the signal integrated over the signal range. It can be interpreted as the cross-correlation of a signal with a set of wavelets of various widths. The wavelet transform essentially scrutinizes a signal by magnifying its local structures, via scale parameter a, at various location b. It technically maps a signal into a two-dimensional function of a and b. By varying a from the largest to the smallest value, the wavelet unravels coherent structures within the signal as it travels along the location dimension b. The process can be graphically depicted by the wavelet transform plot with respect to parameters a and b. In terms of changing average, wavelet transform depicts how weighted averages of a signal vary from one averaging period to the next. In practical applications, changes in averages over various scales might be of more interest than the averages themselves, e.g., changes in yearly average temperature over the desert, and changes in daily average concentration of carbon dioxide over a city. By varying the scale a, we can construct a picture of how averages of a signal over a range of scales are changing from one period of length a to the next. Thus, the wavelet transform plot serves as an exploratory device that can help us to visualize and unravel features of interest.

To recover the original signal from its wavelet transform, an inverse wavelet transform can be employed by integrating the wavelet transform over all scales a's and locations b's as follows:

$$x(t) = \frac{1}{c_\psi} \int_{-\infty}^{\infty} \int_{0}^{\infty} W(a, b) \frac{1}{\sqrt{a}} \psi\left(\frac{t-b}{a}\right) \frac{1}{a^2} dadb. \tag{6.11}$$

In some applications, such as those in geophysics, wavelets with real and imaginary parts might be more appropriate. A common complex wavelet, called the Morlett wavelet, is defined in a simpler form as

$$\psi(t) = \frac{1}{\pi^{1/4}} e^{i2\pi f_0 t} e^{-\frac{t^2}{2}}, \tag{6.12}$$

where f_0 is the central tendency of the mother wavelet (Fig. 6.3). It is technically a complex exponential whose amplitude is modulated by a function proportional to the Gaussian probability density function.

By substituting t with (t-b)/a, the Morlett wavelet becomes

$$\psi\left(\frac{t-b}{a}\right) = \frac{1}{\pi^{1/4}} e^{i2\pi f_0 \left(\frac{t-b}{a}\right)} e^{-\frac{1}{2}\left(\frac{t-b}{a}\right)^2} \tag{6.13}$$

Again, the corresponding transform unravels coherent structures of the signal over the ranges of scale a and location b. For certain applications, the complex wavelets are instrumental because the phase of the wavelet transform may contain useful information.

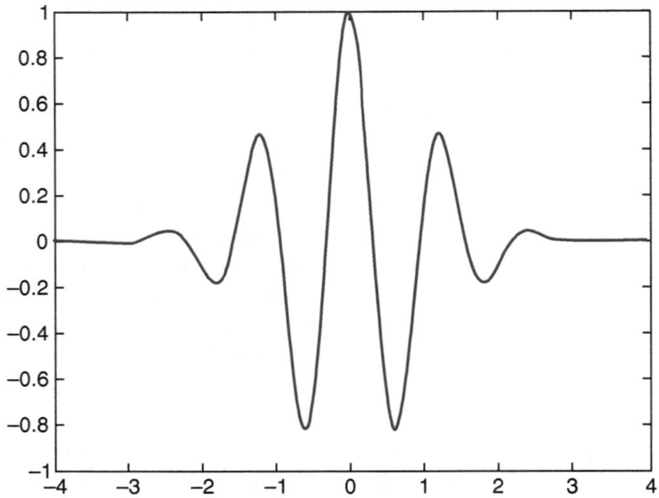

Fig. 6.3 The Morlett wavelet

6.2.2.2 The Discrete Wavelet Transform

In practice, we often need to discretize the wavelet by the appropriate discretization of the parameters a and b to perceive more efficiently and practically the key features of a signal with a finite number of a and b values. To get rid of the redundancy contained in the continuous wavelet transform, a common method is to employ the logarithmic discretization of the a scale and move in discrete proportional steps to each b location. By setting $a = {a_0}^m$ and $b = nb_0{a_0}^m$, where a_0 and b_0 are pre-specified scaling and location steps, respectively, and m and n are the respective controls of dilation and translation, the corresponding wavelet is expressed as

$$\psi_{m,n}(t) = \frac{1}{\sqrt{{a_0}^m}}\psi\left(\frac{t - nb_0{a_0}^m}{{a_0}^m}\right). \tag{6.14}$$

In general, the wavelet transform of a continuous signal with the discrete wavelet in (6.14) becomes

$$\mathrm{W}(m, n) = \int_{-\infty}^{\infty} x(t){a_0}^{-m/2}\psi({a_0}^{-m}t - nb_0)dt. \tag{6.15}$$

Applying the commonly used dyadic grid, i.e., setting a to be of the form 2^{j-1}, j=1,2,3, and within a given dydadic scale 2^{j-1}, select times b that are separated by multiples of 2^j. For $a_0 = 2$ and $b_0 = 1$, the wavelet in (6.14) is the well-known power-of-two dyadic wavelet

$$\psi_{m,n}(t) = \frac{1}{\sqrt{2^m}}\psi\left(\frac{t - n2^m}{2^m}\right). \tag{6.16}$$

The corresponding discrete wavelet transform becomes

$$\mathrm{W}(m, n) = \int_{-\infty}^{\infty} x(t)\, \psi_{m,n}(t)dt. \tag{6.17}$$

The discrete wavelet transform can hence be applied directly to a time series obtained from a discrete set of points in time. Employing the dydadic grid wavelet, the original signal can be reconstructed in terms of the wavelet coefficients, $W(m, n)$, via the following inverse discrete wavelet transform

$$x(t) = \sum_{m=-\infty}^{\infty} \sum_{n=-\infty}^{\infty} \mathrm{W}(m, n)\, \psi_{m,n}(t)dt. \tag{6.18}$$

6.2.3 Wavelet Transforms in High Dimensions

In the mining of spatial structures and processes, our interest often centers on the discovery of local and global distributions of certain spatial phenomena. An effective way to detect such spatial distributions is to convolute a spatial signal with a wavelet and let the resulting wavelet transform to unfold the local relationships through the scale and translation parameters a and b.

The two-dimensional Mexican hat for the (t_1,t_2)-coordinate space is defined as

$$\psi(t) = \left(2 - \left|\mathbf{t}^2\right|\right)e^{-|\mathbf{t}|^2/2}, \tag{6.19}$$

where $\boldsymbol{t} = (t_1,t_2)$ is the spatial coordinate vector with $\mathbf{t} = \sqrt{t_1{}^2 + t_2{}^2}$. With specified parameter a and $\mathbf{b}$ (a vector), the corresponding wavelet transform becomes

$$\mathrm{W}(a, \mathbf{b}) = \frac{1}{a}\int_{-\infty}^{\infty} x(\mathbf{t})\, \psi\left(\frac{\mathbf{t} - \mathbf{b}}{a}\right)d\mathbf{t} \tag{6.20}$$

where $\mathbf{b} = (b_1, b_2)$is the coordinate vector, $1/a$ for energy conservation, and x($\boldsymbol{t}$) can be any geographical measure such as elevation or temperature.

The associated inverse wavelet transform is

$$x(\mathbf{t}) = \frac{1}{c_\psi}\int_{-\infty}^{\infty}\int_{0}^{\infty} \mathrm{W}(a, \mathbf{b})\frac{1}{a}\psi\left(\frac{\mathbf{t} - \mathbf{b}}{a}\right)\frac{1}{a^3}d\mathbf{b}. \tag{6.21}$$

To generalize, the k-dimensional wavelet is expressed as

$$\psi(a, \mathbf{b}) = a^{-k/2}\psi\left(\frac{\mathbf{t} - \mathbf{b}}{a}\right), \tag{6.22}$$

where ***t*** and ***b*** are the k-dimensional vectors, and $a^{-k/2}$ is for energy conservation. The k-dimensional wavelet transform is thus

$$\mathrm{W}(a, \mathbf{b}) = \int_{-\infty}^{\infty} x(\mathbf{t})\, \psi(a, \mathbf{b}) d\mathbf{t}. \tag{6.23}$$

The corresponding inverse wavelet transform is

$$x(\mathbf{t}) = \frac{1}{c_\psi} \int_{-\infty}^{\infty} \int_{0}^{\infty} \mathrm{W}(a, \mathbf{b})\psi(a, \mathbf{b})a^{-(k+1)} da d\mathbf{b} \tag{6.24}$$

6.2.4 Other Data Mining Tasks by Wavelet Transforms

As discussed in Sect. 5.1, fractals are objects that exhibit self-similarity, exact or statistical, over scales. Such property makes wavelet transform a natural mechanism for the examination of fractal objects (Bacry et al. 1993; Fisher 1995). The determination of the scaling properties of the fractional Brownian motion by wavelet transform analysis is a typical example. Scaling of multifractals by wavelet-based characterization has also been made in recent years (Riedi et al. 1999).

6.2.5 Wavelet Analysis of Runoff Changes in the Middle and Upper Reaches of the Yellow River in China

Among other applications, wavelet transforms have been employed in a number of hydrologic analyzes such as the streamflow characterization (Labat et al. 2005), variability of runoffs (Labat et al. 2000), hydrological variations (Andreo et al. 2006), effect of El Nino on streamflow in the Yantze river (Jevrejeva et al. 2003; Zhang et al. 2007), and watersheds characterization (Gaucherel 2002). Since uncertainty is involved in stream development and time scale of high and low flows are unknown, wavelet analysis, without requiring any pre-specified timing of cycles and bursts, enables us to examine the hydrologic dynamics at all times and scales (day, month, year and millennium). It can detect the characteristics of runoff changes under different time scales.

Since the 1980s, frequent occurrences of dry-up episodes have been experienced in the lower reach of the Yellow River in China. Since the hydrologic pattern of the lower reach is directly affected by runoff changes in the middle and upper reaches, it is essential to study the dynamics of the latter in order to obtain a clearer explanation of the former. In their study of runoff changes in the upper and middle reaches of the Yellow River, Jiang et al. (2003) apply the compactly supported spline wavelet with dyadic grid to unravel high and low flows of that region over multi-time scales. The time series are monthly runoffs of four hydrolic stations: Guide, Lanzhou, Hekouzhen, and Sanmenxia, collected between 1919 and 1997 (Fig. 6.4). Wavelet coefficients for scale 1 (1 month)- scale 10 (512 months) are obtained. The lowest resolution is 42.5 years. Fig. 6.5a–d plot the wavelet coefficients against scale and time for Guide, Lanzhou, Hekouzhen, and Sanmenxia stations, respectively. The curves are values of the wavelet coefficients. Positive values are depicted by solid lines indicating peak flow periods (H), and negative values are depicted by dotted lines indicating the low flow periods (L). In Fig. 6.5a, the cycles at Guide are clearly unraveled. They are cycles of approximately 2^9months (42 years), 2^6–2^8months (5–21 years), and 2^3–2^5months (1–3 years). The upper part of Fig. 6.5a clearly shows the 42-year cycles with four consecutive

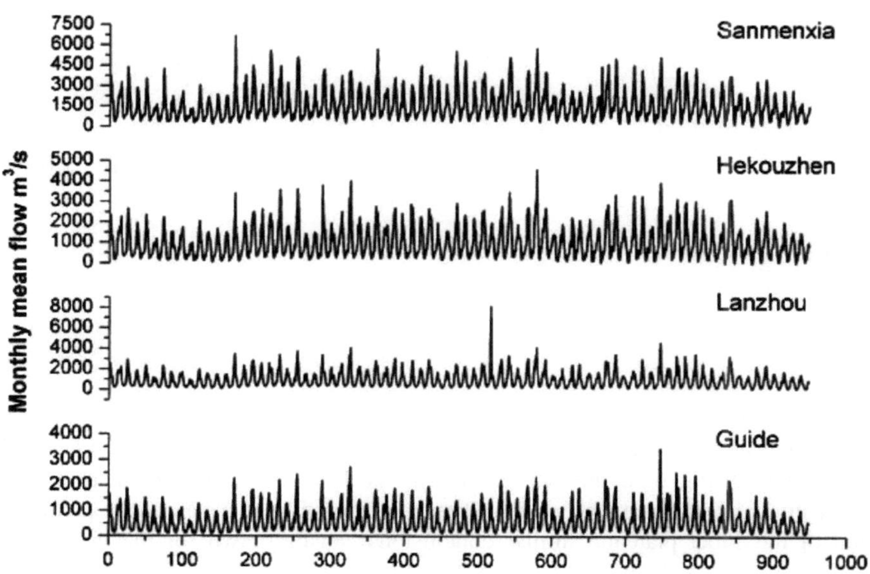

Fig. 6.4 Number of months from July, 1919

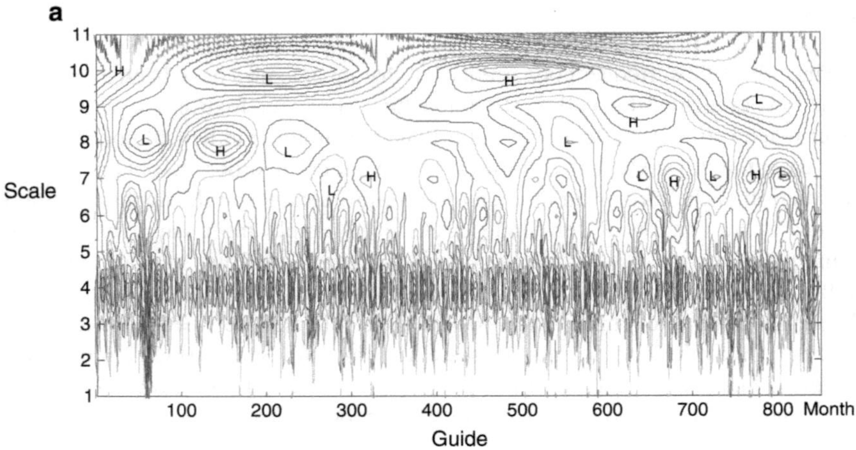

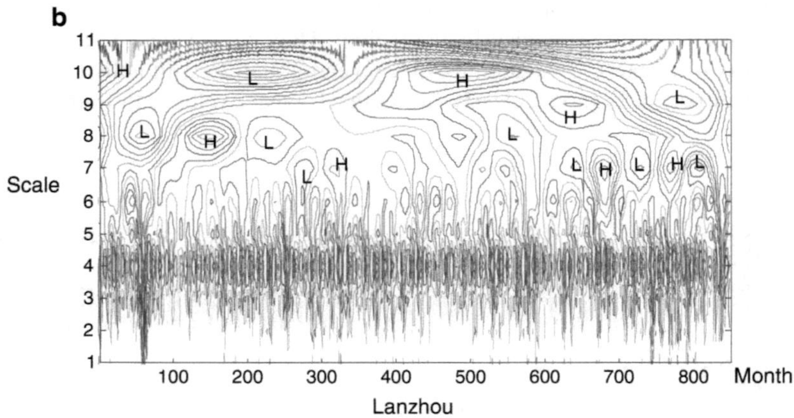

Fig. 6.5 (Continued)

centers of H, L, H and L from 1919 to 1997. The center part of Fig. 6.5b, on the other hand, indicates mainly the 5–21 year cycles with a series of H's and L's. The bottom part of Fig. 6.5a, unravels essentially the 1–3 year cycles which correspond to the annual changes of the runoff. They coincide with the 3-year cycle precipitation of the Tibet plateau. Similar conclusions can be drawn for the three other stations from Fig. 6.5b–d (This is in fact the answer to the question raised with repect to Fig. 1.6 in Sect. 1.5 in Chap. 1). The unraveled runoff dynamics are found to be significantly related to climatic changes and human activities of the region over the years.

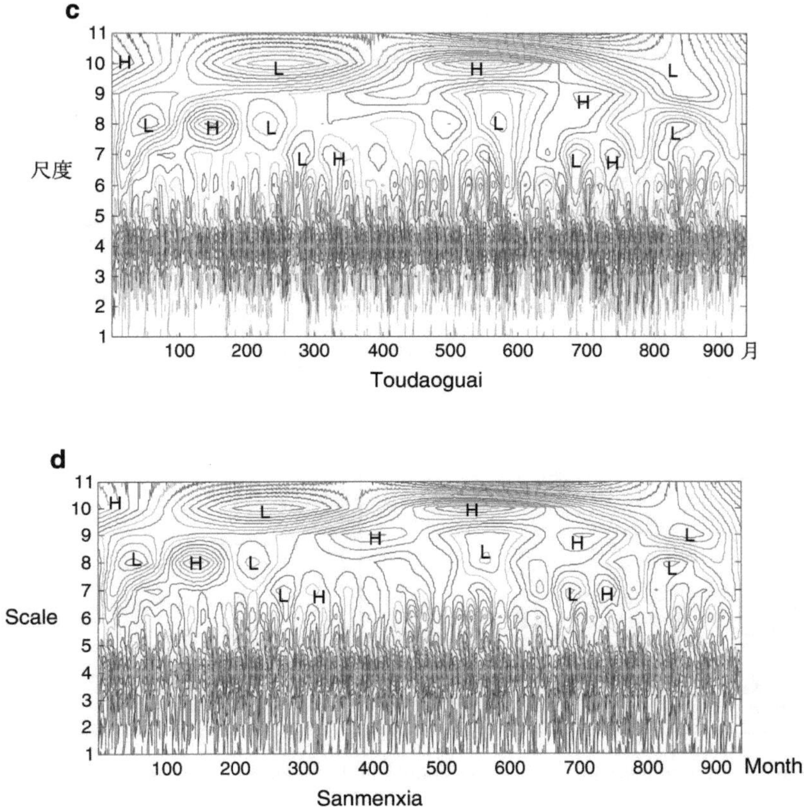

Fig. 6.5 Wavelet coefficient maps of runoff changes

6.2.6 Wavelet Analysis of Runoff Changes of the Yangtze River Basin

The Yangtze River (Changjiang), the longest river in China and the third longest river in the world, lies between 91°E and 122°E and 25°N and 35°N. It has a drainage area of 1,808,500 km^2 and the mean annual discharge of 23,400 m^3s^{-1} measured at Hankou Station. The Yangtze River Basin is located in the monsoon region of East Asia subtropical zone, and has a mean annual precipitation of about 1,090 mm. Frequent flood hazards have exerted tremendous impacts on the socio-economic development and human life in the Yangtze River basin. To understand runoff changes, the wavelet approach has been applied to analyze the periodicity of hydrological extremes (Zhang et al. 2006a,b) and the detection of connections

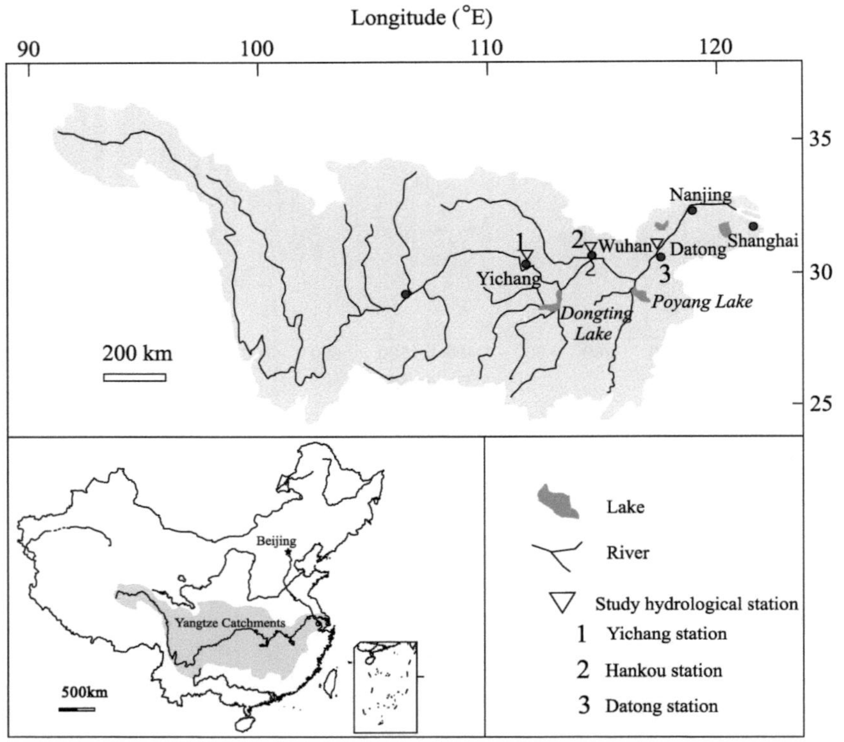

Fig. 6.6 Location of hydrological guaging stations in the yangtze river basin

between ENSO events and annual maximum streamflow (Zhang et al. 2007) in the Yangtze River basin. With respect to periodicity of annual maximum water level and streamflow, Zhang et al. (2006) employ the Mexican hat wavelet to probe into the temporal-frequency space of annual maximum water level and streamflow series of three hydrological guaging stations: Yichang, Hankou and Datong stations. The locations of these three stations are depicted in Fig. 6.6. Similar patterns of annual maximum streamflow and water level are formed at the upper and middle reaches of the river. The periods of water level changes decreased over time, especially at downstream. It means that the occurrence frequency of annual maximum water level became higher over time (Fig. 6.7). The finding could facilitate flood mitigation in the Yangtze River.

The El Niño/Southern Oscillation (ENSO) represents the dominant coupled ocean-atmosphere mode of the tropical Pacific. On inter-annual time scales, the significant part of the global climatic changes can be linked to ENSO (Trenberth et al. 1998). The ENSO extreme phases are usually linked with major episodes of floods and droughts in many locations of the world. Zhang et al. (2007) use the continuous wavelet transform (CWT), cross wavelet and wavelet coherence methods to explore connections between hydrological extremes in the Yangtze

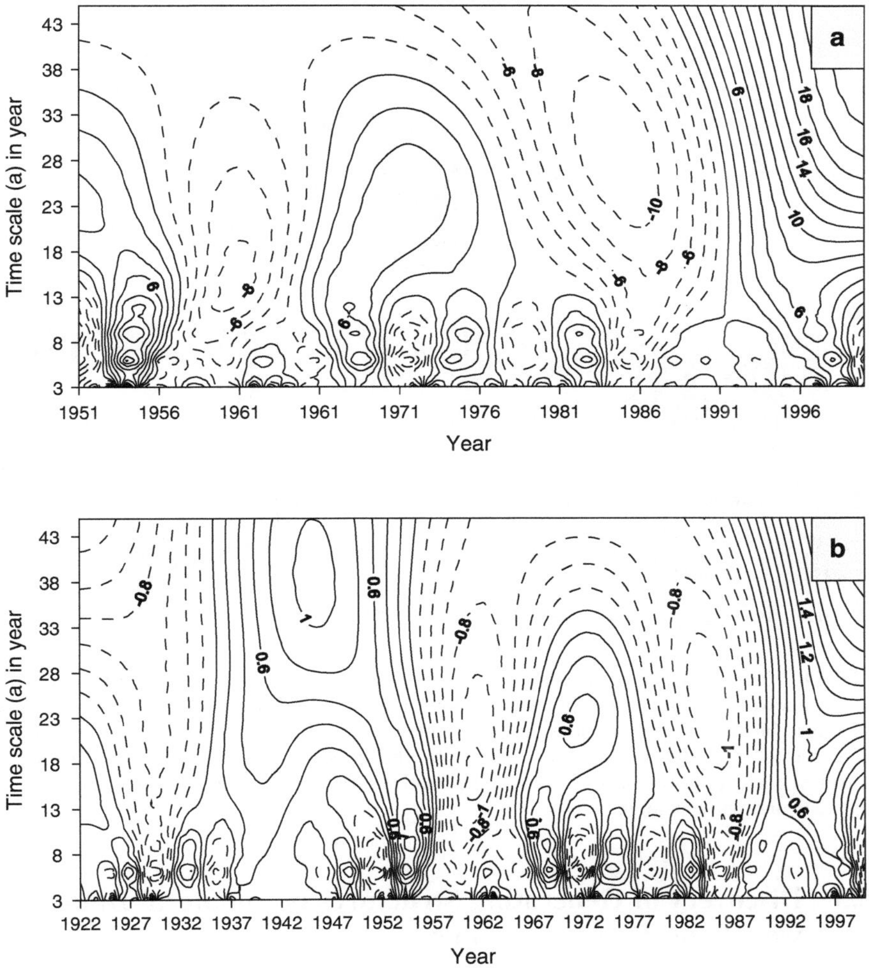

Fig. 6.7 Wavelet analysis of the annual maximum streamflow (**a**) and annual maximum water level (**b**) of the Datong station

River basin and the ENSO events. Different phase relations are identified between the annual maximum streamflow of Yangtze River and the El Niño/Southern Oscillation (ENSO) in the lower, middle and upper Yangtze River basin. In-phase relations of annual maximum streamflow in the lower Yangtze River and anti-phase relations in the upper Yangtze River (Figs. 6.8 and 6.9) were found. Ambiguous phase relations, however, were identified in the middle Yangtze River. Wavelet techniques successfully reveal the connections between global climatic signals and hydrological extremes in the Yangtze River basin. Further understanding of the underlying physical mechanisms responsible for such spatial and temporal variability of hydrological extremes across the Yangtze River basin is thus necessary.

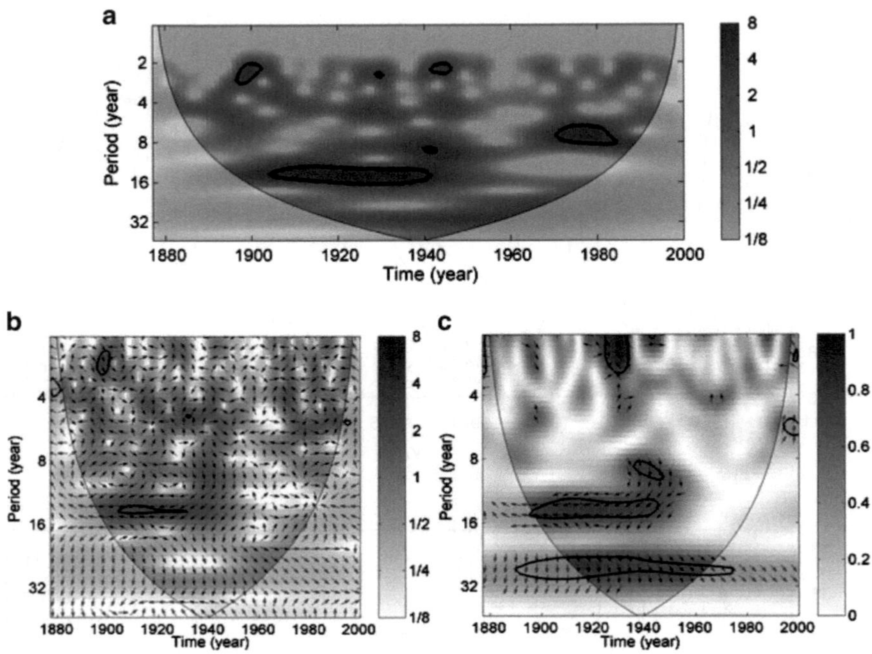

Fig. 6.8 Wavelet analysis of annual maximum streamflow of Datong Station. (**a**) Continuous wavelet power spectrum of the normalized annual maximum streamflow series of Datong station. The thick black contour designates the 95% confidence level against red noise and the cone of influence (COI) is shown as a lighter shade. (**b**) The cross wavelet transform. (**c**) The squared wavelet coherence result. Arrows indicate the relative phase relationship (with in-phase pointing right and anti-phase pointing left)

6.3 Discovery of Generating Structures of Temporal Data with Long-Range Dependence

6.3.1 A Brief Note on Multiple Scaling and Intermittency of Temporal Data

As discussed in Sect. 6.1, time series depicting complex systems are often multifractal with long-range dependence. Variation of air quality over time is a typical example. Long-range dependence implies the presence of stochastic trends in the time series. Thus effective management of air quality requires an understanding of the trends hidden in monitoring data and patterns of high pollution episodes. The correspondence between anthropogenic trends and the long-range dependence (LRD) component in air quality data has been studied extensively in the literature (see, for example, Anh et al. 1997a,b). Existing works on air pollution mainly pay attention to the second-order statistics of the time series (such as their covariance

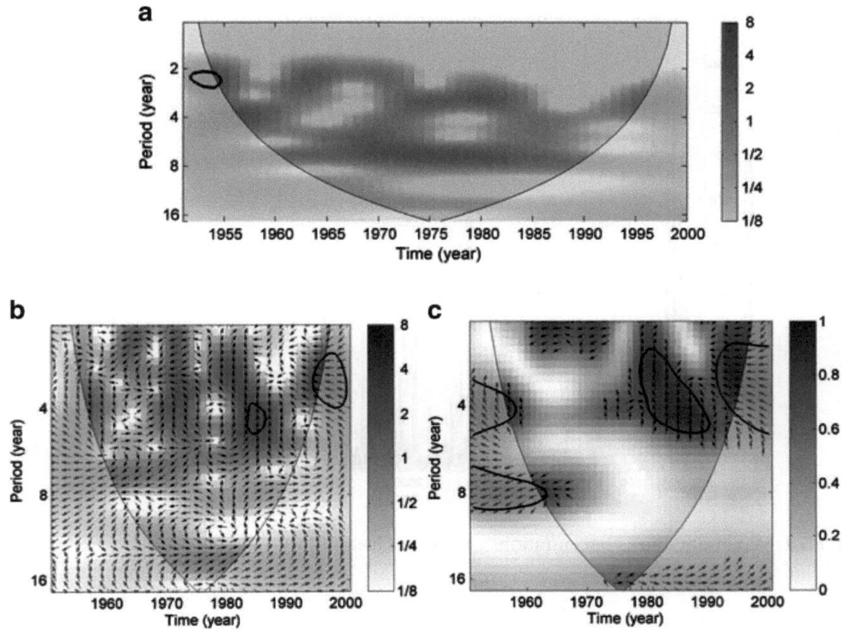

Fig. 6.9 Wavelet analysis of annual maximum streamflow of Yichang Station (**a**) Continuous wavelet power spectrum of the normalized annual maximum streamflow series of Yichang station. The thick black contour designates the 95% confidence level against red noise and the cone of influence (COI) is shown as a lighter shade. (**b**) The cross wavelet transform. (**c**) The squared wavelet coherence result. Arrows indicate the relative phase relationship (with in-phase pointing right and anti-phase pointing left)

structure or spectral density). It is known in many recent studies that turbulent processes display multiple scaling (Meneveau and Sreenivasan 1991; Frisch 1995) that cannot be adequately captured by models based on second-order properties. A description of this behavior requires the consideration of their high-order moments, and a suitable framework of this description is the multifractal formalism. An employment of multifractal models to represent air quality data with intermittency is given in Anh et al. (1999b). An introduction of the model which quite appropriately represents intermittency in air quality data is discussed in here.

6.3.2 Multifractal Approach to the Identification of Intermittency in Time Series Data

Figures 6.10 and 6.11, respectively, depict the maximum daily concentration of SO_2 and NO recorded at the Queen Mary Hospital monitoring station in Hong Kong with 3,650 and 970 points-in-time observations. Our purpose is to unravel the

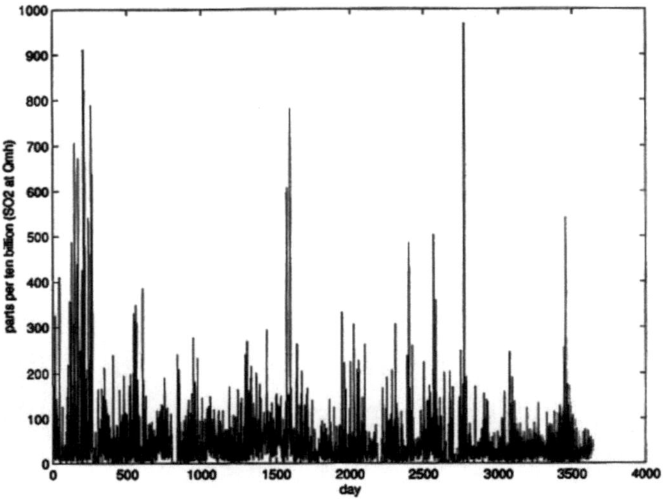

Fig. 6.10 Maximum daily concentrations of SO_2 at Queen Mary Hospital

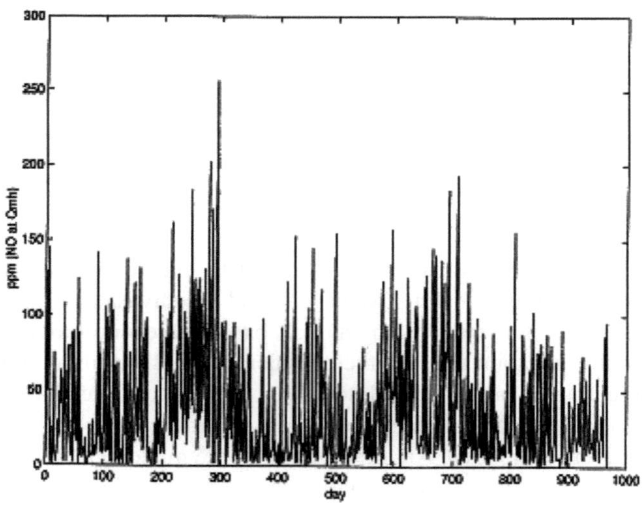

Fig. 6.11 Maximum daily concentrations of NO at queen mary hospital

model which best fits the structure that generates the bursts appearance of these time series.

Let $\{Y(t)\ ,\ t \in R\}$ be a stationary stochastic process with

$$Y(t) \geq 0\,,\ \forall t \quad \text{and} \quad EY(t) = 1. \tag{6.25}$$

Define

$$Y(t\,;\,r) = \frac{1}{r}\int_{t-\frac{r}{2}}^{t+\frac{r}{2}} Y(s)\,ds\,,\ r > 0. \tag{6.26}$$

$Y(t;\,r)$ is a smoothing (coarse graining) of $Y(t)$ at scale $r > 0$ smoothing window of size r. We assume that $Y(t;\,r)$ is an intermittent process (with spiky/bursts appearance, see Chap. 8 of Frisch (1995)). The scaling behavior of $Y(t;\,r)$ can be described by

$$\sum (Y(t;r))^q \sim r^{\tau(q)}\,,\ q \in R, \tag{6.27}$$

as $r \to 0$, where the sum is taken over all disjoint intervals of length r. The function $\tau(q)$ is known as the mass exponent of $Y(t)$ and is related to the generalized Renyi dimension, $D(q)$, by

$$D(q) = \frac{\tau(q)}{q-1}. \tag{6.28}$$

Hentschel and Procaccia (1983) have shown that D_0 is the fractal dimension of the support of $Y(t)$, D_1 is the information dimension and D_2 is the correlation dimension.

An intermittency model for $Y(t)$ is essentially a parameterization of $\tau(q)$. Among the intermittency models, Borgas (1992) conclude that the binomial p-model appears to be the most satisfactory model and best represents the measurements. Along this line of thinking, we assume that $Y(t)$ is generated by a multiplicative cascade with a binomial generator characterized by a probability p, $0 < p \leq 1/2$. In other words, we consider an interval E of unit length and construct a Cantor set $F = \bigcap_{i=0}^{\infty} E_i$ on this interval, where $E_0 = E$, E_k contains 2^k subintervals of length 2^{-k} obtained by dividing each subinterval of E_{k-1} into two halves.

We next define a positive measure μ on F. Let $0 < p \leq 1/2$ be given and consider a unit mass on E_0 (i.e., $\mu_0 = 1$). We can randomly split this unit mass between the two intervals of E_1 so that one has mass p and the other has mass $1-p$. This defines μ_1, which has a constant value of $2p$ on one interval and a constant value of $2(1-p)$ on the other interval. Continuing in this way, so that the mass on each interval of E_k is divided randomly into the proportions p and $1-p$ between its two subintervals in E_{k+1}. This defines a sequence $\{\mu_k\}$, which is a positive martingale and hence converges weakly to a limiting mass distribution μ on F (Kahane 1991).

The basic assumption is that $Y(t)$ is generated by such iterative process, resulting in a multiplicative cascade. Its scaling/intermittency is described by (6.27). Next, we need to determine the function $\tau(q)$. Each generation of the cascade is defined by E_k and μ_k. For each $0 \leq j \leq k$, a number $\binom{k}{j}$ of the 2^k intervals of E_k have mass $p^k(1-p)^{k-j}$, where $\binom{k}{j} = \frac{k!}{j!(k-j)!}$. By the binomial theorem, we have

$$\sum \left(Y(k\,;\,2^{-k})\right)^q = \sum_{j=0}^{k} \binom{k}{j} p^{qj}(1-p)^{q(k-j)} = (p^q + (1-p)^q)^k. \tag{6.29}$$

Let $2^{-k} = r$, i.e., $k = -(\log r / \log 2)$. It follows from (6.29) that

$$\log \sum (Y(k\,;\,r))^q = k \log(p^q + (1-p)^q) = -\log r \frac{\log(p^q + (1-p)^q)}{\log 2}$$
$$= \log\left(r^{-\log_2(p^q+(1-p)^q)}\right). \tag{6.30}$$

From (6.27) to (6.30), we obtain

$$\tau(q) = \lim_{r\to 0} \frac{\log \sum (Y(k\,;\,r))^q}{\log r} = -\log_2(p^q + (1-p)^q), \tag{6.31}$$

and hence

$$D_q = -\frac{\log_2(p^q + (1-p)^q)}{q-1}. \tag{6.32}$$

For the convenience of data fitting, a related exponent can be introduced by defining

$$E(Y(t\,;\,r))^q \sim r^{1-q+\tau(q)}, \; q \geq 0. \tag{6.33}$$

(Monin and Yaglom 1975, p. 534). Define

$$K(q) = -\tau(q) + q - 1. \tag{6.34}$$

Then, for the binomial cascade described above,

$$K(q) = \log_2(p^q + (1-p)^q) + q - 1. \tag{6.35}$$

It follows directly from (6.33) that $K(0) = 0$. Since $EY(t\,;\,r) = 1$, by definition we also have $K(1) = 0$. Considering a sufficiently small r, we obtain from (6.33) to (6.34) that

$$K(q) = \frac{\log E(Y^q)}{\log r}. \tag{6.36}$$

It is shown in Anh et al. (1999a) that $K(q)$ is a convex function, and $K(q) < 0$ *iff* $E(Y^q) < 1, 0 < q < 1$. These are useful results when (6.35) is employed for data fitting.

In order to see directly whether $Y(t)$ is monofractal or multifractal, it is more convenient to consider another scaling exponent defined by

$$E|Y(t) - Y(t-r)|^q \sim r^{\zeta(q)}, \; q \geq 0, \tag{6.37}$$

as $r \to 0$. It can be observed that $\zeta(0) = 0$, and no other exponent is known a priori (in contrast with $K(q)$, where there are two a priori exponents: $K(0) = 0$ and

$K(1) = 0$). It is shown in Anh et al. (1999a) that $\zeta(q)$ is concave. If $Y(t)$ is bounded, the function $\zeta(q)$ is shown to be monotonically non-decreasing (Marshak et al. 1994). These results imply that, if $Y(t)$ is a monofractal, its scaling will simply be given by $\zeta(q) = q\alpha$, where α is a constant, for all q. In particular, for fractional Brownian motion with Hurst index H, a typical monofractal (Falconer 1985; Mandelbrot 1985), we have $\zeta(q) = qH$. This becomes a convenient tool to test whether or not $Y(t)$ is a monofractal.

6.3.3 Experimental Study on Intermittency of Air Quality Data Series

The model in (6.34) is applied to six air quality data series to identify and study their intermittency. Three data series, *QmhSO*2, *VpkSO*2 and *VicSO*2, respectively, record the maximum daily concentrations of sulfur dioxide at three monitoring stations located at Queen Mary Hospital, Victoria Peak and Victoria Road of Hong Kong. The other three series, *QmhNO*, *VpkNO* and *VicNO*, depict the maximum daily concentrations of nitrogen oxide at these respective locations. The SO_2 series cover the period 1986–1995, consisting of 3,650 observations, while the *NO* series only cover the period 1993–1995, consisting of 970 observations. As examples, *QmhSO*2 and *QmhNO* are plotted in Figs. 6.10 and 6.11, respectively. It can be observed that both normalized series display intermittency which is quite distinct from the yearly cycle.

In order to unravel whether the data series contain any long-range dependence, their periodograms (sample spectra) are computed. Figs. 6.12 and 6.13 depict the

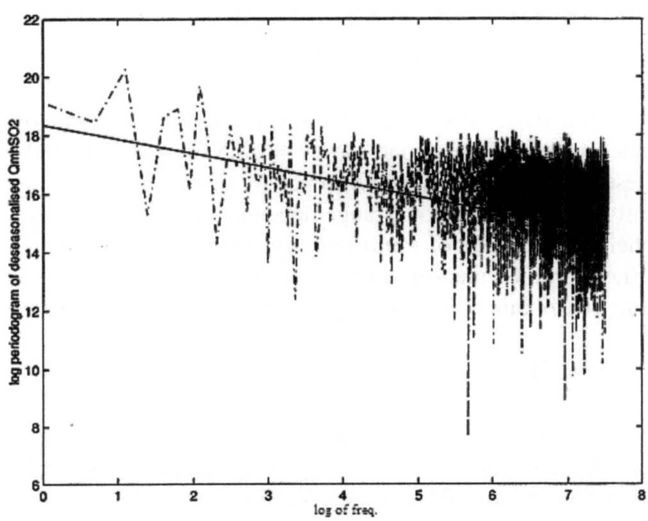

Fig. 6.12 log periodogram and fitted model (continuous line) of the $QmhSO_2$ series

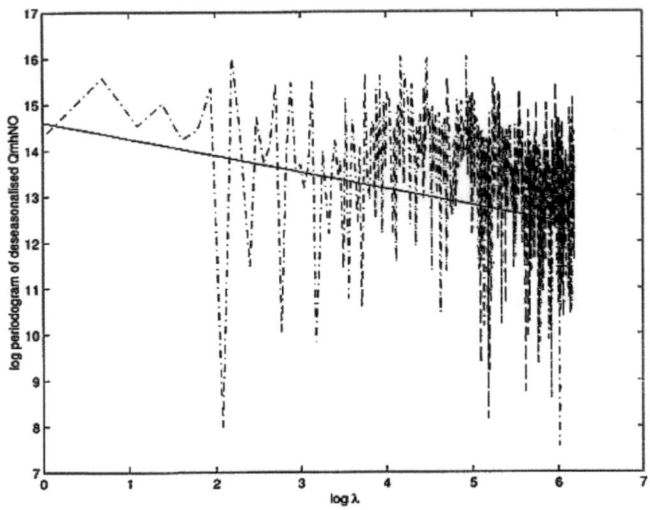

Fig. 6.13 log periodogram and fitted model (continuous line) of the *QmhNO* series

log periodograms against the log frequency for *QmhSO*2 and *QmhNO*, respectively. The spectral density of these data series is assumed to have the form

$$f(\lambda) \sim \frac{1}{|\lambda|^{2\gamma}} \; as \; \lambda \to 0, \tag{6.38}$$

and the long-range dependency exponent γ can be estimated by the corresponding regression model

$$\log f(\lambda) = 2\gamma \log(|\lambda|) + u, \tag{6.39}$$

where u is the white noise. The exponent γ is estimated by least squares for a range of frequencies λ near 0. The estimates are $\hat{\gamma} = 0.25$ for *QmhSO*2 and $\hat{\gamma} = 0.18$ for *QmhNO*. These estimates imply the existence of a singularity at frequency 0, and hence confirm the presence of long-range dependency in the time series.

To decipher the presence of multifractility, the $\zeta(q)$ curves are computed for $q = 0\ (0.1)$ (i.e., for 100 values of q from $_0$ to 10). That is, for each value of q, we, using least squares, obtain a value of $\zeta(q)$ from the slope of the log regression

$$\log E|Y(t) - Y(t-r)|^q = \zeta(q) \log r + u\,, \quad u \sim WN, \tag{6.40}$$

for $r = 1\,,\,2\,,\,\ldots\,,\,10$. Figure 6.14 depicts the $\zeta(q)$ curves for the SO_2 series, while those for the *NO* series are plotted in Fig. 6.15 compared to the $\zeta(q)$ curve of fractional Brownian motion (fBm) which is a linear function of q, the $\zeta(q)$ curves for the SO_2 and *NO* series are all nonlinear (concave), indicating clearly that these data series are multifractal.

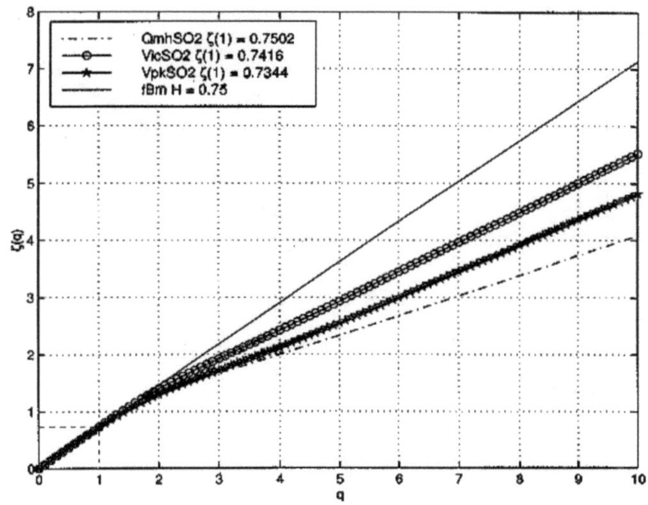

Fig. 6.14 The $\varsigma(q)$ curves for the SO_2 series and fractional Brownian motion

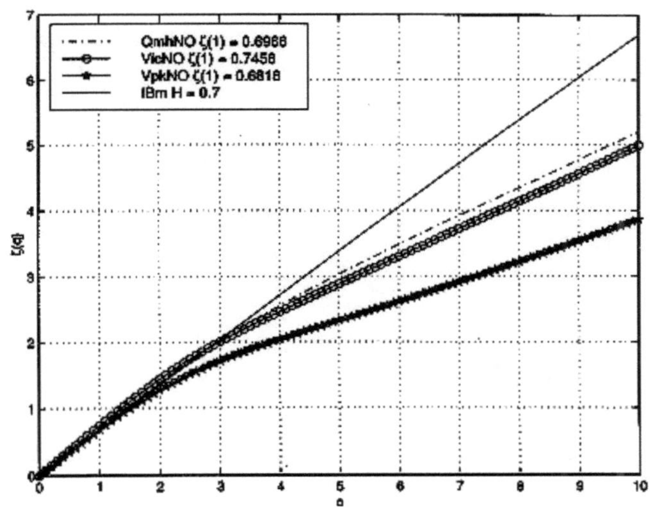

Fig. 6.15 The $\varsigma(q)$ curves for the NO series and fractional Brownian motion

By fitting the model in (6.35) to the data, this multifractality/intermittency can be estimated. Applying (6.27) and (6.33), that is, via the log-regression

$$\log \sum (Y(t;r))^q = \tau(q) \log r + u, \ u \sim WN, \tag{6.41}$$

the $K(q)$ curves are computed (Figs. 6.16 and 6.17), we can observe that these $K(q)$ curves pass through the points $(0, 0)$ and $(1, 0)$, and have a convex shape as

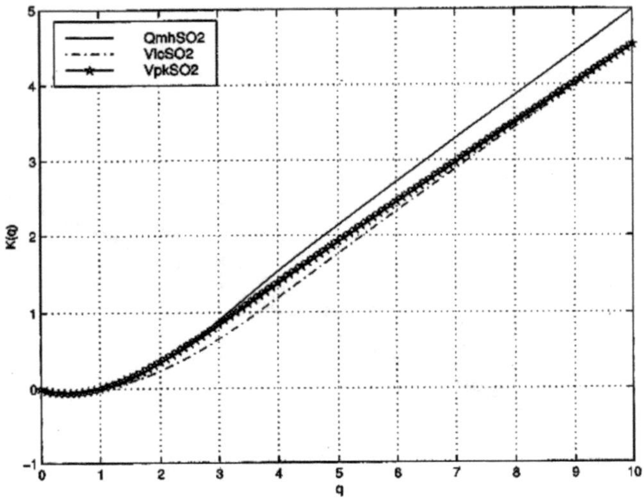

Fig. 6.16 The $K(q)$ curves for the SO_2 series

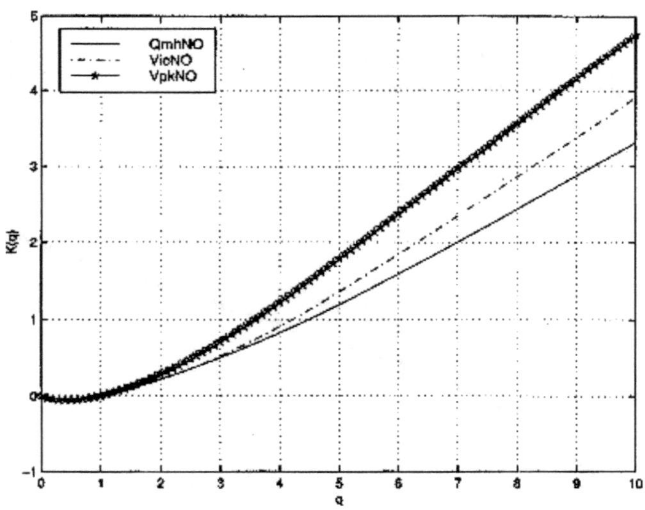

Fig. 6.17 The $K(q)$ curves for the NO series

predicted by the theory. Figure 6.16 indicates that the SO_2 concentrations at Queen Mary Hospital are more intermittent and volatile (higher $K(q)$ curve and smaller p (see also estimates in Table 6.1)) than those at Victoria Peak and Victoria Road (This is actually the answer to the problem raised with respect to Fig. 1.5 in Sect. 1.5 in Chap. 1), while the situation for NO concentrations is in a reverse order.

By comparing the $K(q)$ curves obtained from the data with the curves computed from (6.35), the estimates of the p value can be obtained (Table 6.1). As examples,

Table 6.1 Estimate of p

Series	*QmhSO2*	*VpkSO2*	*VicSO2*	*QmhNO*	*VpkNO*	*VicNO*
p	0.240	0.260	0.270	0.327	0.260	0.300

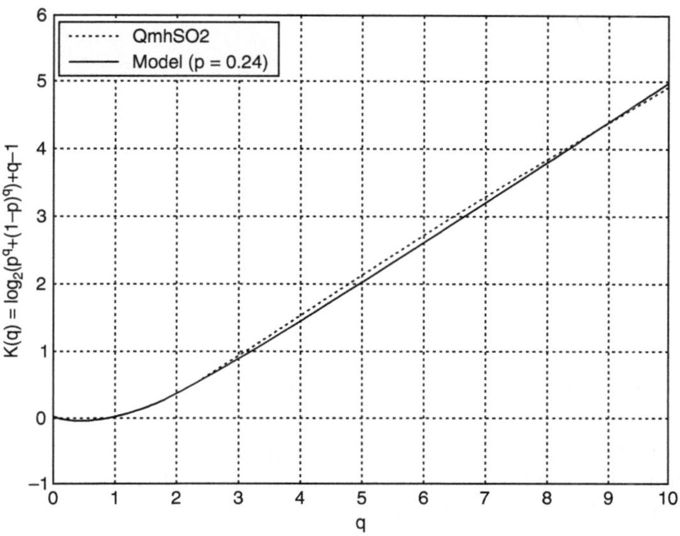

Fig. 6.18 The $K(q)$ curves and fitted model for the $QmhSO_2$ series

the accuracy of the model fitting is depicted in Fig. 6.18 for *QmhSO*2 and Fig. 6.19 for *QmhNO*. The above table of estimates of p provides a useful tool for comparison and classification of the extent of intermittency at different locations of the air shed.

The whole experimental study also shows that complex non-stationary time series data can be substantially compressed and concisely represented. It also paves the road for the analysis of co-integration between multifractal processes, e.g., causality between air pollution and meteorological (and/or socioeconomic) parameters. Further down the road, models of episode prediction can also be developed.

6.4 Finding the Measure Representation of Time Series with Intermittency

6.4.1 Multiplicative Cascade as a Characterization of the Time Series Data

Extending on the analysis discussed in Sect. 6.3, Anh et al. (2005a) further provide a characterization of these data based on their measure representation. This is given in the form of the probability density function of the measure. They have shown that the

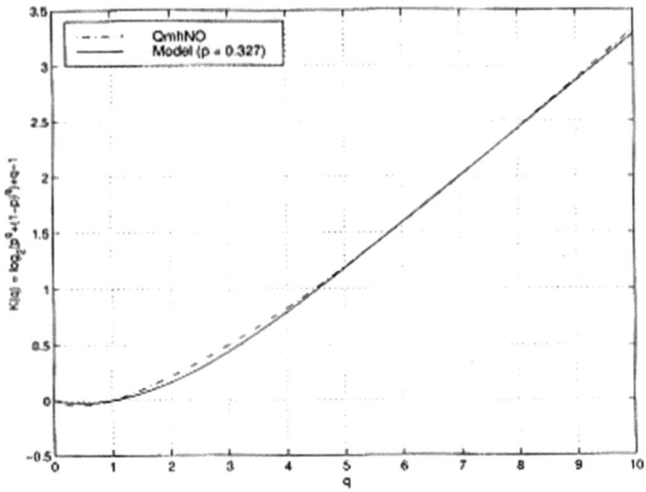

Fig. 6.19 The $K(q)$ curves and fitted model for the *QmhNO* series

stationary stochastic process $\{Y(t), t \in \mathbf{R}\}$ is the limit of a multiplicative cascade with generator W. And the logarithm of W has an infinitely divisible distribution.

It should be observed that the probability density function of a generator W is uniquely determined by the set $\{K(q), q = 0, 1, 2, \ldots\}$ in the multifractal analysis of the cascade. Based on Novikov (1994), if the function $K(q)$ has analytic continuation into the complex plane, then the characteristic function of *ln W*has an infinitely divisible distribution. Anh et al. (2005a) give the most general form of the $K(q)$ curve of the positive stochastic process $\{Y(t), 0 \le t \le 1\}$.

In practice, fitting such $K(q)$ curve to data requires a proper choice of the probability density function of the corresponding measure. According to Novikov (1994), Anh et al. (2005a) show that the Gamma density function, that is,

$$f(x) = Ax^{\alpha-1}\exp(-x/\sigma) \tag{6.42}$$

where A, α, σ are positive constants, provides a very good fit for the $K(q)$ curves of the time series:

$$K(q) = \begin{cases} \kappa\left(q - \frac{(q\sigma+1)^{1-\alpha}-1}{(\sigma+1)^{1-\alpha}-1}\right), & \alpha \neq 1, \\ \kappa\left(q - \frac{\ln(q\sigma+1)}{\ln(\sigma+1)}\right), & \alpha = 1, \end{cases} \tag{6.43}$$

where $\kappa = 1 - b/\ln 2$. The mean and variance are, respectively, $\alpha\sigma$ and $\alpha\sigma^2$.

6.4.2 Experimental Results

The above model has been applied to characterize the air quality data in Hong Kong. These consist of seven SO_2 series, three NO series and three NO_2 series.

The SO_2 series, denoted by *QMH* SO_2, *VPK* SO_2, *VIC* SO_2, *ABD* SO_2, *ALC* SO_2, *CHK* SO_2 and *WFE* SO_2, record the average daily concentrations of sulfur dioxide at Queen Mary Hospital, Victoria Peak, Victoria Road, Aberdeen, Ap Lei Chau, Chung Hom Kok and Wah Fu Estate, respectively. The *NO* series, denoted by *QMH NO*, *VPK NO*, *VIC NO*, and the NO_2 series, denoted by *QMH* NO_2, *VPK* NO_2, *VIC* NO_2, give the average daily concentrations of nitrogen oxide and nitrogen dioxide at Queen Mary Hospital, Victoria Peak and Victoria Road, respectively. The SO_2 series covers the period 1986–1995 (consisting of 3,650 observations), while the *NO* and NO_2 series record the situation from May 1993 to the end of 1995 (consisting of 970 observations). As examples, *QMH* SO_2, *QMH NO*, *QMH* NO_2 are plotted in Figs. 6.20–6.22, respectively. It can be observed that all three series display intermittency, particularly pronounced in the SO_2 series. This intermittency is quite distinct from the yearly cycle.

The long-range dependence and multifractality in these data are unraveled via spectral and multifractal analyzes in Anh et al. (1999a), and the probability distributions of these data are discovered in Anh et al. (2005a). In the latter study, $K_d(q)$ is employed to denote the value of $K(q)$ computed from the data using its definition, and the error is defined as:

$$error = \sum_{j=1}^{J}\left(\kappa\left(q_j - \frac{(q_j\sigma+1)^{1-\alpha}-1}{(\sigma+1)^{1-\alpha}-1}\right) - K_d(q_j)\right)^2. \tag{6.44}$$

The value of κ, σ and α are estimated by minimizing the above error with the assumption: $\kappa \geq 0$, σ, $\alpha \leq 20$. After obtaining the values of κ, σ and α, the $K(q)$ curve can then be estimated from (6.43).

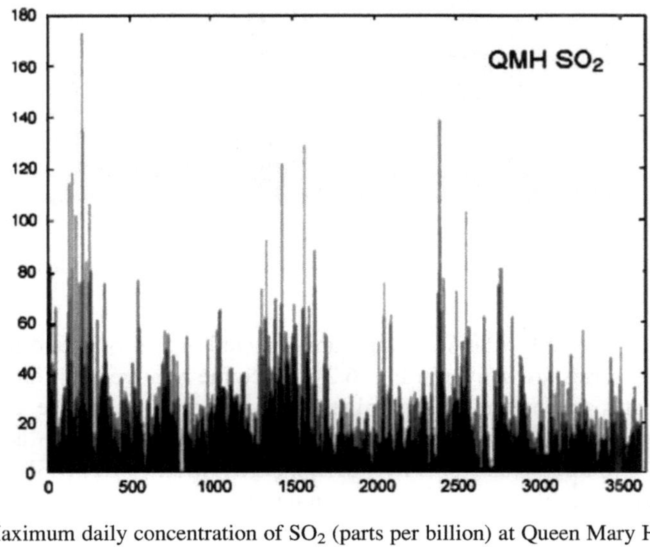

Fig. 6.20 Maximum daily concentration of SO_2 (parts per billion) at Queen Mary Hospital

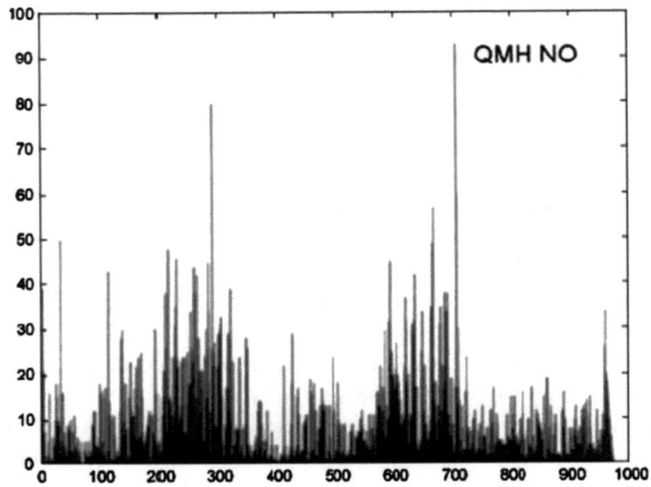

Fig. 6.21 Maximum daily concentration of NO (parts per billion) at Queen Mary Hospital

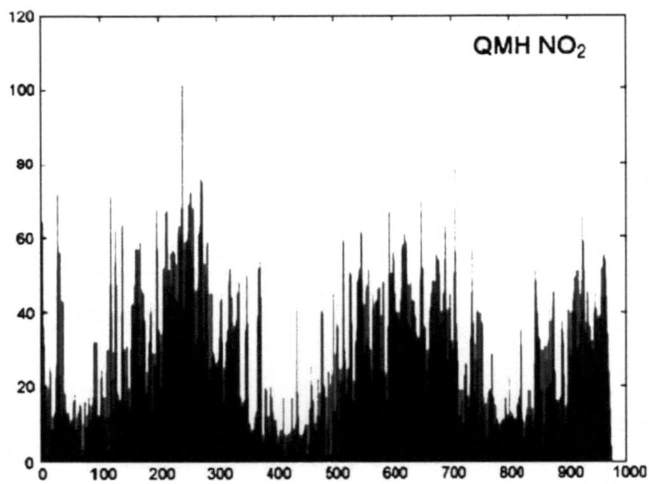

Fig. 6.22 Maximum daily concentration of NO_2 (parts per billion) at Queen Mary Hospital

The $K_d(q)$ curves for the SO_2 series are depicted in Fig. 6.23 and those for the NO and NO_2 series are shown in Fig. 6.24. Since the relative position of the $K(q)$ curve indicates the extent of intermittency in the data, it can be employed to discover clusters in the time series. It can be observed that the SO_2 activities can be grouped into three clusters: (*VPK*, *CHK*), (*VIC*, *ABD*, *QMH*, *WFE*) and (*ALC*). Also the activities of NO and NO_2 are quite apparent (see Fig. 6.24). It can be observed from Fig. 6.23 that the intermittency of the SO_2 series is highest at

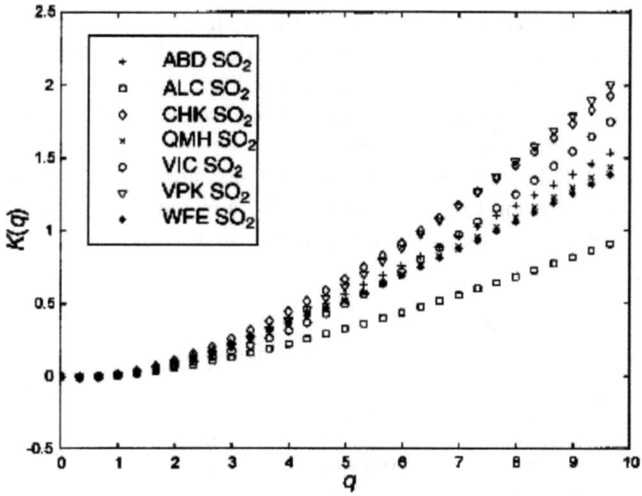

Fig. 6.23 The $K(q)$ curves of seven SO_2 series

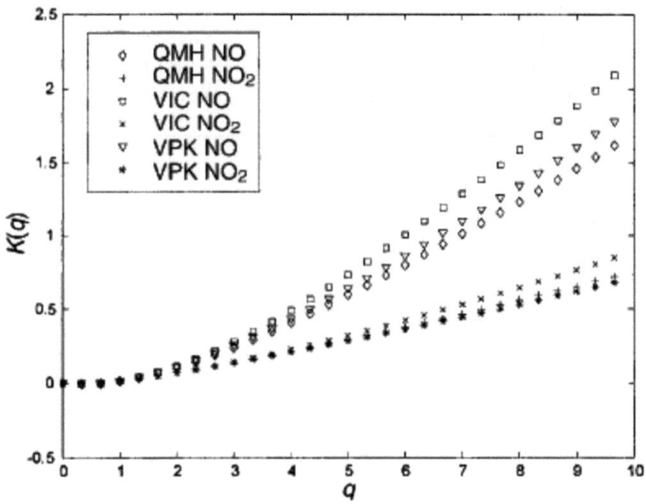

Fig. 6.24 The $K(q)$ curves of three NO series and three NO_2 series

*VPK*and *CHK* because they are furthest away from the pollution source and hence more affected by dispersion rather than by pollution source strength. The former is more variable than the latter. The *ALC* SO_2 series, on the other hand, exhibits the lowest intermittency because of its proximity to the source. The cluster comprising *VIC*,*ABD*, *QMH*and *WFE* lies in between the two and hence exhibits intermediate intermittency.

The difference in NO and NO_2 is related to their origins in the urban environment. NO is the primary pollutant from automobiles and high temperature combustion processes such as power generation. Once omitted, NO is gradually oxidized in the atmosphere and converted to NO_2 and other nitrogen oxide compounds whose concentrations are thus less intermittent.

The statistics of the data fitting based on (6.43) are shown in Table 6.2. The fitting of the SO_2 series in Fig. 6.25 and that of the NO and NO_2 series in Fig. 6.26 clearly show that (6.43) gives a perfect fit to the data.

Table 6.2 Values of quantities κ, σ, α and error of all organisms selected

Pollution	κ	σ	α	Error
ABD SO_2	0.358043	1.612571	0.437936	6.175880×10^{-4}
ALC SO_2	0.412958	3.225351	0.155109	1.516704×10^{-4}
CHK SO_2	0.622163	0.980861	0.334116	7.772050×10^{-4}
QMH SO_2	0.286605	0.565456	0.851996	6.535137×10^{-4}
VIC SO_2	0.740016	0.086973	1.010369	1.172241×10^{-4}
VPK SO_2	0.420615	0.018573	10.754034	1.159886×10^{-4}
WFE SO_2	0.476782	18.919532	0.173357	1.042387×10^{-4}
QMH NO	0.304991	0.868483	0.766563	5.752756×10^{-4}
QMH NO_2	0.110473	8.389144	0.622636	1.293493×10^{-4}
VIC NO	0.347101	0.136793	2.957015	9.016652×10^{-4}
VIC NO_2	0.261564	20.000000	0.196173	1.361025×10^{-3}
VPK NO	0.646551	2.550012	0.214433	1.276346×10^{-3}
VPK NO_2	0.089859	3.717864	1.057626	1.591563×10^{-3}

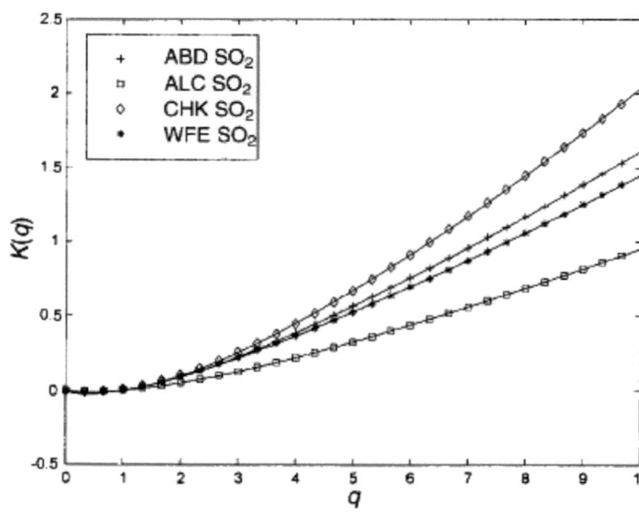

Fig. 6.25 Fitting of the $K(q)$ curves of SO_2 at the sites ABD, ALC, CHK and WFE

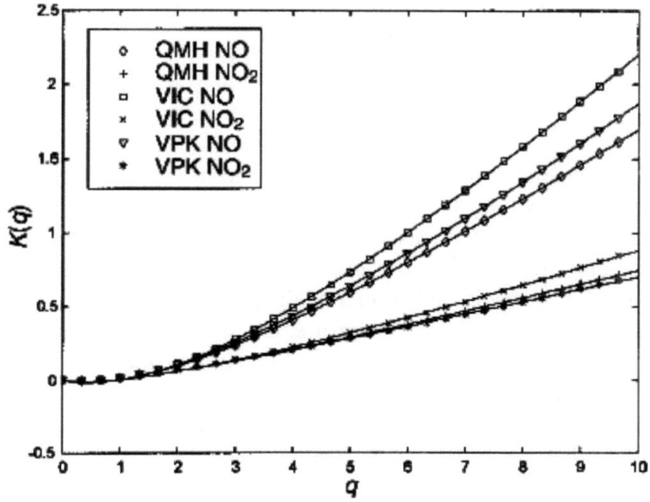

Fig. 6.26 Fitting of the $K(q)$ curves of three NO series and three NO_2 series

6.5 Discovery of Spatial Variability in Time Series Data

6.5.1 Multifractal Analysis of Spatial Variability Over Time

In Sect. 6.4.2, clustering of time series from monitoring stations in different locations is done in a more or less informal way, i.e., by visualizing how various $K(q)$ curves cluster together. Time series at different points in space constitute a cluster if their $K(q)$ curves are adjacent to each other. To discover variation of time series in space, we need a more formal approach. Anh et al. (2005b) show that the clustering of time series can be formally discovered within the multifractal framework. They have formulated a procedure to study spatial variability of rainfall intensity over an area in Southern China bordering the South China Sea.

In their study, each daily rainfall series is considered as a sample path of a multiplicative cascade process generated by a log infinitely divisible distribution. That is, each data series is characterized by the log infinitely divisible generator whose statistical moments can be used in the variability and trend analysis. It should be noted that these moments belong to the generator of a multifractal process, not those obtained directly from the raw data themselves.

The study shows that information based on the first two moments of the above generator, shown to be related to the information dimension and correlation dimension of the generalized Rényi dimensions of the multiplicative cascade, is a suitable tool to yield meaningful results for spatial trend analysis and spatial variability study based on the monitoring data. These two dimensions form a vector representing a rainfall series, and the corresponding Euclidean distance is then used as the basis to perform clustering.

Anh et al. (2005b) shows that the generalized Rényi dimensions of the limit measure μ_∞ of the random cascade is

$$D_q = -\frac{\log_2 E(W^q)}{q-1} + 1, \quad q \neq 1\,, \tag{6.45}$$

where W is the generator of the multiplicative cascade process. As noted in Anh et al. (2005b), the function $K(q)$ is convex with $K(0) = K(1) = 0$. When $K(q)$ is strictly convex for $q \geq 1$, the measure μ_∞ is a multifractal measure. In other words, it contains singularities of possibly many different orders. Under this situation, tradition tools such as mean and spread of the probability measures would not be useful to characterize a multifractal measure. On the other hand, the curves such as D_q and $K(q)$ characterize the singularities of these multifractal measures. We can then use these curves to rank the degree of variability (inherent in singularities) of the rainfall data. In particular, we can use the dimensions D_q for this purpose. It should be noted that, for a fixed q, D_q decreases as the moment of order q of the generator W of the cascade increases.

Among the values of D_q, the special cases $D_0 \equiv$ box-counting dimension, $D_1 \equiv$ information dimension and $D_2 \equiv$ correlation dimension are commonly used. These dimensions also have physical meanings. When the data are normalized to become a density function (with the sum of all values becomes 1), we will have $D_0 = 1$. Thus D_1 and D_2 can be employed to accomplish the task.

The notion of correlation dimension is introduced by Grassberger and Procaccia (1983a, b). Given a sequence of data, $x_1, x_2, x_3, \cdots x_N$, where N is sufficiently large, we embed the sequence into $\mathbf{R}^m$ with time delay $\tau = p\Delta\tau$ as

$$y_i = \left(x_i, x_{i+p}, x_{i+2p}, \cdots, x_{i+(m-1)p}\right), \quad i = 1, 2, \cdots, N_m\,,\, N_m = N - (m-1)p.$$

In this way, we obtain N_m vectors in the embedding space $\mathbf{R}^m$. For any y_i, y_j, we define the distance as

$$r_{ij} = d\left(y_i, y_j\right) = \sum_{l=0}^{m-1} \left|x_{i+lp} - x_{j+lp}\right|. \tag{6.46}$$

If the distance is less than a number r, we say that these two vectors are correlated. The correlation integral is then defined as

$$C_m(r) = \frac{1}{N_m^2} \sum_{i,j=1}^{N_m} H\left(r - r_{ij}\right), \tag{6.47}$$

where H is the Heaviside function

$$H(x) = \begin{cases} 1, & \text{if } x > 0, \\ 0, & \text{if } x \leq 0. \end{cases} \tag{6.48}$$

For an appropriate choice of m and r (not too large), Grassberger and Procaccia (1983b) show that the correlation integral $C_m(r)$ behaves as

$$C_m(r) \quad \propto \quad r^{D_2(m)} .$$

Thus one can define the correlation dimension as

$$D_2 = \lim_{m\to\infty} D_2(m) = \lim_{m\to\infty} \lim_{r\to 0} \frac{\ln C_m(r)}{\ln r} . \tag{6.49}$$

6.5.2 *Detection of Spatial Variability of Rainfall Intensity*

The method discussed in Sect. 6.5.1 is employed to examine spatial variability of rainfall intensity in an area of South China bordering the South China Sea. The area contains many high mountains so that the rainfall field is affected by complex synoptic conditions. Anh et al. (2005b) attempt to find the temporal and spatial trends and the clustering of the rainfall field.

The data set consists of daily rainfall data over the period January 1, 1959 – December 31, 1990 (yielding 11680 observations) at 16 locations of the region (Fig. 6.27).

A typical time series (normalized to have sum equals 1) in the data set is shown in Fig. 6.28. Apart from the pronounced yearly pattern, the time series displays extreme irregularities at many scales of measurement. This latter feature would make conventional time series techniques unsuitable for the analysis of these data.

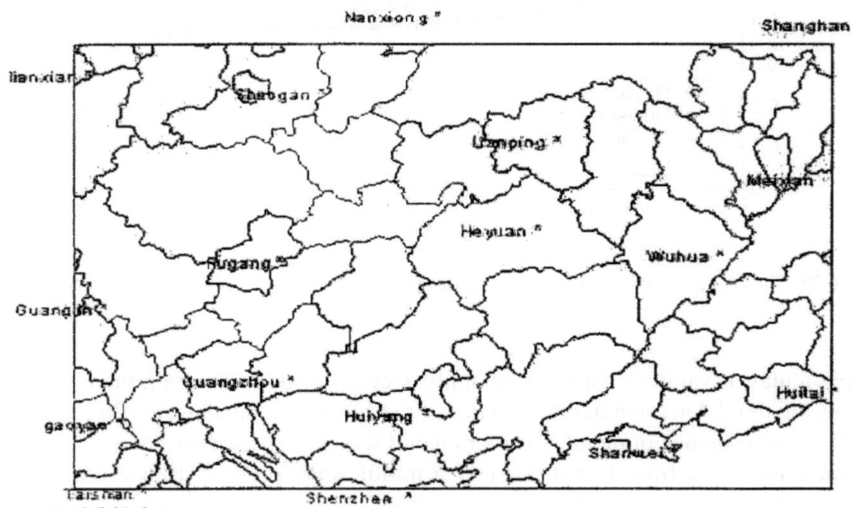

Fig. 6.27 The locations of the 16 stations

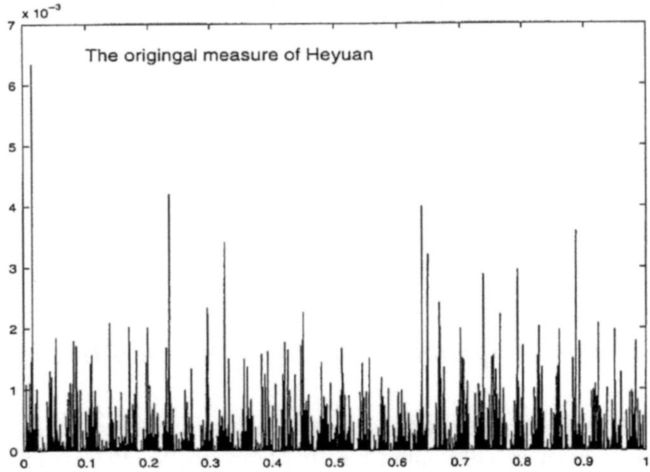

Fig. 6.28 Normalized rainfall data of the Heyuan station

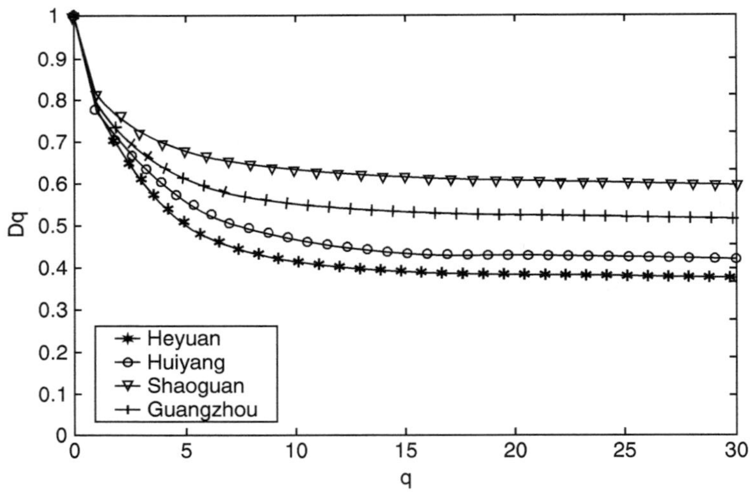

Fig. 6.29 The D_q curves of the 4 stations as examples

In particular, it would be difficult to apply spectral analysis for long memory and stochastic trend detection in the presence of the apparent annual cycle. On the other hand, the persistence of details over a range of scales in these data series exhibits a characteristic of a fractal, and possibly a multifractal, process. Hence the multifractal approach discussed above is appropriate.

The D_q curves for all 16 time series are computed. The curves for Heyuen, Huiyang, Shaoguan and Guangzhou are shown in Fig. 6.29. It is clear from the nonlinearity of these curves that the process is indeed multifractal. This multifractality is also confirmed by the strict convexity of the corresponding $K(q)$ curves.

The values of D_1 and D_2 for all sites over 5-year periods are computed to check for variation of the results over time (Tables 6.3–6.6).

It can be observed that the estimated values of these dimensions are fairly stable over each period, indicating that there is no temporal trend in the rainfall intensity over the 32 year period under study.

The values of D_1 and D_2 for the 16 sites are shown in Table 6.7. The sites are ordered according to the increasing order of D_1. It is interesting to note that the ordering using D_2 yields some similar results, and the plot of the vectors (D_1,D_2) shows a strikingly linear relationship between D_1 and D_2 (Fig. 6.30). It should be noted that, by definition, this kind of relationship does not hold in general.

Table 6.3 D_1, D_2 for every-5-years rainfall data of Gaoyao station

Time period	D_1	D_2
1959–1963	0.6846532	0.6016884
1964–1968	0.6413091	0.5734902
1969–1973	0.6551777	0.5688238
1974–1978	0.6962509	0.6753542
1979–1983	0.6778288	0.6078432
1984–1988	0.6709815	0.4943815

Table 6.4 D_1, D_2 for every-5-years rainfall data of Heyuan station

Time period	D_1	D_2
1959–1963	0.6696920	0.5711966
1964–1968	0.6748458	0.6129775
1969–1973	0.6529554	0.5776837
1974–1978	0.5952509	0.6753542
1979–1983	0.6592873	0.5685446
1984–1988	0.6547381	0.5679712

Table 6.5 D_1, D_2 for every-5-years rainfall data of Huiyang station

Time period	D_1	D_2
1959–1963	0.6314682	0.5369876
1964–1968	0.5325046	0.5389811
1969–1973	0.6379271	0.5549309
1974–1978	0.6179082	0.5262730
1979–1983	0.6554687	0.5579509
1984–1988	0.6320971	0.5021596

Table 6.6 D_1, D_2 for every-5-years rainfall data of Lianping station

Time period	D_1	D_2
1959–1963	0.6988534	0.6375061
1964–1968	0.6705224	0.6269682
1969–1973	0.6822639	0.6201434
1974–1978	0.7105417	0.6501876
1979–1983	0.7036157	0.6445061
1984–1988	0.6761858	0.6201606

Table 6.7 D_1, D_2 of 16 stations using 32 years rainfall data

Station	D_1	D_2
Huilai	0.715512	0.626226
Shanwei	0.726584	0.640939
Shenzhen	0.761043	0.679768
Taishan	0.775577	0.701413
Huiyang	0.778958	0.701244
Heyuan	0.780745	0.690082
Meixian	0.785788	0.71867
Wuhua	0.79708	0.731479
Guangzhou	0.798752	0.728615
Fugang	0.803564	0.731953
Gaoyao	0.811428	0.751742
Lianping	0.811547	0.746411
Shaoguan	0.814573	0.756123
Nanxiong	0.816967	0.759669
Guangling	0.818861	0.757424
Lianxian	0.82804	0.77042

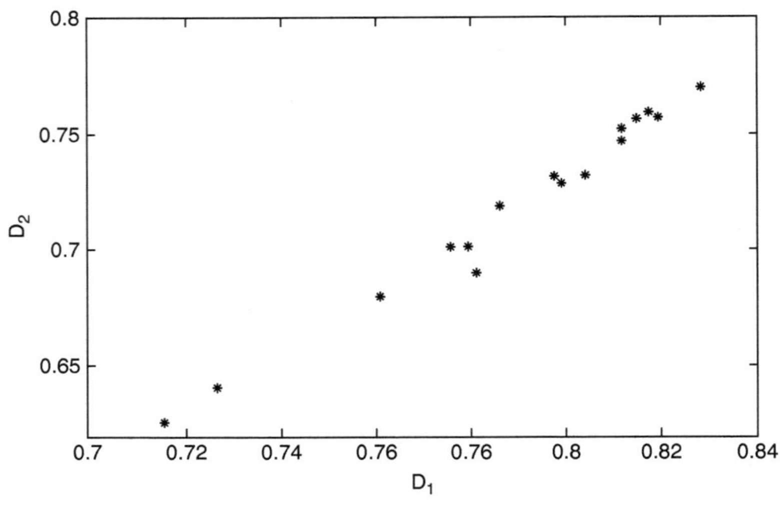

Fig. 6.30 D_1 and D_2 of the 16 stations

6.6 Identification of Multifractality and Spatio-Temperal Long Range Dependence in Multiscaling Remote Sensing

6.6.1 *A Note on Multifractality and Long-Range Dependence in Remote Sensing Data*

The need to resolve many global environmental issues addressed by the International Geosphere-Biosphere Program, Framework Convention for Climate Change, Kyoto Protocol, Biodiversity Convention and other international programs requires the urgent availability of remote sensing and other spatial information at local, regional and global scales. The launch of satellite sensors such as Landsat 7 ETM+, Spot 4 Vegetation, Terra MODIS, Medium Resolution Imaging Spectrometer (MERIS), Global Imager, etc. has partially satisfied this requirement. The remote sensing research community will now need to develop new and suitable methodologies to explore these multi-channel, multi-resolution and multi-temporal data sets to extract useful spatial information for environmental monitoring.

With the improvement of sensor resolution, it has been discovered that many geophysical and geographical phenomena display a certain degree of similarity in geometrical form and complexity over a wide range of scales. This scale invariance is a general and fundamental symmetry principle which must be exploited in the modeling and analysis of these phenomena (Steward et al. 1996; Quattrochi and Goodchild 1997). In particular, Xia and Clarke (1997) provide an extensive review of the applications of the scaling concept in modeling clouds, rain and other atmospheric phenomena; characterization of land surface topography and ocean floor bathymetry; classification of landform regions and geomorphologic processes; modeling of spatial variability of soil properties; analysis of urban land use patterns; simulation of urban growth processes; and digital representation of terrain data. Each topic has developed into a field of active research itself. Amongst the wide variety of studies, land cover, undoubtedly, constitutes the very basic information for monitoring the impact of human activities on the environment (Cihlar 2000).

Wavelets and fractals are major methods employed to describe and model the scaling property of spatial data in recent years (Lam and De Cola 1993). Ranchin and Wald (1993) have demonstrated the use of the wavelet transform in multiresolution analysis of remotely sensed images. Djamdji et al. (1993) employ the wavelet transform for automatic registration of images acquired by the same sensor at different dates or by different sensors with variable spatial resolution. Li and Shao (1994) use wavelet analysis in automatic interpretation of buildings from aerial images. On the other hand, fractal models and techniques have become popular in dealing with a wide range of scaling phenomena. Lovejoy et al. (2001) shows that fractal techniques have been used to model over 20 remotely sensed problems including radar rain and ice surfaces; visible, infrared and passive microwave land reflectivity; topography, etc. Early applications of fractals transform the two-dimensional remotely sensed fields into surfaces in three-dimensional space and treat the surfaces as fractal sets (Peleg et al. 1984; Pentland 1984; Keller et al. 1989; Rees 1995). A typical application is the use of local fractal dimensions for image segmentation and classification (Pentland 1984; Anh et al. 1996).

Although the use of fractal geometry represents a major step forward, this approach suffers a basic limitation that, unlike scale-invariant fractal sets which can be characterized by a single fractal dimension, scale-invariant remotely sensed fields are mostly multifractal, hence require an infinite number of scaling exponents for their characterization. Many recent studies have constituted an empirical basis for multifractal analysis of remote sensing data. These include synthetic aperture radar reflectivity fields by Falco et al. (1996); liquid water distributions in marine stratocumulus by Davis et al. (1996); visible reflectance fields of basaltic volcanoes by Laferrière and Gaonac'h (1999); infrared imagery of volcanic features by Gaonac'h et al. (2000) and Harvey et al. (2000), and phytoplankton and remotely sensed ocean color by Lovejoy et al. (2001). Certain important devices such as generalized fractal dimensions, singularity spectrum, and logarithmic infinitely divisible distributions have been established for many classes of multifracal measures. These methods are useful for the characterization and classification of multifractals, but do not provide a model to simulate sample paths of these multifractals. In a research proposal, Fung et al. (2001) propose to develop suitable models for direct simulation of multifractal random fields. This will give a systematic way to model multiple scaling/multifractality of remote sensing data.

Another key property of spatial processes, which is overlooked in many recent studies, is their spatial long-range dependence (LRD) as discussed in the previous sections. The analysis of LRD is essential to the study of intermittency and stochastic trends in remote sensing data. An adequate model for spatial data therefore must encompass both multifractality and LRD. The proposed project by Fung et al. (2001) will bring these two key aspects of spatial data into a unified framework. It paves the road for the study of cointegration/causality of multifractal processes, e.g., the relationship between desertification and meteorological parameters.

6.6.2 A Proposed Methodology for the Analysis of Multifractality and Long-Range Dependence in Remote Sensing Data

In many applications involving diffusions in a non-homogeneous medium, the correlation function of the corresponding random field often decays to zero at a much slower rate than the usual exponential rate of Markov diffusion and the probability density function has heavier tails, resulting in long-range dependence and anomalous diffusion (Anh and Heyde 1999; Hilfer 2000). There have been a variety of mathematical approaches to tackle specific aspects of anomalous diffusion and related problems. These include

1. Stochastic distributions, the Wick product and Hida-Malliavin calculus
2. Fractal measures, function spaces on fractals and fractional embeddings
3. The Green function solution, Mittag–Leffler functions and fractional calculus.

Approach (1) is an extension of Itô's theory on Markov diffusion. It allows for non-Gaussian multiplicative noise and the resulting solution can be interpreted in the usual strong sense instead of the weak sense of Schwartz distributions (Holden et al. 1996). Approach (2) is based on recent development in Sobolev and Besov spaces of functions defined on fractal sets with appropriate fractal measures. Its sophisticated embedding theorems allow us to draw some concrete results on the forms of fractional diffusion operators and their properties, yielding suitable models for anomalous diffusion (Anh et al. 1999c; Angulo et al. 2000). Approach (3), which is closely related to the continuous time random walk theory, relies on fractional calculus and properties of the Mittag–Leffler functions to tackle fractional diffusion equations and the corresponding Green function solutions (Anh and Leonenko 2000, 2001; Hilfer 2000).

In their proposal, Fung et al. (2001) suggest to adopt approach (3) and consider the situation in which the random fields display long correlation in space and time, such as change of vegetation cover and desertification. They propose to consider the following fractional partial differential equation

$$\begin{gathered}\frac{\partial^\beta C}{\partial t^\beta} = -(I-\Delta)^{\gamma/2}(-\Delta)^{\alpha/2}C(t,u)\,, \\ 0<\beta\leq 1, 0<\alpha<2, \gamma>0, t\in \mathbf{R}_+, u\in \mathbf{R}^2\end{gathered} \tag{6.50}$$

subject to the random initial conditions $C(o\,,\,u) = v(u), u = (x\,,\,y) \in \mathbf{R}^2$, with $v(u)$ being a random field of the form $v(u) = h(\xi(u))$, where the non-random function h $(\cdot)$ and the random field $\xi(u)$ determine the (possibly non-Gaussian) marginal distribution and multifractality of the solution of (6.50). In this equation, $\Delta = \partial^2/\partial x^2 + \partial^2/\partial y^2$ is the Laplacian, I is the identity operator, $(-\Delta)^{\alpha/2}$ and $(I-\Delta)^{\gamma/2}$ are inverses of the Riesz and Bessel potentials, respectively. The time derivative of order $\beta \in (0\,,\,1]$ is defined as

$$\frac{\partial^\beta C}{\partial t^\beta} = \begin{cases} \frac{\partial C}{\partial t}(t\,,\,u)\,, & \text{if } \beta = 1\,, \\ (D_t^\beta C)(t\,,\,u), & \text{if } \beta \in (0\,,\,1)\,, \end{cases} \tag{6.51}$$

where

$$\left(D_t^\beta C\right)(t,u) = \frac{1}{\Gamma(1-\beta)}\left[\frac{\partial}{\partial t}\int_0^t (t-r)^{-\beta}C(\tau,u)d\tau - \frac{C(0,u)}{t^\beta}\right], 0<t\leq T, \tag{6.52}$$

is the fractional derivative in the Caputo–Djrbashian sense (Podlubny 1999). In the Gaussian case, the non-linear function h, with Hermitian rank m, can be expanded in a series of orthogonal Chebyshey–Hermite polynomials (i.e., the chaos expansion, see Holden et al. 1996). Equation (6.26) is a fractional diffusion equation when $0 < \beta \leq 1$ and is a fractional wave equation when $1 < \beta \leq 2$. It is therefore an interpolation between the diffusion and wave equations and is referred to as the fractional diffusion-wave equation. The spatial long-range dependence of $C(t,\ u)$ is obtained from the fractional operator $(-\Delta)^{\alpha/2}$, while its temporal long-range dependence is induced by the fractional time derivation $\partial^\beta C/\partial t^\beta$. Thus, (6.50) encompasses both multifractality and LRD to model multi-scaling remote sensing data. The limiting distributions of the rescaled solutions of the initial value problem in (6.50) have been investigated in Anh and Leonenko (2001) in the Gaussian case.

In order to model the possible multifractality of remote sensing data, we must consider a non-Gaussian situation; for example, we may consider an infinitely divisible (e.g., a-stable or Gamma-correlated) random condition for (6.50). It is proposed to develop a corresponding rescaled solution of (6.50) for this non-Gaussian scenario. Its singularity spectrum, which characterizes the multifractality of the phenomenon, will be obtained. The rescaled solution provides an approximation to the solution of the fractional diffusion-wave equation (6.50).

Equation (6.50) actually provides a tool to visualize changes in a natural phenomenon in fine temporal resolution. Its solution can be written in the moving average form

$$C(t, u) = \int_0^t \int_D G(t, u; s, z)\, v(z)\, dzds, \tag{6.53}$$

where G is the Green function of (6.50). We need to develop a method to simulate sample paths of the initial random field $v(u)$ and fast algorithms to compute the solution of (6.53). It is proposed to consider the simpler case $\beta = 1$ (i.e., there is no temporal long-range dependence). The resulting Green function $G(t, u; s, z)$ then has the expansion

$$G(t, u; s, z) = \begin{cases} \sum_{k \in \mathrm{N}_8} \phi_k(u)\, \phi_k(z) e^{-\lambda_k(t-s)}, & t \geq s, \\ 0, & t < s, \end{cases} \tag{6.54}$$

where $u, z \in (0, L_1) \times (0, L_2) = D \subset \mathbf{R}^2, k = |\mathbf{k}| = (k_1^2 + k_2^2)^{\frac{1}{2}}, \mathrm{N}_* = \{1, 2, 3, \ldots\}$,

$$\phi_k(u) = \left(\frac{4}{V_L}\right)^{\frac{1}{2}} \sin\left(\frac{k_1 \pi x}{L_1}\right) \sin\left(\frac{k_2 \pi x}{L_2}\right) \tag{6.55}$$

are the eigenfunctions of the Laplacian, $V_L = L_1 L_2 \equiv$ area of D, $\lambda_k = \omega_k^{\frac{\alpha}{2}}(1 + \omega_k)^{\frac{\gamma}{2}}$ are the eigenvalues of $(-\Delta)^{\frac{\alpha}{2}}(I - \Delta)^{\frac{\gamma}{2}}$ with $\omega_k = \pi^2\left({k_1}^2/{L_1}^2 + {k_2}^2/{L_2}^2\right)$ being the eigenvalue of $(-\Delta)$ (see Angulo et al. 2000).

The simulations can be done over small squares of 5×5 pixels. The eigenvalues λ_k can be approximated by $\lambda_k \sim k^{\frac{-(\alpha+\lambda)}{2}}$. Hence, we only need to estimate $\alpha + \gamma$ on each 5×5 square. Since the spectral density of the spatial operator of (6.49) is one of the form $c \Big/ \left(|\lambda|^{2\alpha}\left(1 + |\lambda|^2\right)^{\gamma}\right)$, which behaves as $c \Big/ |\lambda|^{2(\alpha+\gamma)}$ when $|\lambda| \to \infty$, we can obtain an estimate of $\alpha + \gamma$ via log-regression. In other words, $\alpha + \gamma$ is $-2x$ slope of the regression of $\ln f(\lambda)$ against $\ln|\lambda|$ on each 5×5 square. It should be noted that every step of this algorithm does not require any optimization. Hence the simulation of (6.50) via (6.53) can be executed in a fast time. By design, this algorithm is useful for segmenting an image and finding its important edges, hence will be particularly relevant for the problem of change detection in remotely sensed imagery.

When the presence of temporal long-range dependence is considered, the Green function of in (6.50) will then be more complex and will involve the Mittage–Leffler function (see Anh and Leonenko 2001). A method to simulate the model in (6.50) via this Green function needs to be investigated. Execution of the simulation over small time steps will provide a visualization of the scene under study. Prediction of future scenarios can also be obtained from (6.53). This prediction provides an extrapolation from one location to another and also over time.

6.7 A Note on the Effect of Trends on the Scaling Behavior of Time Series with Long-Range Dependence

Conventional multifractal analyzes are employed to characterize multifractal properties of normalized and stationary time series. They are generally not suitable for non-stationary time series affected by trends. The existence of trends in time series is actually a common place in many temporal and spatial processes. They might be due to the intrinsic or external conditions generated by some natural or human-made processes. For examples, rainfall and river runoff might have a seasonal periodic trend; population density might have a locational trend; and earthquake occurrence might have spatial and temporal trends. The existence of trends might affect the scaling behavior of processes with long-range dependence. Detrended fluctuation analysis (DFA) is a method for the detection of monofractal scaling properties and the determination of long-range power-law correlation in noisy and non-stationary time series (Peng et al. 1992). It attempts to identify different regimes of a system with respect to its different scaling behaviors in the characterizing time series. Crossover time scales are identified according to the change in the correlation properties of the temporal signal. The effect of trends and exogenous/artificial trends having very little relations with the dynamics of the system has also been studied (Hu et al. 2001).

To account for multifractality and long-range correlation, multifractal detrended fluctuation analysis (MF-DFA) (Kantelhardt et al. 2002), a generalization of DFA, has been formulated to study whether the multifractal nature of the system is solely due to long-range correlation or affected by trends, intrinsic or external, in the time series.

Suppose that $x_k$is a series of length N of compact support. The MF-DFA procedure consists of the following basic steps.

Step 1. Determine the "profile”

$$Y(i) \equiv \sum_{k=1}^{i} [x_k - \langle x \rangle] \ , i = 1 \ , \ \ldots \ , N, \tag{6.56}$$

where $\langle x \rangle$ is the mean of x_k. It should be noted that subtraction of the mean $\langle x \rangle$ is not mandatory, since it would be eliminated by the later detrending in the step 3.

Step 2. Divide the profile *Y(i)* into $N_s \equiv \text{int}\,(N/s)$ non-overlapping segments of equal length s. Since the length N of the series is often not a multiple of the timescale s, a short part at the end of the series may remain. In order not to disregard this part of the series, the same procedure is repeated starting from the opposite end. Hence, $2N_s$ segments are obtained altogether in this step.

Step 3. Calculate the local trend for each of the $2N_s$ segments by a least-squares fit of the series, and determine the variance

$$F^2(s,v) = \frac{1}{s}\sum_{i-1}^{s}\{Y[(v-1)\,s+i] - y_v(i)\}^2 \qquad (6.57)$$

for each segment v, $v = 1\,,\, \ldots\, ,N_s$, and

$$F^2(s,v) = \frac{1}{s}\sum_{i-1}^{s}\{Y[N-(v-1)\,s+i] - y_v(i)\}^2 \qquad (6.58)$$

for $v = N_s + 1\,,\, \ldots\,,\, 2N_s$. Here, $y_v(i)$is the fitting polynomial in segment v. It should be noted that linear, quadratic, cubic or higher order polynomials can be used in the fitting procedure (conventionally called DFA1, DFA2, DFA3, . . .). Since the detrending of time series is through the subtraction of the polynomial fits from the profile, DFA using different orders differ in their capability of eliminating trends in the series. Thus a comparison of the results for different orders of DFA enables us to estimate the type of the polynomial trend in the time series (Chen et al. 2002; Hu et al. 2001). Steps 1 to 3 are standard DFA procedure. To account for multifractality, the following steps need to be appended:

Step 4. Average over all segments to obtain the qth-order fluctuation function, defined as

$$F_q(s) \equiv \left\{\frac{1}{2N_s}\sum_{v=1}^{2N_s}\left[F^2(s,v)\right]^{q/2}\right\}^{1/q}, \qquad (6.59)$$

where $q = 0$ and $sm + 2$. For $q = 2$, the standard DFA procedure is obtained. Generally we are interested in how the generalized q dependent fluctuation functions $F_q(s)$ depend on the timescale s for different values of q. Hence we must repeat steps 2, 3 and 4 for several timescales s.

Step 5. Determine the scaling behavior of the fluctuation functions by analyzing the log-log plots of $F_q(s)$ versus s for each value of q. If the time series x_i are long-range power law correlated, $F_q(s)$ increases, for large values of s, as a power law

$$F_q(s) \propto s^{h(q)}, \qquad (6.60)$$

where $h\,(q)$ is the generalized Hurst exponent. For stationary series, the exponent$h\,(2)$ for small time scales is identical to the well-known Hurst exponent H. For non-stationary time series such as fractional Brownian motion, the corresponding scaling exponent of $F_q(s)$ is identified by $h\,(2) > 1$ (Feder 1988; Peng et al. 1994). In this case, the relationship between the exponents $h\,(2)$ and H will be $H = h\,(2) - 1$. When H equals 0.5, it implies that the series is uncorrelated. When $0.5 < H < 1$, it indicates that the time series has long range correlation. When $0 < H < 0.5$, it suggests that the time series has short memory or long- range anti-correlation.

If the time series is monofractal with compact support, $h(q)$will be independent on q. In this case, the scaling behavior of the variances $F_q(s)$ is identical for all segments v, and the averaging procedure in (6.59) will get the same scaling behavior for all values of q. For positive value of q, the average $F_q(s)$ will be enlarged by high value of $F^2(s,v)$ in each segment v. So $h(q)$ describes the scaling behavior of the segments with large fluctuation. Meanwhile, for the negative values of q, the average $F_q(s)$ will be calculated by the small value of $F^2(s,v)$ in each segment v. Thus, $h(q)$ describes the scaling behavior of the segments with small fluctuation.

Applying MF-DFA to the daily rainfall time series of the Pearl River basin of P.R. China, Yu et al. (2008) find long-range correlation and multifractal behavior in the profile. The second-order fluctuation functions exhibit two distinct points at which one scaling regime switches to another with a new scaling exponent. The two crossover points correspond to time scales around 90 and 365 days, reflecting exactly the seasonal and yearly cycle features in the rainfall series. With appropriate design, MF-DFA procedures should also be able to detect multifractality and long-range correlation in space in general and spatial point processes in particular.

Chapter 7
Summary and Outlooks

7.1 Summary

I have discussed from the conceptual, theoretical and empirical points of view the basic issues of knowledge discovery in spatial and temporal data. The kinds of knowledge that geographers are interested in are essentially spatial structures, processes, and relationships in various domains. The data that we are dealing with are voluminous, multi-scaled, multi-sourced, imperfect, and dynamic. Hidden in these complex databases are conceptually and practically meaningful structures, processes and relationships that might be important to the understanding and sustainable development of the human-land system. Our objective is to develop effective and efficient means to discover the potentially useful spatial knowledge that might otherwise lay unnoticed.

From our discussion, a basic task of knowledge discovery in spatial and temporal data involves the unraveling of structures or processes that might appear as natural clusters in data. I have discussed in Chap. 2 the main objectives and difficulties of such task and methods by which natural clusters can be identified. Therefore, clusters form the spatial knowledge hidden in the database in this kind of spatial data mining task. Geographical research often involves classification of spatial phenomena. Given pre-specified spatial classes, we need to discover from data a surface that can separate the classes which are constituents of the data set. Structure thus discovered serves as class separation surface in the general case. Instead of a separation surface, particularly under complicated situations, classes might be separated by a set of classification rules which can be unraveled from data. In Chaps. 3 and 4, a number of statistical and non-statistical methods for the discovery of separation surfaces and classification rules have been, respectively, examined. In the search for spatial relationships hidden in data, the basic issue is to uncover and determine whether or not a spatial association or a spatial casual relationship is local or global. In Chap. 5, I have examined various methods by which local and global spatial statistics can be established from data and the issue of spatial non-stationarity can be resolved. To be able to study the dynamics of spatial phenomena

Y. Leung, *Knowledge Discovery in Spatial Data*, Advances in Spatial Science,
DOI 10.1007/978-3-642-02664-5_7,

in various time scales, I have discussed the mining of scaling behaviors of spatial processes from data in Chap. 6. The discovery of time varying behaviors from temporal data is investigated on the basis of their self-similarity and long range dependence along a wide spectrum of temporal scales.

To recapitulate, the main objective of this monograph is the development of appropriate methods for the discovery of spatial knowledge in complex spatial data. It has been achieved through several basic tasks that involve the search for natural clusters in data, the mining of separation surfaces or classification rules for data, the unraveling of local and global relationships in data, and the discovery of scaling behaviors of spatial phenomena in temporal data. Methods developed for these tasks target at the special properties of spatial and temporal databases discussed in Chap. 1. There are, however, outstanding issues that have not been examined in this monograph. They constitute directions for further research on the discovery of knowledge in spatial data. In the section follows, I outline in brief the issues involved and plausible approaches to their solutions.

7.2 Directions for Further Research

7.2.1 Discovery of Hierarchical Knowledge Structure from Relational Spatial Data

As discussed throughout this monograph, spatial information system is a highly complex system which contains spatial and aspatial data of various types. The effective and efficient capturing, manipulation, and analysis of data in such systems plays a very important role in research and practice. Over the years, various data models have been formulated to structure and manage spatial data. Systems built on the relational data model are ample in research and applications (Burrough 1986; Vaughan et al. 1988; Maguire et al. 1991; Goodchild 1992; Laurini and Thompson 1992). The majority of knowledge discovery tasks have been performed in such databases. Due to the complex relationships among spatial objects and attributes, it has, however, been argued that object-oriented geographical information system (OOGIS) is more effective than the conventional raster- or vector-based GIS (structured as relational databases) in the modeling of geographical information and knowledge (Danforth and Tomlinson 1988; Worboys et al. 1990; David et al. 1993; Kim et al. 1993; Milne et al. 1993; Gunther and Lamerts 1994; Worboys 1994; Tang et al. 1996; Leung et al. 1999; Garvey et al. 2000). It not only provides rich semantics for spatial information representation and processing, but can also express in a natural and understandable way deep and complex spatial knowledge in a hierarchical manner. Furthermore, it can facilitate spatial reasoning with geographical concepts and commonsense, which is particularly crucial in the building of intelligent systems and in performing data mining and knowledge discovery in spatial databases (Gahegan and Roberts 1988; Manago and Kodratoff 1991; Rafanelli et al. 1995; Han et al. 1998).

The ubiquity of relational spatial databases and the inertia in their usage often deter researchers and practitioners to rebuild or convert their systems to OOGIS. To take advantage of and to better utilize the readily available relational databases, which are relatively poor in semantics and less efficient in spatial reasoning, formulating powerful methods for the discovery of the intrinsic hierarchical knowledge structures hidden in relational GIS is of theoretical and pragmatic importance. Appropriate solutions of this problem will build an important bridge between relational and object-oriented databases, and enhance our ability in the utilization of spatial information. Since objects and the associated attributes form a highly complicated network of relationships traversing the whole hierarchy, discovering relationships and rules in hierarchical spatial information systems is a complex task. To be able to perform effective and efficient data mining and knowledge discovery in such systems will have important bearing on the success of OOGIS in research and practice.

In the analysis of OOGIS, Leung et al. (1999) proposed a generic concept-based OOGIS which organizes spatial information via a hierarchical conceptual model with entity and feature hierarchies. It is rich in spatial semantics and efficient in data management. Within the framework, each concept is characterized by its extension (objects the concept covers) and intension (attributes describing the concept). The structure itself is a lattice with a partial order relation. Based on Wille (1982) and Ganter and Wille (1999), it is in fact a concept lattice whose theory has not been extended to the research in spatial knowledge discovery. Using the Hasse diagram, the structure graph of the concept lattice, generated by the partial order relation, one can find a close correspondence between OOGIS and concept lattice. Each node (concept) in a concept lattice is also characterized by its extension and intension. Thus, the theory of concept lattice is a natural theoretical foundation for OOGIS (Leung et al. 2008c).

Based on the notions of concept lattice and the associated knowledge reduction methods, we can discover hierarchical spatial knowledge structure (in the form of object-oriented format) from relational database. Tables in relational spatial databases can be considered as information systems that can be analyzed by the concept lattice methods. These tables are linked by key fields. We can study the implication relationship between concepts in the same table and use the concept of inclusion (Zhang et al. 1996) and rough set theory to find possible implication relationships between concepts in different tables linked by the key fields. Merge and split for concept lattice can be studied for the derivation of hierarchical knowledge structure that can be expressed by Hasse diagrams in concept lattice.

However, structures unraveled may contain redundant attributes for describing the entities and constructing the spatial concept lattice. By identifying and removing the redundant attributes, we can simplify the knowledge hierarchy but at the same time preserving the structure. In this task, we can develop concept lattice reduction methods to reconstruct the concept hierarchy. The idea is to search for the smallest attribute set necessary for the construction of a spatial concept lattice identical to that unraveled by using all of the attributes. Therefore, knowledge reduction boils down to the search for the minimum concept lattice consistent set, called a reduction. The hierarchical knowledge structures thus discovered possess the properties of data encapsulation, inheritance, generalization, and specialization basic to OOGIS. The generalization operator, for example, can be naturally defined by the structure graph of the concept lattice.

7.2.2 Errors in Spatial Knowledge Discovery

With the development and advancement of technologies in the acquisition of spatial information, data measured at different scales may be obtained from various sources. Geographic information, for example, can be obtained from satellite sensors, map scanning, optoelectronic and radar devices, aircraft, and global positioning system. Applications with low and high spatial and temporal resolutions have found their way in large-scale mapping, monitoring, and spatial decision-making. Nowadays, more and more problems in a large variety of fields are involved with the use of multi-source data. The phenomenon is ubiquitous in present day scientific research and practical applications.

Multi-scale problems are usually associated with multi-source problems since data from different sources (e.g., sensors) generally possess different properties in terms of scales or resolutions. Through fusion (or integration), benefits of information coming from multiple sources and measured in different scales can be integrated, resulting in a new database which is more informative and is of higher quality for spatial data mining and knowledge discovery. It has been demonstrated in satellite-imagery study that by fusing image data from various sources with different resolutions, we can enhance the spectral resolution of the fused images so that more objects can be more accurately identified (Schistad Solberg et al. 1994). Therefore, multi-source and multi-scale (or resolution) data may coexist in a knowledge discovery problem and need to be handled simultaneously and in accordance with their properties.

As argued throughout this monograph, scale is an inherent issue in geographical research in general and remote sensing in particular. An important and practical problem is how to fuse multi-source and multi-scale data into more accurate and revealing information for data mining. The keen interest in scale has led to the development of many methods for its handling in scientific investigations and applications. In GIS and remote sensing, a number of multi-scale methods have been proposed (Quattrochi and Goodchild 1997).

In recent years, extensive research on the development of techniques for multi-sensor data fusion systems has been carried out in many domains (Joshi and Sanderson 1999). Data fusion techniques include the correlation and conditioning of data products, both spatially and temporally, and the fusion and interpretation of data using a hybrid reasoning approach. Schistad Solberg et al. (1996) divide the methods of data fusion into four main categories: statistical (e.g., Laferte et al. 1995; Lee et al. 1987), fuzzy logic (Grégoire and Konieczny 2006), Dempster-Shafer theory of evidence (Rottensteiner et al. 2005), and neural network. Pohl and

Van Genderen (1998) review methods for fusion of multi-sensor image data in remote sensing. Techniques for image fusion include the intensity-hue-saturation technique (Zhang and Hong 2005), wavelet transform (Ulfarsson et al. 2003), multi-scale Kalman filter (Simone et al. 2000), multisensor Kalman filter (Caron et al. 2006), and pyramid-based algorithms (Sadjadi 2005). These methods improve the understanding of heterogeneous and multi-source data, and provide better opportunities for knowledge discovery. The integration of multi-source spatial data measured at different scales is thus very fundamental in theory and practice.

There are, however, two outstanding gaps in the study of information fusion in the literature. They are (1) the lack of study in the fusion of vector-based GIS, and (2) the scarcity of investigation of error propagation in the fusion of multi-source and multi-scale spatial data, particularly in spatial data mining and knowledge discovery. The former gap may possibly be due to the relatively large proportion of studies that are related to remotely sensed imagery. Nevertheless, I regard it equally necessary and important to study fusion of multi-source and multi-scale vector-based data (of course, its fusion with remotely sensed data also) since it is central to geographical research and applications.

The latter gap is even more critical. While information fusion has been rather extensively studied, there is very little research on error analysis and propagation in the fusion of multi-source and multi-scale spatial data. Map overlay, for example, is a fundamental operation in the analysis of geo-referenced data. The fusion of layers of maps from different sources and measured at different scales with different error variances gives rise to the question about how errors propagate throughout the fusion process and how they would affect the accuracy of the final fused product and the knowledge discovered. Failure in accounting for such process will lead to serious consequences in the spatial knowledge discovery tasks. Naturally, one, for example, wants to know how to determine the accuracy of some measures such as the length and area of an object in the fused data. Conceptually, overlay may be viewed as a special case of data fusion in the narrow sense. In general, when the information fusion methods in the literature are implemented, it is essential to know how the accuracy (or reversely the error variances) of the final output is determined by the accuracies (or error variances) of the input data coming from various sources and measured at different scales. Although effective integration of multi-source and multi-scale geo-referenced data has been found to reduce uncertainty, a formal analysis of uncertainty reduction on the theoretical and empirical basis is necessary to make it convincing and reliable (Pohl and van Genderen 1998).

Error analysis and error propagation in single-source and single-scale GIS has been made over the years (Goodchild and Gopal 1989; Goodchild 1991; Veregin 1995; Heuvelink 1998). A general framework for error analysis in measurement-based GIS has been reported in a series of papers by Leung et al. (2004a–d). These works are fundamental to the study of error propagation problems in information fusion in general. From the statistical point of view, the fusion problem may be posed as an estimation problem where the best fusion algorithm minimizes the mean square error between the fusion result and the true situation. Lee and Bajcsy (2004) propose a spatially and temporally varying uncertainty model of acquired and transformed multisensor measurements. The proposed uncertainty model includes errors due to (1) each sensor; (2) transformations of measured values; (3) vector data spatial interpolation; and (4) temporal interpolation. Blum (2005) proposes a robust image fusion approach that provides a mean square error smaller than a given bound. These studies are preludes to the investigation of uncertainty problems in information fusion.

The consistent treatment of uncertainty is fundamental to the correct fusion of different data sources and the discovery of genuine and correct knowledge in such data types. Without knowing the relative weightings that we need to give to each data source, we cannot know how to correctly combine them and how to determine the error contained in the fused output. This line of research should fill the crucial gaps in the fusion of multi-source and multi-scale spatial information by first developing error propagation schemes for basic fusion algorithms in the literature, and then, more importantly, formulating a general theoretical framework (with empirical supports) for error propagation in the fusion of multi-source and multi-scale remotely sensed data (raster-based), vector-based geographical information, and both so that uncertainty in the fused product can be minimized, and the hidden knowledge can be accurately unraveled.

Specifically, we need to establish schemes of error propagation for generic data fusion algorithms so that errors can be tracked and analyzed throughout the data fusion and knowledge discovery process. As a generalization, we need to formulate a general-purpose optimal scheme for error propagation in the fusion of multi-source and multi-scale information (raster-based, vector-based, and both) so that uncertainty in the fused product and knowledge discovered can be traced and minimized (Leung et al. 2006c).

7.2.3 Other Challenges

In addition to the above areas, there are other issues that need to be tackled in order to further advance research on spatial knowledge discovery. Closely related to the problem of unraveling error involving multi-source and multi-scale spatial data discussed in Sect. 7.2.2, knowledge discovery in distributed spatial data pose a research problem on data management and algorithms development. Due to the advances in communication technologies, data are distributed over the wired and wireless networks. How to discover spatial knowledge in a distributed computing environment is thus an area for further research. Algorithms for clustering, classification, associations and relationships need to be developed so that corresponding structures and processes can be efficiently and effectively unraveled in data coming from heterogeneous sites within the network-based computational environment.

Geographical phenomena and processes are dynamic in nature. Their changes over time can be gradual, intermittent or rapid. The discovery of some processes needs to be made on-line or in real time. The tracking of oil spill, typhoons or floods, for example, requires real-time or near-real time database management, and algorithms that can perform on-line or real-time knowledge discovery with multiple high-speed continuous or potentially infinite data streams. How to model an infinite amount of continuous data in order to discover time-varying concepts, patterns, and trends imposes a challenge to spatial data mining. Our ability to perform data mining in active databases is thus crucial to the solution of instant emergent systems such as natural and human-made disasters.

The above, of course, is not an exhaustive account of challenges lying ahead of us. They are, however, major issues that have not been dealt with in a thorough and rigorous manner. One may come up with other issues, such as knowledge discovery in multi-type spatial data, sampling stratregies, etc., that are important and challenging.

7.3 Concluding Remark

As pointed out at the beginning of this monograph, knowledge discovery in spatial and temporal data has been a long tradition in geographical, particularly geographical data analysis, research. We have been searching for patterns, processes and relationships from data way before data mining and knowledge discovery becomes an accepted term and a research area in the academic and professional circles. With the advancement of information and communication technologies, and our ability to amass voluminous and highly complex spatial and temporal data, the importance of knowledge discovery in data has thus been brought to the forefront of our search for knowledge necessary to the solution of problems constantly cropping up in our everyday life. Due to the nature of the spatial and temporal data, we need to develop methods and algorithms that are efficient and effective for the discovery of the kind of knowledge that is of interest to us. This monograph is a modest attempt to give an exposure of such undertaking. I hope that it can stimulate further research in the years to come.

Knowledge discovery is not a blind search process. It should be guided by knowledge also. Knowledge about the phenomena and processes that we are dealing with, knowledge about the data and information that we are facing, and knowledge about the methodologies that we are using and developing are absolutely important to a successful discovery. A thorough understanding of our problems and data is thus essential to such endeavor. The knowledge discovery process is recursive. New knowledge is discovered on the basis of existing knowledge. Knowledge thus discovered is, in turn, employed to enrich and improve the processes and methods to be used for further knowledge discovery. In our ever changing world, new structures and processes will emerge in space and time. Knowledge discovery will continue to be a key issue in research and practice.

Bibliography

Abry P (2003) Scaling and wavelets: an introductory walk. In: Rangarajan G, Ding M (eds) Processes with long-range correlations: theory and applications. Springer, Berlin
Acton ST, Mukherjee DP (2000) Scale space classification using area morphology. J IEEE Trans Image Process 9(4):623–635
Aha DW, Kibler D, Albert MK (1991) Instance-based learning algorithms. Mach Learn 6:37–66
Ahlqvist O (2005) Using uncertain conceptual spaces to translate between land cover categories. Int J Geogr Inform Sci 19:831–857
Ahlqvist O, Keukelaar J, Oukbir K (2000) Rough classification and accuracy assessment. Int J Geogr Inform Sci 14:475–496
Ahlqvist O, Keukelaar J, Oukbir K (2003) Rough and fuzzy geographical data intergration. Int J Geogr Inform Sci 17:223–234
Aldridge CH (1998) A theory of empirical spatial knowledge supporting rough set based knowledge discovery in geographical databases. Ph.D. thesis, University of Otago, Dunedin, New Zealand
Allenby GM, Rossi PE (1994) Modeling household purchase behavior with logistic normal regression. J Am Stat Assoc 89:1218–1231
Amari S (1995) Information geometry of the EM and EM algorithms for neural. Neural Network 8 (9):1379–1409
Amorese D, Lagarde JL, Laville E (1999) A point pattern analysis of the distribution of earthquakes in Normandy (in France). J Bull Seismol Soc Am 89(3):742–749
Anderberg MR (1973) Cluster analysis for applications. Academic, New York
Anderson JA (1982) Logistic discrimination. In: Krishnaiah PR, Kanal L (eds) Hand book of statistics, vol 2. North-Holland, Amsterdam, pp 169–191
Andreo B, Jiménez P, Durán JJ, Carrasco I, Vadillo I, Mangin A (2006) Climatic and hydrological variations during the last 117–166 years in the south of the Iberian Penninsula, for spectral and correlation analyses and continuous wavelet analyses. J Hydrol 324:24–39
Angulo C, Catala A (2000) K_SVCR. A Multi-class support vector machine. In: Lopez de Mantaras R, Plaza E (eds) ECML 2000, LNAI 1810. Springer, Berlin, pp. 31–38
Angulo JM, Ruiz-Medina MD, Anh VV, Grecksch W (2000) Fractional diffusion and fractional heat equation. Appl Prob 32:1077–1099
Anh VV, Heyde CC (Eds) (1999c) Special issue on long-range dependence. J Statist Plann Inf 80:1-292
Anh VV, Leonenko NN (2000) Scaling laws for the fractional diffusion-wave equation with random data. Stat Prob Lett 48:239–252
Anh VV, Leonenko NN (2001) Spectral analysis of fractional kinetic equations with random data. J Stat Phys 104(516):1349–1387

Anh VV, Duc H, Azzi M (1997a) Modelling anthropogenic trends in air quality data. J Air Waste Manag Assoc 47(1):66–71

Anh VV, Duc H, Tieng Q (1997b) Modelling persistence and intermittency in air pollution. In: Power H, Tirabassi T, Brebbia CA (eds) Air pollution V: modelling monitoring and management. Computational Mechanics Publications, Southampton, pp 443–452

Anh VV, Gras F, Tsui HT (1996) Multifractal description of natural scenes. Fractals 4(1):35–43

Anh VV, Heyde CC, Tieng Q (1999a) Stochastic models for fractal processes. J Stat Plann Infer 80 (1/2):123–135

Anh VV, Lam KC, Leung Y, Tieng Q (1999b) Multifractal analysis of Hong Kong air quality data. Environmetrics 10:139–149

Anh VV, Leung Y, Chen D, Yu ZG (2005b) Spatial variability of daily rainfall using multifractal analysis (unpublished paper)

Anh VV, Leung Y, Lam KC, Yu ZG (2005b) Multifractal characterization of Hong Kong air quality data. Environmetrics 16:1–12

Ankerst M, Breuning M, Kriegel HP, Sander J (1999) OPTICS: ordering points to identify the clustering structure. In: Proceedings of the 1999 ACM-SIGMOD international conference on management of data (SIGMOD'99), pp. 49–60

Anselin L (1988) Spatial econometrics: methods and models. Kluwer, Dordrecht

Anselin L (1990) Spatial dependence and spatial structural instability in applied regression analysis. J Reg Sci 30:185–207

Anselin L (1995) Local indicators of spatial association – LISA. Geogr Anal 27:93–115

Anselin L (1998a) Exploratory spatial data analysis in a geocomputational environment. In: Longley P, Brooks A, McDonnell SM, Macmillan B (eds) Geocomputation: a primer. Wiley, Chichester

Anselin L (1998b) Exploratory spatial data analysis in a geocomputational environment. In: Longley P, Brooks A, McDonnell SM, Macmillan B (eds) Geocomputation: a primer. Wiley, Chichester

Anselin L, Griffith DA (1988) Do spatial effects really matter in regression analysis? Paper Reg Sci Assoc 65:11–34

Anselin L, Rey S (1991) Properties of tests for spatial dependence in linear regression models. Geogr Anal 23:112–131

Arabie P, Hubert L, De Soete G (eds) (1996) Clustering and classification. World Scientific, Singapore

Arbia G (1989) Spatial data configuration in statistical analysis of regional economic and related problems. Kluwer, Dordrecht

Arbib MA (ed) (1995) The handbook of Brain Theory and Neural Networks. MIT, Cambridge

Asano T, Katoh N (1996) Variants for the Hough transform for line detection. Comput Geom 6:231–252

Atallah MJ (1992) Parallel techniques for computational geometry. Proc IEEE 80(9):1435–1448

Atkinson PM, Curran PJ (1997) Choosing an appropriate spatial resolution for remote sensing investigations. Photogramm Eng Rem Sens 63(12):1345–1351

Atkinson PM, Tatnall ARL (1997) Neural networks in remote sensing. Int J Remote Sens 18 (4):699–709

Babaud J, Witkin AP, Baudin M, Duda R (1986) Uniqueness of the Gaussian kernel for scale-space filtering. J IEEE Trans Pattern Anal Mach Intell 8:26–33

Bacry E, Muzy J, Arneodo A (1993) Singularity spectrum of fractal signals from wavelet analysis: exact results. J Stat Phys 70(314):635–647

Ball G, Hall D (1976) A clustering technique for summarizing multivariate data. J Behav Sci 12:153–155

Banfield JD, Raftery AE (1993) Model-based Gaussian and non-Gaussian clustering. Biometrtics 49:803–821

Bao S, Henry M (1996) Heterogeneity issues in local measurements of spatial association. Geogr Syst 3:1–13

Barnett V, Lewis T (1994) Outliners in statistical data. Wiley, New York
Basak J, Mahata D (2000) A connectionist model for corner detection in binary and gray images. J IEEE Trans Neural Network 11:1124–1132
Benediktsson JA, Swain PH, Ersoy OK (1990) Neural network approaches versus statistical methods in classification of multi-source remote sensing data. IEEE Trans Geosci Rem Sens 28(4):540–552
Beniston M, Stephenson DB (2004) Extreme climate events and their evolution under changing climatic conditions. Global Planet Change 44:1–9
Bennett RJ (1979) Spatial time series: analysis-forecasting-control. Pion, London
Bentley JL, Clarkson KL, Levine DB (1993) Fast Linear expected-time algorithms for computing maxima and convex hulls. J Algorithmica 9:168–183
Beran J (1992) Statistical methods for data with long-range dependence. Stat Sci 7:404–416
Beran J (1994) Statistics for long-memory processes. Chapman and Hall, New York
Berkson J (1944) Application of the logistic function to bio-assay. J Am Stat Assoc 39:357–365
Bern MW, Karloff HJ, Schieber B (1992) Fast Geometric approximation techniques and geometric embedding problems. J Theor Comput Sci 106:265–281
Bezdek JC (1980) A convergence theorem for the fuzzy ISODATA clustering algorithms. J IEEE Trans Pattern Anal Machine Intell 2:1–8
Bezdek JC, Coray C, Gunderson R, Watson J (1981) Detection and characterization of cluster substructure. I. linear structure: fuzzy C-line. SIAN J Appl Math 40(2):339–357
Bezdek JC, Hathaway RJ, Windham MP (1991) Numerical comparison of the RFCM and AP algorithms for clustering relational data. J Pattern Recogn 24(8):783–791
Bezdek JC, Keller JM, Krishnapuram R, Pal NR (1999) Fuzzy models and algorithms for pattern recognition and image processing. Kluwer, Boston
Bhattacharya U, Parui SK (1997) An improved back-propagation neural network for detection of road-like features in satellite imagery. Int J Rem Sens 18(6):3379–3394
Bischof H, Schneider W, Pinz AJ (1992) Multi-spectral classification of landsat images using neural network. IEEE Trans Geosci Rem Sens 30:482–490
Bishop CM (1995a) Neural networks for pattern recognition. Clarendon Press, Oxford
Bishop CM (1995b) Radial basis functions, neural networks for pattern recognition. Clarendon Press, Oxford
Bittner T, Stell JG (2002) Vagueness and rough location. Geoinformatica 6:99–121
Blatt M, Wiseman S, Domany E (1997) Data clustering using a model granular magnet. J Neural Comput 9:1805–1847
Blum RS (2005) Robust image fusion using a statistical signal processing approach. Inform Fusion 6(2):119–128
Bonsal BR, Zhang X, Vincent LA, Hogg WD (2001) Characteristic of daily and extreme temperatures over Canada. J Clim 14:1959–1976
Boots B, Getis A (1988) Point pattern analysis. Sage Publications, London
Boots B, Tiefelsdorf M (2000) Global and local spatial autocorrelation in bounded regular tessellations. J Geogr Syst 2:319–348
Borgas MS (1992) A comparison of intermittency models in turbulence. Phys Fluid A 4 (9):2055–2061
Box GEP, Jenkins GM (1976) Time series analysis: forecasting and control. Holden-Day, San Francisco, CA
Box GEP, Jenkins GM, Reinsel GC (1994) Time series analysis: forecasting and control. Prentice Hall, Englewood Cliffs, NJ
Breiman L (2001) Random forests. Mach Learn 45:5–32
Breiman L, Friedman JH, Olshen RA, Stone CJ (1984) Classification and regression trees. Wadsworth, California
Brown M, Lewis HG, Gunn SR (1999) Support vector machines for spectral unmixing. IGRASS'99 2:1363–1365

Brown M, Lewis HG, Gunn SR (2000) Linear spectral mixture models and support vector machines for remote sensing. IEEE Trans Geosci Rem Sens 38(5):2346–2360

Brunsdon C, Fotheringham AS, Charlton M (1996) Geographically weighted regression: a method for exploring spatial nonstationarity. Geogr Anal 28:281–298

Brunsdon C, Fotheringham AS, Charlton M (1997) Geographical instability in linear regression modeling – a preliminary investigation. In: New techniques and technologies for statistics II: Proceedings of the Second Bonn Seminar IOS Press, Amsterdam, pp. 149–158

Brunsdon C, Fotheringham AS, Charlton M (1999) Some notes on parametric significance test for geographically weighted regression. J Reg Sci 39(3):497–584

Brunsdon C (1995) Estimating probability surfaces for geographical point data: an adaptive kernel algorithm. Comput Geosci 21:877–894

Bruzzone L, Prieto DF (1999) A technique for the selection of kernel-function parameters in RBF neural networks for classification of remote-sensing images. IEEE Trans Geosci Rem Sens 37 (2):551–559

Bruzzone L, Prieto DF (2000) Automatic analysis of the difference image for unsupervised change detection. IEEE Trans Geosci Rem Sens 38(3):1171–1182

Bruzzone L, Prieto DF (2001) Unsupervised retraining of a maximum likelihood classifier for the analysis of multitemporal remote sensing images. IEEE Trans Geosci Rem Sens 39 (2):456–460

Bruzzone L, Prieto DF, Serpico SB (1999) A neural-statistical approach to multitrmporal and multisource remote-sensing image classification. J IEEE Trans Geosci Rem Sens 37:1350–1359

Burges C, Scholkopf B (1997) Improving the accuracy and speed of support vector machines. In: Mozer M, Jordan M, Petsche T (eds) Neural information processing systems. MIT, Cambridge

Burges CJC (1998) A tutorial on support vector machines for pattern recognition. Data Min Knowl Discov 2:121–167

Burridge P (1980) On the cliff-ord test for spatial correlation. J Roy Stat Soc B 42:107–108

Burrough PA (1986) Principles of geographic information systems for land resources assessment. Oxford University Press, Oxford

Cao Z, Fu Z (1999) Clustering of long- and medium-term seismicity on China mainland (in Chinese with English abstract). J Earthquake 19(4):338–344

Cao Z, Kandel A, Li L (1990) A new model of fuzzy reasoning. Fuzzy Set Syst 36:311–325

Caron F, Duflos E, Pomorski D, Vanheeghe P (2006) GPS/IMU data fusion using multisensor Kalman filtering: introduction of contextual aspects. Inform Fusion 7(2):221–230

Carpenter GA, Grossberg S (1987) ART2: self-organization of stable category recognition codes for analog input pattern. In: Proceedings IEEE international conference neural networks. San Diego, CA. pp. 727–736

Carpenter GA, Grossberg S (1988) The ART of adaptive pattern recognition by a self-organizing neural network. Computer 21:77–88

Carpenter GA, Grossberg S, Reynolds JH (1991) ARTMAP: supervised real time learning and classification of nonstationary data by a self-organising neural network. Neural Networks 4:565–588

Carpenter GA, Grossberg S, Markuzon N, Reynolds JH, Rosen DB (1992) Fuzzy ARTMAP: a neural network architecture for incremental supervised learning of analog multi-dimensional maps. IEEE Trans Neural Network 3:698–713

Casetti E (1972) Generating models by the expansion method: applications to geographical research. J Geogr Anal 4:81–91

Casetti E (1982) Drift analysis of regression analysis: an application to the investigation of fertility development relations. Modeling Simul 13:961–966

Casetti E (1986) The dual expansion method: an application for evaluating the effects of population growth on development. In: IEEE transactions on systems, man and cybernetics SMC-16, pp. 29–39

Casetti E (1997) The expansion method, mathematical modelling, and spatial econometrics. Int Reg Sci Rev 20:9–32
Caudill SB, Acharya RN (1998) Maximum likelihood estimation of a mixture of normal regressions: starting values and singularities. Comm Stat Simulat 27(3):667–674
Celeux G, Govaert G (1992) A classification EM algorithm for clustering and two stochastic versions. J Comput Stat Data Anal 14:315–332
Chakravarthy SV, Ghosh J (1996) Scale-based clustering using the radial basis function network. J IEEE Trans Nural Network 7(5):1250–1261
Chen T, Chen H (1995) Approximation capability to functions of several variables, nonlinear functions, and operators by radial basis function neural networks. IEEE Trans Neural Network 6:904–910
Chen Z, Ivanov PC, Hu K, Stanley HE (2002) Effect of nonstationarities on detrended fluctuation analysis. Phys Rev E 65(4):041107
Cherkassky V, Mulier F (1997) Learning from data: concepts, theory and methods. Wiley, New York
Chmielewski MR, Grzymala-Busse JW (1996) Global discretization of continuous attributes as preprocessing for machine learning. Int J Approx Reason 15:319–331
Cihlar J (2000) Land cover mapping of large areas from satellites: status and research priorities. Int J Remote Sens 21(6):1093–1114
Civco DL (1993) Artificial neural networks for land cover classification and mapping. Int J Geogr Inform Syst 7:173–186
Cleveland WS (1979) Robust locally weighted regression and smoothing scatter-plots. J Am Stat Assoc 74:829–836
Cleveland WS, Devlin SJ (1988) Locally weighted regression: an approach to regression analysis by local fitting. J Am Stat Assoc 83:596–610
Cleveland WS, Devlin SJ, Grosse E (1988) Regression by local fitting: methods, properties and computational algorithms. J Econom 37:87–114
Cliff AD, Ord JK (1972) Testing for spatial autocorrelation among regression residuals. Geogr Anal 4:267–284
Cliff AD, Ord JK (1973) Spatial autocorrelation. Pion, London
Collett D (1991) Modelling binary data. Chapman and Hall, London
Cordy CB, Griffith DA (1993) Efficiency of least squares estimators in the presence of spatial autocorrelation. Commun Stat Simul Comput 22:1161–1179
Coren S, Ward L, Enns J (1994) Sensation and perception. Harcourt Brace College Publishers, Fort Worth, TX
Cortes C, Vapnik VN (1995) Support vector networks. Mach Learn 20:273–297
Cortijo FJ, LA PDE, Blanca N (1998) Improving classical contextual classification. Int J Remote Sens 19(8):1591–1613
Costanzo CM, Hubert L, Golledge RG (1983) A higher moment for spatial statistics. Geogr Anal 15:347–351
Cote S, Tatnall ARL (1997) The Hopfield neural network as a tool for feature tracking and recognition from satellite sensor images. Int J Remote Sens 18(4):871–885
Couloigner I, Ranchin T (2000) Mapping of urban areas: A multiresolution modeling approach for semi-automatic extraction of streets. Photogramm Eng Rem Sens 66(7):867–874
Cox KR (1969) The voting decision in a spatial context. Prog Geogr 1:81–117
Cressie NAC (1993) Statistical for spatial data. Wiley, New York
Cristianini N, Shawe-Taylor J (2000) An introduction to support vector machines: and other kernel-based learning methods. Cambridge University Press, Cambridge
Dacey MF (1960a) The spacing of river towns. Ann Assoc Am Geogr 50:59–61
Dacey MF (1960b) The spacing of river towns. Ann Assoc Am Geogr 50:59–60
Danforth S, Tomlinson C (1988) Type theories and object-oriented programming. Comput Surv 20 (1):29–72

Danuser G, Stricker M (1998) Parametric model fitting: from inlier characterization to outlier detection. IEEE Trans Pattern Anal Mach Intell 20(2):263–280
Das Gupta S (1980) Discriminant analysis. In: Krishnaiah PR, Kanal L, Fisher RA (eds) An appreciation. Springer, New York, pp 161–170
Dattareya GR, Kanal LN (1990) Estimation of mixing probabilities in multi-class finite mixtures. J IEEE Trans Syst Man Cyvern 20:149–158
Daubechies I (1992) Ten lectures on wavelets. Society for industrial and applied mathematics, Philadelphia, Pennsylvania, pp. 357
Dave RN (1991) Characterization and detection of noise in clustering. J Pattern Recogn Lett 12:657–664
Dave RN, Krishnapuram R (1997a) Robust clustering methods: a unified view. J IEEE Trans Fuzzy Syst 5:270–293
Dave RN, Krishnapuram R (1997b) Robust clustering methods: a unified view. IEEE Trans Fuzzy Syst 5(2):270–293
David B, Raynal L, Schorter G (1993) GeO_2: Why objects in a geographical DBMS? In: Abel D, Ooi BC (eds) Advances in spatial databases: Proceedings of Third International Symposium, SSD'93. Lecture notes in computer Science, vol 692. Springer, Berlin, pp 264–276
Davis A, Marshak A, Wiscombe W, Cahalan R (1996) Scale in invariance of liquid water distributions in marine stratocumulus. J Atmos Sci 53:1538–1558
De Veaux RD (1989) Mixtures of linear regressions. Comput Stat Data Anal 8:227–245
DeGaetano AT (1996) Recent trends in maximum and minimum temperature threshold exceedences in the northeastern United States. J Clim 9:1646–1660
DeGaetano AT, Allen RJ (2002) Trends in twentieth-century temperature extremes across the United States. J Clim 15:3188–3205
Dempster AP, Laird NM, Rubin DB (1977) Maximum likelihood estimation from incomplete data via EM algorithm. J Roy Stat Soc B 39:1–38
Derin H (1987) Estimating components of univariate Gaussian mixtures using Prony's methods. J IEEE Trans Pattern Anal Mach Intell 9:142–148
Di K, Li DL, Li DY (1998) A mathematical morphology based algorithm for discovering clusters in spatial databases. J Image Graph 3(3):173–178
Djamdji J-P, Bijaoui A, Maniere R (1993) Geometrical registration of images: the multiresolution approach. Photogr Eng Rem Sens 59:645
Dubes RO, Jain AK (1976) Clusterinf techniques: the user's dilemma. J Pattern Recogn 8:247–260
Dubois D, Prade H (1980) Fuzzy sets and systems: theory and applications. Academic, Orlando
Duda RO, Hart PF (1974) Pattern classification and scene analysis. Wiley, New York
Efron B (1975) The efficiency of logistic regression compared to normal discriminant analysis. J Am Stat Assoc 70(352):892–898
Eiumnoh A, Shrestha RP (2000) Application of DEM data to landsat image classification: evaluation in a tropical wet-dry landscape of Thailand. Photogramm Eng Rem Sens 66 (3):297–304
Ester M, Kriegel HP, Xu X (1995) Knowledge discovery in large spatial databases: focusing techniques for efficient class identification. In Proceedings of the fourth international Symposium on large spatial databases (SSD'95), pp. 67–82
Ester M, Kriegel HP, Sander J, Xu X (1996) A density-based algorithm for discovering clusters in large spatial databases with noise. In: Proceedings of the second international conference on knowledge discovery and data mining, Portland, Oregon, pp.324-331
Ester M, Kriegel HP, Sander J, Xu X (1998) Clustering for Mining in Large Spatial Databases. J Spec Issue Data Min, KI-J 12(1):18–24
Everitt BS (1993) Cluster analysis. Halsted Press, New York
Falco T, Francis F, Lovejoy S, Schertzer D, Kerman B, Drinkwater M (1996) Scale invariance and universal multifractals in sea ice synthetic aperature radar reflectivity fields. IEEE Trans Geosci Rem Sens 34:906–914
Falconer KJ (1985) The geometry of fractal sets. Cambridge University Press, Cambridge

Fayyad U, Stolorz P (1997) Data mining and KDD: promise and challenges. Future Gener Comput Syst 13:99–115
Feder J (1988) Fractals. Plenum Press, New York
Feldman DS (1993) Fuzzy network synthesis with genetic algorithms. In: Forrest S (ed) Proceedings of the 5th International conference on genetic algorithms. Morgan Kaufmann, San Mateo, CA, pp 312–317
Ferro CJS, Warner TA (2002) Scale and texture in digital image classification. Photogramm Eng Rem Sens 68(1):51–63
Fischer MM, Getis A (eds) (1997) Recent developments in spatial analysis: spatial statistics, behavioural modeling, and computational intelligence. Springer, Berlin
Fischer MM, Leung Y (1998) A genetic-algorithms based evolutionary computational neural network for modelling spatial interaction data. Ann Reg Sci 32:295–298
Fischer MM, Leung Y (eds) (2001) GeoComputational modelling. Techniques and applications. Springer, Berlin
Fisher RA (1936) The use of multiple measurements in taxonomic problems. Ann Eugen 7:179–188
Fisher Y (ed) (1995) Fractal image compression, theory and application. Springer, New York
Fitzpatrick DB (1976) An analysis of bank credit card profit. J Bank Res 7:199–205
Fix E, Hodges JL (1951) Discriminatory analysis – nonparametric discrimination: consistency properties. Report No. 4, Project no. 21-29-004. USAF School of Aviation Medicine, Randolph Field, Texas. Reprinted in Int Stat Rev 57(1989):238–247
Flygare AM (1997) A comparison of contextual classification methods using Landsat TM. Int J Remote Sens 18(18):3835–3842
Foody GM (1995a) Land cover classification using and artificial neural network with ancillary information. Int J Geogr Inform Syst 9:527–542
Foody GM (1995b) Using prior knowledge in artificial neural network classification with a minimal training set. Int J Remote Sens 16:301–312
Foster SA, Gorr WL (1986) An adaptive filter for estimating spatially-varying parameters: application to modelling police hours spent in response to calls for service. Manag Sci 32:878–889
Fotheringham AS (1997) Trends in quantitative methods I: stressing the local. Prog Hum Geog 21:88–96
Fotheringham AS, Brunsdon C (1999) Local forms of spatial analysis. Geogr Anal 31:340–358
Fotheringham AS, Pitts TC (1995) Directional variation in distance-decay. Environ Plann A 27:715–729
Fotheringham AS, Brunsdon C, Charlton M (2002) Geographically weighted regression – the Analysis of spatially varying relationships. Wiley, Chichester
Fotheringham AS, Charlton M, Brunsdon C (1997a) Measuring spatial variations in relationships with geographically weighted regression. In: Fischer MM, Getis A (eds) Recent developments in spatial analysis. Springer, London, pp 60–82
Fotheringham AS, Charlton M, Brunsdon C (1997b) Two techniques for exploring non-stationarity in geographical data. Geogr Syst 4:59–82
Friedman JH (1977) A recursive partitioning decision rule for nonparametric classification. IEEE Trans Comput C-26:404–408
Friedman JH (1994) An overview of predictive learning and function approximation. In: Cherkassky V, Friedman JH, Wechsler H (eds) Statistics to neural networks: Theory and pattern recognition applications. Springer, Berlin, pp 1–61
Frigui H, Krishnapuram R (1999) A robust competitive clustering algorithm with applications in computer vision. J IEEE Trans Pattern Anal Mach Intell 21(5):450–465
Frisch U (1995a) Turbulence. Cambridge University Press, Cambridge
Frisch U (1995b) Turbulence. The legacy of A. Kolmogorov. Cambridge University Press, Cambridge
Fu L (1994) Neural networks in computer intelligence. McGraw-Hill, New York

Fu Z (1997) Research on the earthquake activity mechanics in China's mainland. Earthquake Press, Beijing, pp 124–128

Fu Z, Jiang L (1994) Strong earthquake clustering in the Fenwei and North China Plain seismic zones (in Chinese with English Abstract). J Earthquake Res China 10(2):160–167

Fukunaga K, Hayes RR (1989) Estimation of classifier performance. IEEE Trans Pattern Anal Mach Intell 11:1087–1101

Fung T (2003) Landscape dynamics in the Maipo Ramsar Wetland site. In Roy PS (ed) Geoinformatics for tropical ecosystems. Asian association of remote sensing. Bishen Singh Mahendra Pal Singh, Dehradun, India pp. 539–553

Fung T, Leung Y, Xu ZB (2007) A vision-based approach to remote sensing image classification (a research project funded by the Hong Kong Research Grants Council)

Fung T, Leung Y, Anh VV, Marafa LM (2001) A multifractal approach for modeling, visualization and prediction of land cover changes with remote sensing data (proposal of a research project)

Gahegan MN, Roberts SA (1988) An intelligent, object-oriented geographical information system. Int J Geogr Inform Syst 2(2):101–110

Ganter B, Wille R (1999) Formal concept analysis: mathematical foundations. Springer, Berlin

Gao Y, Leung Y, Xu ZB (1996) A new genetic algorithm with no genetic operators (unpublished paper)

Gaonac'h H, Lovejoy S, Schertzer D (2003) Multifractal analysis of infrared imagery of active thermal features at Kilauea volcano. J Geophys Res 24(11):2323–2344

Garvey M, Jackson MS, Roberts M (2000) An object-oriented GIS. In: Proceedings of Net. Object Days 2000, 9–12 October 2000, Erfurt, Germany, pp. 604–613

Gaucherel C (2002) Use of wavelet transform for temporal characterization of remote watersheds. J Hydrol 269:101–121

Geary RC (1954) The contiguity ratio and statistical mapping. Inc Stat 5:115–145

Geisser S (1982) Beyesian discrimination. In: Krishnaiah PR, Kanal L (eds) Hand book of statistics, vol 2. North-Holland, Amsterdam, pp 101–120

Getis A, Ord JK (1992) The analysis of spatial association by use of distance statistics. Geogr Anal 24:189–206

Girosi F (1994) Regulation theory, radial basis functions, and networks. In: Cherkassky V, Friedman JH (eds) From statistics to neural networks – Theory and pattern recognition applications. Springer, Germany, pp 166–187

Glymour C, Madigan D, Pregibon D, Symth P (1997) Statistical themes and lessons for data mining. Data Min Knowl Disc 1:11–28

Goldberg DE (1989) Genetic algorithms in search optimization and machine learning. Addison-Wesley, New York

Golin M, Sedgewick (1988) Analysis of a simple yet efficient convex hull algorithm. In: Proceedings 4th Annual Symposium on Computational Geometry, pp. 153–163

Gomm JB, Yu D (2000) Selecting radial basis function network centers with recursive orthogonal least squares training. IEEE Trans Neural Network 11(2):306–314

Gong P (1996) Integrated analysis of spatial data from multiple sources: using evidential reasoning and artificial neural network techniques for geological mapping. Photogramm Eng Rem Sens 62(5):513–523

Gong P, Pu R, Chen J (1996) Mapping ecological land systems and classification uncertainties from digital elevation and forest-cover data using neural networks. Photogramm Eng Rem Sens 62(11):1249–1260

Goodchild MF (1991) Issues of quality and uncertainty. In: Muller JC (ed) Advances in cartography. Elsevier, London, pp 113–139

Goodchild MF (1992a) Geographical data modeling. Comput Geosci 18:401–408

Goodchild MF (1992b) Geographical data modelling. Comput Geosci 18:401–408

Goodchild MF, Gopal S (eds) (1989) Accuracy of spatial databases. Taylor and Francis, London

Gopal S, Fischer MM (1997) Fuzzy ARTMAP – A neural classifier for multi-spectral image classification. In: Fischer MM, Getis A (eds) Recent developments in spatial analysis. Berlin, Spinger, pp 306–335
Gorden RL (1977) Unidimensional scaling of social variables: concepts and procedures. The Free Press, New York
Gorr WL, Olligschlaeger AM (1994) Weighted spatial adaptive filtering: Monte Carlo studies and application to illicit drug market modeling. Geogr Anal 26:67–87
Gowda KC, Diday E (1992) Symbolic clustering using a new similarity measure. J IEEE Trans Syst Man Cybern 22:368–378
Graham RL (1972) An efficient algorithm for determining the convex hull of a finite planar set. Inform Process Lett 1:132–133
Granger CW (1980) Long memory relationships and the aggregation of dynamic models. Econometrics 14:227–238
Grassberger P, Procaccia I (1983a) Measuring the strangeness of strange attractors. Phys D 9:189–208
Grassberger P, Procaccia I (1983b) Characterisation of strange attractors. Phys Rev Lett 50:346–349
Grégoire E, Konieczny S (2006) Logic-based approaches to information fusion. Inform Fusion 7 (1):4–18
Griffith DA (1988) Advanced spatial statistics. Kluwer, Dordrecht
Grossberg S (1976) Adaptive pattern classification and universal recording. I: Parallel development and coding neural feature detectors. Biol Cybern 23:121–134
Guadagni P, Little J (1983) A logit model of brand choice calibrated on scanner data. Market Sci 2:203–238
Guha S, Rastogi R, Shim K (1998) CURE: An efficient clustering algorithm for large databases. In: Proceeding of the 1998 ACM-SIGMOD international conference management of data (SIGMOD'98), pp. 73–84
Guibas L, Salesin D, Stolfi J (1993) Constructing strongly convex approximate hulls with inaccurate primitives. J Algorithmica 9:534–560
Gunther O, Lamerts J (1994) Object-oriented techniques for the management of geographic and environmental data. Comput J 37:16–25
Guttman L (1968) A general nonmetric technique for finding the smallest the smallest coordinate space for a configuration of points. Psychometrika 33:469–506
Han JW, Nishio S, Kawano H, Wang W (1998) Generalization-based data mining in object-oriented databases using an object cube model. Data Knowl Eng 25:55–97
Hand DJ (1982) Kernel discriminant analysis. Research Studies Press, Letchworth
Hand DJ (1986) Recent advances in error rate estimation. Pattern Recogn Lett 4:335–346
Hand DJ (1998) Data mining: statistics and more? Am Stat 52(2):112–118
Hand DJ, Henley WE (1997) Statistical classification methods in consumer credit scoring. J Roy Stat Soc Series A 160:523–541
Hart PE (1968) The condensed nearest neighbour rule. IEEE Trans Inform Theor 14:515–516
Harvey DA, Gaonac'h H, Lovejoy S, Schertzer D (2002) Multifractal characterization of remotely sensed volcanic features: a case study from Kiluea volcano, Hawaii. Fractals 10(3):265–274
Hastie TJ, Tibshirani RJ (1996) Discriminant analysis by Gaussian mixtures. J R Stat Soc B 58:155–176
Hastie TJ, Tibshirani RJ (1990) Generalized additive models. Chapman and Hall, London
Hathaway RJ, Bezdek JC (1993) Switching regression models and fuzzy clustering. IEEE Trans Fuzzy Syst 1(3):195–204
Hathaway RJ, Bezdek JC (1994) NERF c-means non-Euclidean relational fuzzy clustering. J Pattern Recogn 27(3):429–437
Hathaway RJ, Bezdek JC, Davenport JW (1994) On relational data versions of c-means algorithms. J Pattern Recogn Lett 17:607–612

Hathaway RJ, Davenport JW, Bezdek JC (1989) Relational duals of c-means clustering algorithms. J Pattern Recogn 22(2):205–212

Hawkins D (1980) Identification of outliners. Chapman and Hall, Boca Raton, FL

Hearst MA, Scholkopf B, Dumais S (1998) Trends and controversies-support vector machines. IEEE Intell Syst 13(4):18–28

Heermann PD, Khazenie N (1992) Classification of multi-spectral remote sensing data using a back-propagation neural network. IEEE Trans Geosci Rem Sens 30(1):81–88

Heino R, Brazdil R, Forland E, Tuomenvirta H, Alexandersson H, Beniston M, Pfister C, Rebetez M, Rosenhagen G, Rosner S, Wibig J (1999) Progress in the study of climate extremes in northern and central Europe. Clim Change 42:151–181

Hentschel HGE, Procaccia I (1983) The infinite number of generalized dimensions of fractals and strange attractors. Phys D 8:435–444

Hepple LW (1998) Exact testing for spatial autocorrelation among regression residuals. Environ Plann A 30:85–109

Hermes L, Frieauff D, Puzicha J, Buhmann JM (1999) Support vector machines for land usage classification in Landsat TM imagery. In: Proceeding of the IEEE international geoscience and remote sensing symposium, vol. 1. Hamburg, pp. 348–350

Heuvelink GBM (1998) Error propagation in environmental modelling with GIS. Taylor and Francis, London

Hilfer R (2000) Fractional time evolution. In: Hilfer R (ed) Fractional calculus in physics. World Scientific, Singapore, pp 87–130

Hinneburg A, Keim DA (1998) An efficient approach to clustering in large multimedia databases with noise. In: Proceedings 1998 international conference knowledge discovery and data mining (KDD'98), pp. 58–65

Hocking RR (1996) Methods and applications of linear models. Wiley, New York

Holden M, Øksendal B, Ubøe J, Zhang TS (1996) Stochastic partial differential equations. A modelling, white noise functional approach. Birkhäuser, Boston

Holland JH (1975) Adaptation in natural and artificial systems. University of Michigan Press, Ann Arbor

Holmes QA, Nuesch DR, Schuchman RA (1984) Textural analysis and real-time classification of sea-ice types using digital SAR data. IEEE Trans Geosci Remote Sens GE-22:113–120

Holsheimer M, Kersten M (1995) A perspective on databases and data mining. Proceedings of 1st international conference on knowledge discovery and data mining, pp. 150–155

Honda K, Togo N, Fujii T, Ichihashi H (2002) Linear fuzzy clustering based on least absolute deviations. In: Proceedings of 2002 IEEE international conference of fuzzy systems, pp. 1444–1449

Hopfield JJ (1982) Neural networks and physical systems with emergent collective computational abilities. Proc Natl Acad Sci USA 81:3088–3092

Hosking JRM, Pednault EPD, Sudan M (1997) A statistical perspective on data mining. Future Gener Comput Syst 13:117–134

Hosmer DW, Lemeshow S (1989) Applied logistic regression. Wiley, New York

Hsu CN, Knoblock CA (1995) Estimating the robustness of discovered knowledge. In: Proceedings of the first international conference on knowledge discovery and data mining. Canada, Aug. 20–21, pp. 156–161

Hu K, Ivanov PC, Chen Z, Carpena P, Eugene Stanley H (2001) Effect of trends on detrended fluctuation analysis. Phys Rev E 64(1):011114

Huang Y, Leung Y (2002) Analysing regional industralisation in jiangsu province using geographically weighted regression. J Geogr Syst 4(2):233–249

Hubert L (1974) Approximate evaluation technique for the single-link and complete-link hierarchical clustering procedure. J Am Stat Assoc 69:968

Hummel R, Moniot R (1989) Reconstructions form zero crossings in scale space. IEEE Trans Acoust Speech Signal Process 37(12):245–295

Hwang YK, Ahuja N (1993) Cross motion planning – A survey. ACM Comput Surv 24 (3):219–291
Imhof JP (1961) Computing the distribution of quadratic forms in normal variables. Biometrika 48:419–426
Imielinski T, Mannila H (1996) A database perspective on knowledge discovery. Commun ACM 39:58–64
Ishibuchi H, Nozaki K, Yamamoto N, Tanaka H (1995) Selecting fuzzy if-then rules for classification problems using genetic algorithms. IEEE Trans Fuzzy Syst 3(3):260–270
Jain AK, Dubes RO (1998) Algorithms for clustering data. Prentice Hall, Englewood Cliffs, NJ
Jenson JR (1996) Introductory to digital image processing: a remote sensing perspective. Prentice Hall, Upper Saddle River, NJ
Jenson JR, Langari R (1999) Fuzzy logic: intelligence, control and information. Prentice Hall, Upper Saddle River, NJ
Jevrejeva S, Moore JC, Grinsted A (2003) Influence of the arctic oscillation and El Niño-Southern Oscillation (ENSO) on ice conditions in the baltic sea: the wavelet approach. J Geophys Res 108(D21):4677. doi:10.1029/2003JD003417
Ji M (2003) Using fuzzy sets to improve cluster labeling in unsupervised classification. Int J Rem Sens 24:657–671
Ji Q, Haralick RM (1998) Breakpoint detection using covariance propagation. IEEE Trans Pattern Anal Mach Intell 20(8):845–951
Jiang M, Ma Z (1985) A comparison between the third and fourth seismic periods in north china. J Earthquake 6:5–11
Jiang XH, Liu CM, Huang Q (2003) Multiple time scales analysis and cause of runoff changes of the upper and middle reaches of the Yellow River. Journal of Natural Resources 18(2):142–147 (in Chinese)
John GH (1995) Robust decision tree: removing outliers from databases. In: Proceedings of the first international conference on knowledge discovery and data mining. Canada, Aug. 20–21, 1995. pp. 174–179
John GH, Langley P (1995) Estimating continuous distributions in Bayesian classifiers In Proceedings of the 11th conference on uncertainty in artificial intelligence. Morgan Kaufmann, San Mateo, CA
Johnson RA, Wichern DW (1992) Applied multivariate statistical analysis, 3rd edn. Prentice-Hall, Englewood Cliffs, NJ
Johnson SC (1967) Hierarchical clustering scheme. J Psychometrika 32:241
Johnston RJ (1973) Spatial patterns and influences on voting in multi-candidate elections: the Christchurch City county elections, 1968. Urban Stud 10:69–81
Jones JP III, Casetti E (1992) Applications of the expansion method. Routledge, London
Jones RH, Stewart RC (1997) A method for determining significant structures in a cloud of earthquake. J Geophys Res 102:8245–8254
Kagan YY (1999) Is earthquake seismology a hard, quantitative science? J Pure Appl Geophys 155:233–258
Kagan YY, Jackson DD (1991) Long-term earthquake clustering. Geophys J Int 104:117–133
Kahane J-P (1991) Produits de poids aléatoires indépendants et applications. In: Bélair J, Dubuc S (eds) Fractal geometry and analysis. Kluwer, Dordrecht, pp 277–324
Kanellopoulos I, Wilkinson GG (1997) Strategies and best practice for neural network image classification. Int J Rem Sens 18(4):711–725
Kantelhardt JW, Zschiegner SA, Koscienlny-Bunde E, Halvin S, Bunde A, Stanley HE (2002) Multifractal detrended fluctuation analysis of nonstationary time series. Phys A: Stat Mech Appl 316(1–4):87–114
Kantz H, Schreider T (2004) Nonlinear time series analysis. Cambridge University Press, Cambridge
Karr L (1991) Design of an adaptive fuzzy logic controller using a genetic algorithms. In: Belew RK, Booker LB (eds) Proceedings of the 4th International Conference on Genetic Algorithms. Morgan Kaufmann, San Mateo, CA, pp 450–457

Karypis G, Han EH, Kumar V (1999) CHAMELEON: a hierarchical clustering algorithm using dynamic modeling. Computer 32:68–75

Kaufman L, Rousseeuw PJ (1990) Finding groups in data: an introduction to cluster analysis. Wiley, New York

Keerthi SS, Shevade SK, Bhattacharya U, Murthy KRK (2000) A fast iterative nearest point algorithm for support vector machine classifier design. IEEE Trans Neural Network 11 (1):124–136

Keller JM, Chen S, Crownover RM (1989) Texture description and segmentation through fractal geometry. Comput Graph Image Process 45:150–166

Kendall MG (1987) Kendall advanced theory of statistics, 5th edn. Charles Griffin, London

Kim E, Park M, Ji S, Park M (1997) A new approach to fuzzy modeling. IEEE Trans Fuzzy Syst 5(3):328–337

Kim W, Garza J, Keskin A (1993) Spatial data management in database systems: research directions. In: Abel D, Ooi BC (eds) Advances in spatial databases: third International Symposium, SSD'93. Lecture notes in computer science, vol 692. Springer, Berlin, pp 1–13

Kirpatrick S, Gelatt CD, Vecchi MP (1983) Optimization by simulated annealing. J Sci 200:671–680

Knoke JD (1986) The robust estimation of classification error rates. Comput Math Appl 12A:253–260

Koenderink JJ (1984) The structure of images. Biol Cybern 50:363–370

Koerts J, Abrahamse API (1968) On the power of the BLUS procedure. J Am Stat Assoc 63:1227–1236

Kohavi R (1996) Scaling up the accuracy of naive-Bayes classifiers: a deceision-tree hybrid. In: Proceedings of the second international conference on knowledge discovery and data mining. Morgan Kaufman, San Mateo CA

Kohonen T (1982) Clustering, taxonomy, and topological maps of patterns. In: Proceedings of the 6th international conference of pattern recognition. Munich, Germany, pp. 114–128

Kohonen T (1988) Self-organization and associative memory. Springer, Berlin

Kontkanen P, Myllymaki P, Silander T, Tirri H (1998) BAYDA: software for Bayesian classification and feature selection. Proceedings of the fourth intelnational conference on knowledge discovery and data mining. AAAI Press, Menlo Park, CA, pp. 254–258

Kosko B (1992) Neural networks and fuzzy systems. Prentice-Hall, Englewood Cliffs, NJ

Krämer W, Donninger C (1987) Spatial autocorrelation among errors and the relative efficiency of OLS in the linear regression model. J Am Stat Assoc 82:577–579

Krishnapuram R, Keller JM (1993) A possibilistic approach to clustering. J IEEE Trans Fuzzy Syst 1:98–110

Kryszkiewicz M (2001) Comparative study of alternative types of knowledge reduction in inconsistent systems. Int J Intell Syst 16:105–120

Krzanowski WJ (1977) The performances of Fisher's linear discriminant function under non-optimal conditions. Technometrics 19:191–199

Kulkarni AD (1994) Artificial neural networks for image understanding. Van Nostrand Reinhold, New York

Kulldorf M, Feuer E, Miller B, Freedman L (1997a) Breast cancer in Northeastern United States: a geographical analysis. Am J Epidemiol 146:161–170

Kulldorf M, Feuer E, Miller B, Freedman L (1997b) Brest cancer in Northeastern United States: a geographical analysis. Am J Epidemiol 146:161–170

Kurosu T, Yokoyama S, Fujita M (2001) Land use classification with textual analysis and the aggregation technique using multi-temporal JERS-1 L-band SAR images. Int J Remote Sens 22(4):595–613

Labat D, Ababou R, Mangin A (2000) Rainfall-runoff relations for karstic springs. Part II: continuous wavelet and discrete orthogonal multi-resolution analyses. J Hydrol 238:149–178

Labat D, Ronchail J, Guyot JL (2005) Recent advances in wavelet analyses: Part 2–Amazon, Parana, Orinoco and Congo discharges time scale variability. J Hydrol 314:289–311

Laferrière A, Gaonac'h H (1999) Multifractal properties of visible reflectance fields from basaltic volcanoes. J Geophys Res 104:5115–5126

Laferte J-M, Heitz F, Perez P, Fabre E (1995) Hierarchical statistical models for the fusion of multiresolution image data. In: Proceedings of 5th international conference on computer vision, 20–23 June 1995, pp. 908–913

Lam NSN, De Cola L (eds) (1993) Fractals in geography. Prentice Hall, Englewood Cliffs, NJ

Langley P, Sage S (1994) Induction of selective Bayesian classifers. In: Proceedings of the 10th conference on uncertainty in artificial intelligence. Morgan Kaufmann, Seattle, WA,

Lau K, Leung PL, Tse KA (1998) Nonlinear programming approach to metric unidimensional scaling. J Classif 15:2–14

Lau KN, Yang CH, Post GV (1996) Stochastic preference modeling within a switching regression framework. Comput Oper Res 23(12):1163–1169

Laurini R, Thompson D (1992) Fundamentals of spatial information systems. Academic, New York

Lawrence KD, Arthur JL (eds) (1990) Robust Regression. Marcel Dekker, New York

Lawson AB (2001) Statistical methods in spatial epidemiology. Wiley, Chichester

Lee SC, Bajcsy P (2004) Multisensor raster and vector data fusion based on uncertainty modeling, In: ICIP '04. 2004 international conference on image processing, 5:3355–3358

Lee T, Richards JA, Swain PH (1987) Probabilistic and evidential approaches for multisource data analysis. IEEE Trans Geosci Rem Sens 25(3):283–293

Lepage R, Rouhana RG, St-onge B, Noumeir R, Desjardins R (2000) Cellular neural network for automated detection of geological lineaments on radarsat images. J IEEE Trans Geosci Rem Sens 38:1224–1233

Leung Y (1982) Approximate characterization of some fundamental concepts of spatial analysis. Geogr Anal Int J Theor Geogr 14:19–40

Leung Y (1984) Towards a flexible framework for regionalization. Environ Plann A 16:1613–1632

Leung Y (1987) On the imprecision of boundaries. Geogr Anal Int J Theor Geogr 19:125–151

Leung Y (1988a) Spatial analysis and planning under imprecision. Elsevier, Amsterdam

Leung Y (1988b) Interregional equilibrium and fuzzy linear programming: 1. Environ Plan A 20:25–40

Leung Y (1988c) Interregional equilibrium and fuzzy linear programming: 2. Environ Plan A 20:219–230

Leung Y (1994) Inference with spatial knowledge: an artificial neural network approach. Geogr Syst 1:103–121

Leung Y (1997) Intelligent spatial decision support systems. Springer, Berlin

Leung Y (1999) Fuzzy sets approach to spatial analysis. In: Zimmermann H-J (ed) Practical applications of fuzzy technologies, the handbooks of fuzzy sets series. Kluwer, Norwell, MA, pp 267–300

Leung Y (2001) Neural and evolutionary computation methods for spatial classification and knowledge acquisition. In: Fisher MM, Leung Y (eds) GeoComputational modelling: techniques and applications. Springer, Berlin, pp 71–108

Leung Y, Leung KS (1993a) An intelligent expert system shell for knowledge-based geographical information systems: 1. the tools. Int J Geogr Inform Syst 7:189–199

Leung Y, Leung KS (1993b) An intelligent expert system shell for knowledge-based geographical information systems: 2. some applications. Int J Geogr Inform Syst 7:201–213

Leung Y, Li DY (2003) Maximal consistent block technique for rule acquisition in incomplete information systems. Inform Sci 153:85–106

Leung Y, Ma JH (2006c) An optimization model for error propagation in the fusion of multi-source and multi-scale spatial information (unpublished paper)

Leung Y, Gao Y, Xu ZB (1997a) Degree of population diversity – A perspective on premature convergence in genetic algorithms and its Markov chain analysis. IEEE Trans Neural Network 8:1165–1176

Leung Y, Gao Y, Zhang WX (2001b) A genetic-based method for training fuzzy systems. In: Proceedings of the 10th IEEE international conference on fuzzy systems – meeting the ground challenge: machines that serve people, organized by the institute of electrical and electronics engineers. Australia, Melbourne

Leung Y, Ge Y, Ma JH (2004e) Clustering of remote sensing data by unidimentional scaling (unpublished paper)

Leung Y, Leung KS, He JZ (1999) A generic concept-based object-oriented geographical information system. Int J Geogr Inform Sci 13(5):475–498

Leung Y, Leung KS, Lau CK (1997b) A development shell for intelligent spatial decision support systems: 1. Concepts and tools. Geogr Syst 4:19–37

Leung Y, Leung KS, Ma JH (2003a) Data mining for bank databases (unpublished paper)

Leung Y, Leung KS, Mei CL (2003b) Data mining for credit card promotion in the banking sector (unpublished paper)

Leung Y, Leung KS, Yuan XJ (2003c) Discovery of promotion strategies for banking services by classification trees (unpublished paper)

Leung Y, Li G, Xu ZB (1998) A genetic algorithm for multiple destination routing problem. IEEE Trans Evol Comput 2(4):150–161

Leung Y, Luo JC, Zhou CH (2002a) A knowledge-integrated radial basis function model for the classification of multispectral remote sensing images (unpublished paper)

Leung Y, Ma JH, Goodchild MF (2004b) A general framework for error analysis in measurement-based GIS, Part 1: The basic measurement-error model and related concepts. J Geogr Syst 6:325–354

Leung Y, Ma JH, Goodchild MF (2004c) A general framework for error analysis in measurement-based GIS, Part 2: The algebra-based probability model for point-in-polygon analysis. J Geogr Syst 6:355–380

Leung Y, Ma JH, Goodchild MF (2004d) A general framework for error analysis in measurement-based GIS, Part 3: Error analysis in intersections and overlays. J Geogr Syst 6:381–402

Leung Y, Ma JH, Goodchild MF (2004e) A general framework for error analysis in measurement-based GIS, Part 4: Error analysis in length and area measurements. J Geogr Syst 6:403–428

Leung Y, Ma JH, Zhang WX (2001b) A New method for mining regression classes in Large data sets. IEEE Trans Pattern Anal Mach Intell 23(1):5–21

Leung Y, Ma JM, Zhang WX, Qiu GF (2008c) Discovery of hierarchical knowledge structure in geographical information systems – the concept lattice approach (unpublished paper)

Leung Y, Mei CL, Wang N (2008b) A semi-parametric spatially varying coefficient model and its local-linearity-based estimation: a generalization of the mixed GWR model (unpublished paper)

Leung Y, Mei CL, Zhang WX (2000a) Statistical tests for spatial non-stationarity based on geographically weighted regression model. Environ Plann A 32:9–32

Leung Y, Mei CL, Zhang WX (2000b) Testing for spatial autocorrection among the residuals of the geographically weighted regression. Environ Plann A 32:871–890

Leung Y, Mei CL, Zhang WX (2003d) Statistical test for local patterns of spatial association. Environ Plann A 35:725–744

Leung Y, Wu WZ, Zhang WX (2006a) Knowledge acquisition in incomplete information systems: a rough set approach. Eur J Oper Res 168:164–180

Leung Y, Zhang JS, Xu ZB (2000c) Clustering by scale-space filtering. IEEE Trans Pattern Anal Mach Intell 22(12):1396–1410

Leung Y, Zhang JS, Xu ZB (1997c) Neural networks for convex hull computation. IEEE Trans Neural Network 8(3):606–611

Leung Y, Zhang JS, Xu ZB (2000d) Clustering by scale-space filtering. IEEE Trans Pattern Anal Mach Intell 22:1396–1410

Leung Y, Fischer MM, Wu WZ, Mi JS (2008c) A rough set approach for the discovery of classification rules in interval-valued information systems. Int J Approx Reason 47:233–246

Leung Y, Fung T, Mi JS, Wu WZ (2007) A rough set approach to the discovery of classification rules in spatial data. Int J Geogr Inform Sci 21:1033–1058

Leung Y, Leung KS, Zhao SP, Lau CK (1997d) A development shell for intelligent spatial decision support systems: 2. An application in flood simulation and damage assessment. Geogr Syst 4:39–57

Leung Y, Luo JC, Ma JH, Ming DP (2006b) A new method for feature mining in remotely sensed image. Geoinformatica 10:295–312

Leung Y, Luo JC, Zhou CH, Ma JH (2002b) Support vector machine for spatial feature extraction and classification of high resolution remote sensing images (unpublished paper)

Li D, Shao J (1994) Wavelet theory and its application in image edge detection. Int J Photogr Rem Sens 49:4

Li DR, Wang SL, Li DY (2006) Spatial data mining theories and applications. Science Publisher, Beijing

Linneman HV (1996a) An econometric study of international trade flows. North-Holland, Amesterdam

Linneman HV (1996b) An econometric study of international trade flows. North-Holland, Amsterdam

Lippmann RP (1994) Neural networks, Bayesion a posteriori probabilities, and pattern classification. In: Cherkassky V, Friedman JH (eds) From statistics to neural networks— theory and pattern recognition applications. Germany, Springer, pp 83–104

Loh WY, Vanichsetakul N (1988) Tree-structured classification via generalized discriminant analysis (with discussion). J Am Stat Assoc 83:715–728

Lovejoy S, Schertzer D, Tessies Y, Gaonac'h H (2001) Multifractals and resolution-dependent remote sensing algorithm: the example of ocean colour. Int J Rem Sens 22:1191–1234

Luo JC, Leung Y, Zheng J, Ma JH (2004) An elliptical basis function for the classification of remote sensing images. J Geogr Syst 6:219–236

Ma Z, Jiang M (1987) Strong earthquake periods and episodes in China (in Chinese with English Abstract). J Earthquake Res China 3(1):47–51

MacQueen J (1967) Some methods for classification and analysis of multivariate observations. In: Proceedings of 5th Berkeley Symposium on Mathematical Statistics and Probability, 1, pp. 281–297

Maguire DJ, Goodchild MF, Rhind D (1991) Geographical information systems: principles and applications, Vol. 1: Principles; Vol.2: Applications. Longman, Harlow

Mak M, Kung S (2000) Estimation of elliptical basis function parameters by the EM algorithm with application to speaker verification. IEEE Trans Neural Network 11(4):961–969

Man Y, Gath I (1994) Detection and separation of ring-shaped clusters using fuzzy clustering. J IEEE Trans Pattern Anal Mach Intell 16:855–861

Manago M, Kodratoff Y (1991) Induction of decision trees from complex structured data. In: Piatetsky-Shapiro G, Frawley WJ (eds) Knowledge discovery in databases. AAAI Press, Menlo Park, CA, pp 289–306

Mandelbrot BB (1974) Intermittent turbulence in self-similar cascades: divergence of high moments and dimension of the carrier. J Fluid Mech 62:331–358

Mandelbrot BB (1985) Self-affine fractals and factal dimension. Phys Scripta 32:257–260

Mandelbrot BB (1999a) Multifractals and 1/f noise: wild self-affinity in physics. Springer, New York

Mandelbrot BB (1999b) Renormalization and fixed points in finance, since 1962. Phys A 263 (1):477–487

Mandelbrot BB, Van Ness LW (1968) Fractional Brownian motions, fractional noises and applications. SIAM Rev 10:422–437

Mannan B, Roy J, Ray AK (1998) Fuzzy ARTMAP supervised classification of multi-spectral remotely-sensed images. Int J Rem Sens 19(4):767–774

Maragos P (1989) Pattern spectrum and multiscale shape representation. J IEEE Trans Pattern Anal Mach Intell 11(7):701–716

Mather PM (1999) Land cover classification revisited. In: Atkinson PM, Tate NJ (eds) Advances in remote sensing and GIS analysis. Wiley, London, pp 7–16

McIver JP, Carmines EG (1981) Unidimensional scaling. Sage Publications, Beverly Hills, CA

McLachlan GJ (1992) Discriminant analysis and statistical pattern recognition. Wiley, New York

McLachlan GJ, Basford KE (1988) Mixture models: inference and applications to clustering. Marcel Dekker, New York

McLachlan GJ, Krishnan T (1997) The EM algorithm and extensions. Wiley, London

Medsker LR (1994) Hybrid neural network and expert systems. Kluwer, Dordrecht

Mei CL, He SY, Fang KT (2004) A note on the mixed geographically weighted regression model. J Reg Sci 44:143–157

Menard SW (1995) Applied logistic regression analysis. Sage Publication, Thousand Oaks, CA

Meneveau C, Scrrnivasan KR (1991) The multifractal nature of turbulent energy dissipation. J Fluid Mech 224:429–484

Meng DY, Xu ZB (2006) Visual learning theory (unpublished paper)

Meng DY, Xu ZB, Leung Y, Fung T (2008) The strong convergence of visual method and its applications in disease diagnosis. Paper presented at the 3rd international conference on pattern recognition in bioinformatics, Melbourne Australia

Mi JS, Wu WZ, Zhang WX (2004) Approaches to knowledge reduction based on variable precision rough sets model. Inform Sci 159:255–272

Miline P, Milton S, Smith JL (1993) Geographical object-oriented databases-a case study. Int J Geogr Inform Syst 7:39–55

Miller D, Rose K (1996) Hierarchical, unsupercised learning with growing via phase transitions. J Neural Comput 8:425–450

Miller HJ, Han J (2001) Geographic data mining and knowledge discovery. Taylor and Francis, London

Milne P, Milton S, Smith JL (1993) Geographical object-oriented databases – a case study. Int J Geogr Inform Syst 7:39–55

Mola F, Siciliano R (1997) A fast splitting procedure for classification trees. Stat Comput 7:208–216

Monin AS, Yaglom AM (1975) Statistical fluid mechanism, vol 2. MIT, Cambridge, MA

Moody J, Darken CJ (1989) Fast learning in network of locally-turned processing units. Neural Comput 1:281–294

Moran PAP (1950) Notes on continuous stochastic phenomena. Biometrika 37:17–23

Murai H, Omatu S (1997) Remote sensing image analysis using a neural network and knowledge-based processing. Int J Rem Sens 18(4):811–828

Murray AT, Estivill-Castro V (1998) Cluster discovery techniques for exploratory spatial data analysis. J Int J Geogr Inform Sci 12:431–443

Neter J, Kutner MH, Nachtsheim CJ, Wasserman W (1996) Applied Linear Regression Models, 4th edn. Irwin, Chicago

Ng R, Han J (1994) Efficient and effective clustering method for spatial data mining. In: Proceedings of 1994 international conference on very large data bases (VLDB'94), pp. 144–155

Nicolin B, Cabler R (1987) A knowledge-based system for the analysis of aerial images. IEEE Trans Geosci Rem Sens 25(3):317–329

Novikov EA (1994) Infinitely divisible distributions in turbulence. Phys Rev E 50(5):3303–3305

O'hara Hines RJ (1997) An application of retrospective sampling in the analysis of a very large clustered data set. J Stat Comput Simul 59:63–81

Ohashi Y (1984) Fuzzy clustering and robust estimation. In: 9th Meet. SAS Users Group International FL, Hollywood Beach

Openshaw S (1993) Exploratory space-time-attribute pattern analysis. In: Fotheringham AS, Rogerson PA (eds) Spatial Analysis and GIS. Taylor and Francis, London, pp 147–163

Openshaw S, Charlton M, Wymer C, Craft AW (1987) A mark I geographical analysis machine for the automated analysis of point data sets. Int J Geogr Inform Syst 1:359–377

Ord JK, Getis A (1995) Local spatial autocorrelation statistics distributional issues and an application. Geogr Anal 27:286–306

Ord JK, Getis A (2001) Testing for local spatial autocorrelation in the presence of global autocorrelation. J Reg Sci 41:411–432

Osuna E, Freund R, Girosi F (1997) Training support vector machines: an application to face detection. In: Proceedings of CVPR '97. Puerto Rico

Pao YH (1989) Adaptive pattern recognition and neural networks. Addison-Wesley, Reading, MA

Paola JD, Schowengerdt RA (1995) A review and analysis of back-propagation neural networks for classification of remotely-sensed multi-spectral imagery. Int J Rem Sens 16:3033–3058

Park D, Kandel A, Langholz G (1994) Genetic-based new fuzzy reasoning models with application to fuzzy control. IEEE Trans Syst Man Cybern 24(1):39–47

Park KR, Lee C (1996) Scale-space using mathematical morphology. J IEEE Trans Pattern Anal Mach Intell 18(11):1121–1126

Pawlak Z (1982) Rough sets. Int J Inform Comput Sci 11:341–356

Pawlak Z (1991) Rough sets: theoretical aspects of reasoning about data. Kluwer, Boston

Pearl J (1988) Probabilistic reasoning in intelligent systems: networks of plausible inference. Morgan Kaufmann, San Mateo, CA

Pearson ES (1959) Note on an approximation to the distribution of non-central χ^2. Biometrika 46:364

Peddle DR (1995) Knowledge formulation for supervised evidential classification. Photogramm Eng Rem Sens 61(4):409–417

Peleg S, Naor J, Hartley R, Avnir D (1984) Multiple resolution texture analysis and classification. IEEE PAMI 6:518–523

Peng CK, Buldyrev SV, Havlin S, Simmons M, Stanley HE, Goldberger AL (1994) Mosaic organization of DNA nucleotides. Phys Rev E 49(2):1685

Pentland A (1984) Fractal based description of natural scenses. IEEE Trans PAMI 6:661–674

Pernell C, Themlin J, Renders J, Acheroy M (1995) Optimization of fuzzy expert systems using genetic algorithms and neural networks. IEEE Trans Fuzzy Syst 3(3):300–312

Piramuthu S (1999) Feature selection for financial credit-risk evaluation. Inform J Comput 11 (3):258–266

Pliner V (1984) A class of metic scaling models. J Autom Rem Contr 47:560–567

Pliner V (1996) Metric unidimensional scaling and global optimization. J Classif 13:3–18

Podlubny I (1999) Fractional differential equations. Academic, San Diego, MA

Pohl C, Van Genderen JL (1998) Multisensor image fusion in remote sensing: concepts, methods, and applications. Int J Remote Sens 19:823–854

Polkowski L, Skowron A (eds) (1998) Rough sets in knowledge discovery 1: methodology and applications, 2: Applications. Physica-Verlag, Heidelberg

Polkowski L, Tsumoto S, Lin TY (2000) Rough set methods and applications. Physica-Verlag, Heidelberg

Postaire JG, Zhang RD, Botte CL (1993) Cluster Analysis by Binary Morphology. J IEEE Trans Pattern Anal Mach Intell 15(2):170–180

Powell MJD (1987) Radial basis functions for multivariable interpolation: a review. In: Mason JC, Cox MG (eds) Algorithms for Approximation of Functions and Data. Oxford University Press, Oxford, pp 143–167

Preparata FP (1979) An optimal real-time algorithm for planar convex hull. Commun ACM 22:402–405

Preparata FP, Hong SJ (1977) Convex hulls of finite sets of points in two and three dimensions. Commun ACM 20:87–93

Preparata FP, Shamos MI (1985) Computational geometry: an introduction. Springer, Berlin

Prieto L, Herrera RG, Díaz J, Hernández E, Teso T (2004) Minimum extreme temperatures over Peninsular Spain. Global Planet Change 44:19–71

Qian WH, Lin X (2004) Regional trends in recent temperature indices in China. Clim Res 27:119–134

Qin CZ, Leung Y, Zhang JS (2006) Identification of seismic activities through visualization and scale-space filtering. In: Ruan D, D'hondt P, Fantoni PF, De Cock M, Nachtegael M, Kerre EE (eds) Applied artificial intelligence, processings of the 7th international FLINS conference. World Scientific, New Jersey, pp 643–650

Quagliarella D, Periaux J, Poloni C, Winter G (eds) (1998) Genetic algorithms and evolution strategies in engineering and computer science. Wiley, England

Quandt RE, Ramsey JB (1978) Estimating mixtures of normal distributions and switching regressions. J Am Stat Assoc 73(364):730–738

Quattrochi DA, Goodchild MF (eds) (1997) Scale in remote sensing and GIS. CRC Lewis, Boca Raton, FL

Quinlan JR (1986) Induction of decision trees. Mach Learn 1:81–106

Rafanelli M, Ferri F, Maceratini R, Sindoni G (1995) An object oriented decision support system for the planning of health resource allocation. Comput Methods Programs Biomed 48(1–2):163–168

Ramoni M, Sebastiani P (1996) Robust learning with missing data. Technical Report Kmi-TR-28, Knowledge Media Institute, The Open University

Ranchin T, Wald L (1993) The wavelet transform for the analysis of remotely sensed data. Int J Rem Sens 14:615

Rangarajan G, Ding M (eds) (2003) Processes with long-range correlations: theory and applications. Springer, Berlin

Redner RA, Walker HF (1984) Mixture densities, maximum likelihood and the EM algorithm. SIAM Rev 26(2):195–239

Rees WG (1995) Characterization of imaging of fractal topography. In: Wilkinson G (Ed.) Fractals in geoscience and remote sensing. Luxembourg: Office for Official Publications of the European Communities, pp. 298–325

Richards JA, Jia XP (1998) Remote sensing digital image analysis: an introduction. Springer, New York

Riedi RH, Crouse MS, Ribeiro VJ, Baraniuk RG (1999) A multifractal wavelet model with application to network traffic. IEEE Trans Inform Theor 45(3):992–1019

Ripley BD (1996) Patter recognition and neural networks. Cambridge University Press, Cambridge

Roberts SJ (1997) parametric and non-parametric unsupervised clustering analysis. J Pattern Recogn 30(2):261–272

Rose K, Gurewitz E, Fox G (1990) A deterministic annealing approach to clustering. J Pattern Recogn Lett 11:589–594

Rosenblatt F (1958) The Perceptron: a probabilistic model for information storage and organization in the brain. Psychol Rev 65:386–408

Rottensteiner FJ, Trinder SC, Kubik K (2005) Using the Dempster–Shafer method for the fusion of LIDAR data and multi-spectral images for building detection. Inform Fusion 6(4):283–300

Rousseeuw PJ, Leroy AM (1987) Robust regression and outlier detection. Wiley, New York

Royston JP (1982) An extension of Shapiro and Wilk's W test for normality to large samples. Appl Stat 31(2):115–124

Rubinstein YD, Hastie TJ (1997) Discriminative vs informative learning. In: Proceedings of 3rd international conference on knowledge discovery and data mining. AAAI Press, Menlo Park, CA, pp. 49–53

Rudolph G (1994) Convergence properties of canonical genetic algorithms. IEEE Trans Neural Network 5(1):96–101

Rumelhart DE, McClelland JL, the PDP Research Group (1986) Parallel distributed processing: exploration in the microstructure of cognition, vol 1. MIT, Cambridge

Sadjadi F (2005) Comparative image fusion analysais, 2nd Joint IEEE International Workshop on Object Tracking and Classification in and Beyond the Visible Spectrum (OTCBVS) Program. San Diego, CA, USA, June 20, 2005 (http://www.cse.ohio-state.edu/OTCBVS/05/OTCBVS-05-FINALPAPERS/W01_13.pdf)

SAS Institute Inc (1995) Logistic regression examples using the SAS Systerm, Version 6, vol 1. SAS Institute, Cary, NC

Schistad Solberg AH, Jain AK, Taxt T (1994) Multisource classification of remotely sensed data: fusion of Landsat TM and SARimages. IEEE Trans Geosci Rem Sens 32(4):768–778

Schistad Solberg AH, Taxt T, Jain AK (1996) A Markov random field model for classification of multisource satellite imagery. IEEE Trans Geosci Rem Sens 34:100–112

Scholkopf B, Burges CJC, Smola AJ (1999) Advances in kernel methods: support vector learning. MIT, Cambridge

Scholkopf B, Sung KK, Burges CJC, Girosi F, Niyogi P, Poggio T, Vapnik VN (1997) Comparing support vector machines with Gaussian Kernels to radial basis function classifiers. IEEE Trans Signal Process 45(11):2758–2765

Sen S, Dave RN (1998) Clustering of relational data containing noise and outliers. In: Proceedings of 1998 IEEE international conference on fuzzy systems 2, pp. 1411–1416

Serpico SB, Bruzzone L, Roli F (1996) An experimental comparison of neural and statistical non-parametric algorithms for supervised classification of remote-sensing images. Pattern Recogn Lett 17:1331–1341

Shafer G (1976) A mathematical theory of evidence. Princeton, Princeton University Press

Sheikholeslami G, Chatterjee S, Zhang A (1998) WaveCluster: a multi-resolution clustering approach for very large spatial databases. In: Proceedings of 1998 international conference on very large data bases (VLDB'98), pp. 428–439

Simantiraki E (1996) Unidimensional scaling: a linear programming approach minimizing absolute deviations. J Classif 13:19–25

Simone G, Morabito FC, Farina A (2000) Radar image fusion by multiscale Kalman filtering. FUSION 2000. In: Proceedings of the 3rd international conference on information fusion, 10–13 July 2000, Vol. 2, pp WED3/10 - WED3/17

Simpson EH (1951) The interpretation of interaction in contingency tables. J Roy Stat Soc B 13:238–241

Skowron A, Rauszer C (1992) The discernibility matrices and functions in information systems. In: Slowinski R (ed) Intelligent decision support-Handbook of applications and advances of the rough sets theory. Kluwer, London, pp 331–362

Smith CAB (1947) Some examples of discrimination. Ann Eugen 13:272–282

Sokal RR, Oden NL, Thomson BA (1998) Local spatial autocorrelation in a biological model. Geogr Anal 30:331–354

Stanfill C, Waltz D (1986) Toward memory-based reasoning. Commun ACM 29:1213–1228

Stell JG, Worboys MF (1998) Stratified map spaces: a formal basis for multiresolution spatial databases. In: Poiker TK, Chisman N (eds) SDH'98 proceedings 8th international symposium on spatial data handling. International Geographical Union, pp. 180–189

Steward JB, Angman ET, Feddes TA, Kerr Y (eds) (1966) Scaling up in hydrology using remote sensing. Wiley, London

Stewart CV (1995) MINPRAN: a new robust estimator for computer vision. IEEE Trans Pattern Anal Mach Intell 17(10):925–938

Stuart A, Ord JK (1994) Kendall's advanced theory of statistics, vol 1, 6th edn, Distribution theory. Edward Arnold, London

Sundararajan N, Saratchandran P, Lu Y (1999) Radial basis function neural networks with sequential learning. World Scientific, Singapore

Tadjudin S, Landgrebe DA (2000) Robust parameter estimation for mixture model. IEEE Trans Geosci Rem Sens 38(1):439–445

Tang AY, Adams TM, Usery EL (1996) A spatial data model design for feature-based geographical information systems. Int J Geogr Inform Syst 10:643–659

Taven P, Grubmuller H, Huhnel H (1990) Self-organization of associative memory and pattern classification: recurrent signal processing on topological feature maps. J Biol Cybern 64:95–105

Thomas LC, Crook JN, Edelman DB (eds) (1992) Credit scoring and credit control. Clarendon Press, Oxford

Tiefelsdorf M (1998) Some practical applications of Moran's I's exact conditional distribution. Pap Reg Sci 77:101–129

Tiefelsdorf M (2000) Modelling spatial processes: the identification and analysis of spatial relationships in regression residuals by means of Moran's I. Springer, Berlin

Tiefelsdorf M, Boots B (1995) The exact distribution of Moran's I. Environ Plann A 27:985–999

Tiefelsdorf M, Boots B (1996) Letter to the editor: the exact distribution of Moran's I. Environ Plann A 28:1900

Tiefelsdorf M, Boots B (1997) A note on the extremities of local Moran's I_is and their impact on global Moran's I_i. Geogr Anal 29:248–257

Tinkler KJ (1971) Statistical analysis of tectonic patterns in areal volcanism: the Bunyaraguru volcanic field in Western Uganda. Math Geol 3:335–355

Titterington DM, Smith AFM, Makov UE (1987) Statistical analysis of finite mixture distributions. Wiley, New York

Tong H (1990) Non-linear time series: a dynamical system approach. Oxford University Press, New York

Ulfarsson MO, Benediktsson JA, Sveinsson JR (2003) Data fusion and feature extraction in the wavelet domain. Int J Remote Sens 24:3933–3945

Unwin A (1996) Exploratory spatial analysis and local statistics. Computation Stat 11:387–400

Vapnik VN (1995) The nature of statistical learning theory. Springer, New York

Vapnik VN (1998) Statistical learning theory. Wiley, London

Vapnik VN (1999) An overview of statistical learning theory. IEEE Trans Neural Network 10 (5):988–999

Vaughan RA, Antony R, Kirby RP (1988) Geographical information systems and remote sensing for local resource planning. Remote Sensing Products and Publications, Dundee

Vergin H (1995) Developing and testing of an error propagation model for GIS overlay operations. Int J Geogr Inform Syst 9:595–619

Waldemark J (1997) An automated procedure for cluster analysis of multivariate satellite data. J Int J Neural System 8(1):3–15

Wand MP, Jones MC (1995) Kernel smoothing. Chapman and Hall, London

Wang F (1991) Integrating GIS's and remote sensing image analysis systems by unifying knowledge representation schemes. IEEE Trans Geosci Rem Sens 29(4):656–663

Wang F, Newkirk R (1998) A knowledge-based systems for highway network extraction. IEEE Trans Geosci Rem Sens 26(5):525–531

Wang L, Souza PG (2004) Integration of pixel-based and object-based classification for mapping mangroves with IKONOS imagery. Int J Remote Sens 25(24):5655–5668

Wang M, Leung Y, Zhou CH, Pei T, Luo JC (2006) A mathematical morphology based scale space method for the mining of linear features in geographic data. Data Min Knowl Discov 12:97–118

Wang N, Mei CL, Yan XD (2005) Analysis of spatial relationship between mean and extreme temperatures in China with geographically weighted regression technique (unpublished paper)

Wang N, Mei CL, Yan XD (2008) Local linear estimation of spatially varying coefficient models: an improvement on the geographically weighted regression technique. Environ Plann A 40 (4):986–1005

Wang SL, Li D, Shi WZ, Wang XZ (2002) Geo-rough space. Geo-Spatial Inform Sci 6:54–61

Wang SL, Wang XZ, Shi WZ (2001) Development of a data mining method for land control. Geo-Spatial Inform Sci 4:68–76

Wang W, Yang J, Muntz R (1997) STING: A Statistical Information Grid Approach to Spatial Data Mining. In: Proceedings of 1997 interface conference on very large data bases (VLDB'97), pp. 186–195

Wennmyr E (1989) A convex hull algorithm for neural networks. J IEEE Trans Circuits Syst 36:1478–1484

Wilhelm A, Steck R (1998) Exploring spatial data by using interactive graphics and local statistics. Statistician 47:423–430

Wilkinson GG, Folving S, Kanellopoulos I, McCormick N, Fullerton K, Megier J (1995) Forest mapping from multi-source satellite data using neural network classifiers - an experiment in Portugal. Rem Sens Rev 12:83–106

Wille R (1982) Restructuring lattice theory: an approach based on hierarchies of concepts. In: Rival I (ed) Ordered Sets. Reidel, Dordrecht, pp 445–470

Wilson R, Spann M (1990) A new approach to clustering. J Pattern Recogn 23(12):1413–1425

Witkin AP (1983) Scale space filtering. In: Proceedings of International Joint Conference on Artificial Intelligence, Karlsruhe, pp. 1019–1022

Witkin AP (1984) Scale space filtering: a new approach to multi-scale description. In: Ullman S, Richards W (eds) Image understanding. Norwood, NJ, Ablex

Wolfer J, Roberge J, Grace T (1994) Robust multi-spectral road classification in Landsat thematic mapper imagery. In Proceedings of World Congress on Neural Networks – San Diego, I260–I268

Wolke R, Schwetlik H (1988) Iteratively reweighted least squares: algorithms, convergence and numerical comparisons. SIAM J Sci Stat Comput 9:907–921

Wong HS, Guan L (2001) A neural learning approach for adaptive image restoration using a fuzzy model-based network architecture. J IEEE Trans Neural Network 12:516–531

Wong YF (1993) Clustering data by melting. J Neural Comput 5(1):89–104

Worboys MF (1992) A generic model for planar geographical objects. Int J Geogr Inform Syst 6:353–372

Worboys MF (1994) Object-oriented approaches to geo-referenced information. Int J Geogr Inform Syst 8:385–399

Worboys MF (1998a) Computation with imprecise geographical data. Comput Environ Urban Syst 22:85–106

Worboys MF (1998b) Imprecision in finite resolution spatial data. GeoInformatica 2:257–279

Worboys MF, Hearnshaw HM, Maguire DJ (1990) Object-oriented data modeling for spatial databases. Int J Geogr Inform Syst 4(4):369–383

Wu WZ, Zhang M, Li HZ, Mi JS (2005) Knowledge reduction in random information systems via Dempster-Shafer theory of evidence. Inform Sci 174:143–164

Xu ZB, Leung Y (2004) How neural networks can be made more effective and efficient: a view of learning theory (unpublished paper)

Xu ZB, Leung Y, He XW (1994) Asymmetric bidirectional associative memories. IEEE Trans Syst Man Cybern 24:1558–1564

Yan Z, Yang C, Jones P (2001) Influence of inhomogeneity on the estimation of mean and extreme temperature trends in Beijing and Shanghai. Adv Atmos Sci 18:309–321

Yao X (1999) Evolving artificial neural networks. In: Proceedings of the IEEE 89, IEEE, pp. 1423–1447

Yasdi R (1996) Combining rough sets learning and neural learning: method to deal with uncertain and imprecise information. Neuralcomputing 7:61–84

Yu JG, Leung Y, Chen YQ, Zhang Q (2008) Multifractal analyses of daily rainfall in the Pearl River delta of China (unpublished paper)

Zadeh LA (1994) Fuzzy logical and soft computing: issues, contentions and perspectives. In: Proceedings of 3rd international conference on fuzzy logical, neural networks and soft computing. Fuzzy Logic Systems Institute, Japan, pp. 1–2

Zhai PM, Pan XH (2003) Trends in temperature extremes during 1951–1999 in China. Geophys Res Lett 30(17):1913

Zhai PM, Sun AJ, Ren FM, Liu XN, Gao B, Zhang Q (1999) Changes of climate extremes in China. Clim Change 42:203–218

Zhang D, Lutz T (1989) Structural control of igneous complexes and kimberlites: a new statistical method. J Tectonophys 159:137–148

Zhang JS, Leung Y (2001) A method for robust fuzzy relational clustering (unpublished paper)

Zhang Q, Xu CY, Becker S, Jiang T (2006a) Sediment and runoff changes in the Yangtze past 50 years. J Hydrol 331:511–523

Zhang TS, Ramakrishnan R, Livny M (1996) BIRCH: an efficient data clustering method foe very large databases. In: Proceedings of 1996 ACM-SIGMOD International Conference on Management of Data (SIGMOD'96), pp. 103–114

Zhang WX, Mi JS (2004) Incomplete information system and its optimal selections. Comput Math Appl 48:691–698

Zhang WX, Leung Y, Wu WZ (2003a) Information systems and knowledge discovery. Science Press, Beijing in Chinese

Zhang WX, Mi JS, Wu WZ (2003b) Approaches to knowledge reductions in inconsistent systems. Int J Intell Syst 18:989–1000

Zhang WX, Xu ZB, Leung Y, Leung KS (2006b) Theory of inclusion. Fuzzy Math Syst 10(4):1–9 in Chinese

Zhang Y (2000) A method for continuous extraction of multispectrally classified urban rivers. Photogramm Eng Rem Sens 66(8):991–999

Zhang Y, Hong G (2005) An IHS and wavelet integrated approach to improve pan-sharpening visual quality of natural colour IKONOS and QuickBird images. Inform Fusion 6(3):225–234

Zhou W (1999) Verification of the non-parametric characteristics of back-propagation neural networks for image classification. IEEE Trans Geosci Rem Sens 37(2):771–779

Zhuang X, Huang Y, Zhao Y (1996) Gaussian mixture density modeling, decomposition, and applications. IEEE Trans Image Process 5(9):1293–1301

Zhuang X, Wang T, Zhang P (1992) A highly robust estimator through partially likehood function modeling and its application in computer vision. J IEEE Trans Pattern Anal Mach Intell 14:19–35

Zupan B, Bohanec M (1997) A database decomposition approach to data mining and machine discovery. In: Proceedings of 1st international conference on knowledge discovery and data mining, pp. 299–303

Author Index

Subject Index

MIX
Papier aus verantwortungsvollen Quellen
Paper from responsible sources
FSC® C105338

If you have any concerns about our products,
you can contact us on
ProductSafety@springernature.com

In case Publisher is established outside the EU,
the EU authorized representative is:
Springer Nature Customer Service Center GmbH
Europaplatz 3, 69115 Heidelberg, Germany

Printed by Libri Plureos GmbH
in Hamburg, Germany